Tuna Wars

Steven Adolf

Tuna Wars

Powers Around the Fish We Love
to Conserve

 Springer

Steven Adolf
Amsterdam, The Netherlands

Translated by Anna Asbury and Suzanne Heukensfeldt Jansen

N ederlands
letterenfonds
dutch foundation
for literature

WWF

This publication has been made possible with financial support from the Dutch Foundation for Literature and as well as a financial contribution of the WWF Netherlands (Wereld Natuur Fonds-Nederland).

ISBN 978-3-030-20643-7   ISBN 978-3-030-20641-3   (eBook)
https://doi.org/10.1007/978-3-030-20641-3

This Springer imprint is published by the registered company Springer Nature Switzerland AG.
The registered company address is: Gewerbestrasse 11, 6330 Cham, Switzerland

Acknowledgements

This book could not have been written without the support and knowledge of a great number of people around the globe who are involved in the world of tuna. I owe a lot to the research for this book to the support, enthusiasm and sharing of scientific knowledge of Simon Bush and the discussions and comments of Paul van Zwieten, who both lead the BESTTuna research project of Wageningen University. Sietze Vellema gave me the necessary insights into the global value chain theories. In particular I owe a lot to Henk Brus and Maurice Brownjohn for their comments and their sheer endless knowledge of the tuna industry and sustainability issues. Alain Fonteneau has been a great help in finding my way in the science of biology and economics and the history of bluefin tuna statistics, just like his Japanese counterpart Makoto Peter Miyake and their Spanish colleague José Luis Cort. Stefaan Depypere was a great support in explaining the European fisheries policies and a great help with ICCAT. For the archeological part of the tuna trail, I owe a lot to Clive Finlayson, Darío Bernal and Ángel Muñoz. From the ONG and researcher side I want to thank Sergi Tudela, Bubba Cook, Roberto Mielgo, Rolf Groeneveld, Giuseppe di Carlo, Carel Drijver, Reinier Hille Ris Lambers, Raul García, Elsa Lee, Oliver Knowles, Ric O'Barry, Sari Tolvanen, Ephraim Batungbacal, Hans Nieuwenhuis, Bill Holden, Amanda Nickson, Paulus Tak, Gerald Leape, Andy Sharpless, Daniel Pauly, Alice Miller, Victor Restrepo, Christien Absil, Boris Worms, Francisco Blaha, Wietse van der Werf, Jean-Marc Fromentin, Johan Verreth, Jonah van Beijnen, Ted Bestor, Elisabeth Havice, Kees Lankester, Kate Barclay, Liam Campling, Quentin Hanich and Raphael Vassallo. From the industrial fisheries and tuna governance institutions, I am very grateful to Transform Aqorau, Julio Morón, Javier Garat, Juan Manuel Vieites, Juan Corrales Garavilla, Manuel Calvo, David Martínez, Chris Hue, Francisco Buencamino, Willem Huisman, Narin Niruttinanon, Harry Koster, Gelare Nader, Floor Kuijt, Diego Crespo, Jakob Doorn, Gorjan Nikolk, Glenn Hurry and Herman Wisse. Suzanne Jansen and Anna Asbury did a great job in translating the main part of the manuscript in fluent English. Thanks Inge den Boer for her help in the final editing. I would like to honour the memory of my cousin and good friend P.W. Goedhart who helped me a lot with the biological details of fish. The memory of Luisa Isabel Álvarez de Toledo is of a woman who herself was a force of nature when it came to defend the legacy of tuna.

Finally, my love goes to Luis, my family and friends who have suffered for my Tuna Wars for much too long.

This book is translated with funding of the Dutch Foundation of Literature and support of the Dutch WNF.

Part of the research of this book was conducted under the Wageningen Interdisciplinary Research Fund (INREF) supported by BESTTuna Programme.

Steven Adolf gratefully acknowledges the support of the David and Lucile Packard Foundation that helped funding part of the research under Grant 2012-38353 and 2014-39668.

Tuna Timeline

72,000 BC	Early humans wait for tuna on the beaches of South Africa and eat them
50,000 BC	Gibraltar Cave Neanderthals also put tuna on the menu and paint a hashtag
1500 BC	Development of Almadraba fisheries and large-scale tuna production in the Phoenician cities of the Eastern Mediterranean
1000 BC	Phoenicians go West; Almadraba spreads out in Mediterranean settlements in Carthage, Gades (Cadiz) and beyond
264–241 BC	First Punic-Romans War, Romans take over Sicilian tuna fisheries
241–218 BC	Barcas dynasty established in South of Spain, establish large-scale tuna fisheries
218–201 BC	Second Punic-Roman War, Hannibal Barcas Cross the Alps with elephant army
149–146 BC	Third Punic-Roman War, Fall of Carthage, industrial tuna fisheries become Roman
146 BC–300 AD	Roman supremacy in large-scale fisheries and trade salted tuna and garum
300	The Fall of the Western Roman Empire, large-scale fisheries and trade decline
500–1000	Dark Age large-scale tuna fisheries
1294	Guzmán 'el Buno', conqueror of Tarifa, is rewarded with tuna fisheries right of Spanish south coast
1369–1445	Guzmán family grow into biggest tuna magnate-nobility of Spain, Dukes of tuna
1568	Spanish reach the Solomon Islands
1571	The Battle of Lepanto, Almadraba flourishes under Dukes of Medina Sidonia
1577	People admire tuna in Coenen's Fish Book
1588	Spanish Armada sails off under the command of tuna Duke of Medina Sidonia
1600 around	Cervantes writes tuna novel

1609	Grotius publishes *The Freedom of the Seas*
1641	Failed Coup d'état Tuna Duke, decline in Tuna Fisheries
1757	'Tuna Saint' Martín Sarmiento writes research and advice to promote sustainable tuna fisheries
1813–1815	Napoleonic Wars, start development of jars and cans for conservation
1816	End of tuna privilege Spanish Tuna Dukes
1840	Invention of sushi in Japan, including raw tuna
1880–1900	Introduction canned sardines, later tuna
1914–1918	World I, demand of canned tuna
1920	Plague of bluefin tuna damages fisheries Norway
1928	Start of Spanish Consorcio Nacional Almadrabero
1930	Japanese tuna fisheries expanding through pole and line
1935	Ceasing activities almadraba, strong growth US tuna canning industry
1940–1945	WO II, canned tuna demand
1942–1944	Start of Philippine tuna fisheries under Japanese occupation
1946–1952	Plague bluefin tuna, Trebeurden, France
	French Bask fishermen copy US purse seiner fisheries
1950	Record catch Norwegian fleet, southern bluefin tuna fisheries on the rise, the start of FAO tuna statistics
1952	General McArthur lifts Japanese fishing ban
	Japanese tuna fleet come back, longliners in Pacific, Indian Ocean, Atlantic
	Japanese tuna fisheries expansion followed by Korea, Taiwan, China
	American canning brands explore Philippine tuna fisheries
	Spanish and French Tuna fleet spread out West African Coast
1954	Gordon-Schaefer bioeconomic fisheries model for computing sustainable fisheries yield
1960	Top year catch of Southern bluefin south of Indonesia before catch nosedives
1968	'The Tragedy of the Commons' by Garrett Hardin
1970s	New Japanese techniques for freezing fish
	Port Lincoln growing tuna fisheries
	Japan discovers US east coast for bluefin tuna
	Japanese traders explore Philippine yellowfin for sushi
	Introduction of the FAD, known as payao in the Philippines
1980s	Japan enters in Mediterranean bluefin fishery
	Massive slaughter of dolphins in East Pacific yellowfin tuna fisheries
	Purse seine becomes main gear in tuna fisheries
	US tuna fleet moves from Eastern to Western and Central Pacific

1981	French tuna fleet enters the Indian Ocean, followed by Spanish fleet
1982	PNA, the start of cartel tuna producing Island states
	UN Convention on the Law of the Sea UNCLOS)
1985	BFT disappears from North Sea
1988	Boycott Big Three US brands by Dolphin Coalition
	Start of Dolphin Safe label
1990	American Big Three declare to no longer use dolphin unsafe tuna
1990s	Expansion purse seine bluefin tuna and tuna ranching Mediterranean, export to Japan rockets
1992	UN drift net ban
1996	Birth MSC certification label
2000	Start of Flipper Wars, Conquest Pacific
2003	Indonesia and the Philippines biggest tuna fishing nations in Western Central Pacific
2004	First Kinki Tuna, farmed tuna on the market
	PNA starts Vessel Day Scheme to limit tuna fisheries
2005	Illegal fisheries a major threat to sustainable tuna fisheries, Pirate Wars Horn of Africa
2006	Decline in Mediterranean bluefin tuna fisheries
2007	Birth of Kobe plot diagram for sustainable tuna fisheries
2009	Start of battle of Fisheries Science
2009	Flipper War, Battle US-Mexico, Mexican tuna fisheries take Dolphin Safe to WTO
	Pirate War: hijacking Bask purse seiner Alakrana
2010	Red Alert for stock crash BFT Mediterranean, failed attempt to get bluefin tuna on CITES list endangered species
	Use of FADs in tuna fisheries intensifies
	PNA cartel starts own certified sustainable, Battle for Pacifical
	Birth of European Union carding system to fight against IUU fisheries
2011	Endangered Southern bluefin tuna on IUCN list
	Vulnerable Pacific bluefin tuna on IUCN list
	Arab Spring Tunisia, the Battle of Sirte, exile Ben Ali, death of Gaddafi, collapse of illegal fisheries in Libya
2012	Recovery bluefin tuna stock Mediterranean
2013	Pacifical Brand sustainable skipjack enters market
2014	Shrimp slavery scandal Thailand
2015	Indonesia arrests Philippine crew, bombarding 'illegal' tuna boats
	The Tuna Conspiracy War: Big Three brands taken to court for price fixing

2015–2016	Global Tuna catch reaches 5 mln tons (FAO), worth $42 billion (PEW)
2016	Battle of Tuna Corruption: Mozambique $850 mln tuna bond scam
2017	ICCAT set higher Total Allowable Catch bluefin tuna
	Philippine tuna boats attack fishers in PNG waters
2018	Indonesia sinks illegal tuna boats
	Battle of Tuna Corruption: Europol 'Operation Tarantelo' against illegal bluefin tuna

Contents

About the Author

Steven Adolf (1959) is a researcher and consultant, and a writer and journalist based in the Netherlands. He worked as an editor and correspondent in Spain, Portugal and Morocco for the Dutch daily papers Het Financieele Dagblad, De Volkskrant and NRC Handelsblad, the Dutch weekly magazine Elsevier, and the Belgian newspapers De Morgen and The Standard. He is the author of bestselling country portraits of Spain and Morocco and a book on Spanish immigration in the Netherlands. As a researching economist he is currently involved in a project on sustainable tuna fisheries at Wageningen University.

Photo: Ngoc Binh Tran

Abbreviations

CCSBT	Commission for the Conservation of Southern Bluefin Tuna
DWFF	Distant Water Fishing Fleet
DWFN	Distant Water Fishing Nations
EEZ	Exclusive Economic Zone
EU	European Union
FADs	Fish Aggregating Devices
FFA	Forum Fisheries Agency
IATTC	Inter-American Tropical Tuna Commission
ICCAT	International Commission for the Conservation of Atlantic Tunas
IO	Indian Ocean
IOTC	Indian Ocean Tuna Commission
IQ	Individual Quotas
ITQ	Individual Transferable Quotas
MEY	Maximum Economic Yield
MSC	Marine Stewardship Council
MSY	Maximum Sustainable Yield
PNA	Parties to the Nauru Agreement
RFMOs	Regional Fisheries Management Organisations
SOFIA	State of World Fisheries and Aquaculture
TAC	Total Allowable Catch
TAE	Total Allocated Effort
VDS	Vessel Day Scheme
WCPFC	Western and Central Pacific Fisheries Commission
WCPO	Western and Central Pacific

List of Figures

Introduction

1

*It's because we all came from the sea. (…) We have salt in
our blood, in our sweat, in our tears. We are tied to the ocean.
And when we go back to the sea (…) we are going back from
whence we came.*

John F. Kennedy, 1962

The Tuna Trail

The tale of tuna is a journey without borders, through space and time, crossing distant oceans between every continent. It is a journey from prehistory and the formation of civilisation to today's globalised society. It tells a story from excess to scarcity and the need to design and implement policies that guarantee the fishing for future generations. It is above all a journey of discovery into one of the most amazing fish in our seas, one that has profoundly influenced society and can teach us to reflect on how a sustainable global economy should proceed.

My tuna journey began quite simply, with the opening of a can. We occasionally had tuna at my parents' home, but it was generally reserved for festive occasions, where the fish was served on toast. Large tuna salads weren't something we were used to in the Netherlands of the second half of the twentieth century. It took a while before the tuna sandwich crossed the Atlantic from distant America. The Dutch catch and trade fish, but aren't big fish eaters themselves, the exception being soused herring, a national symbol, eaten standing at stalls, accompanied by freshly chopped onions and pickled gherkins.

Nevertheless, some trace of tuna must have been dormant within me even then, like a latent addiction of which I was barely aware. It was because of the sea, of course; not the muddy grey North Sea, but the sun-drenched, clear azure water of the Mediterranean. In the early 1970s I sat glued to the television screen every Sunday evening watching *L'Odissea* [210], an Italian, French, German and Yugoslavian

© Springer Nature Switzerland AG 2019

S. Adolf, *Tuna Wars*, https://doi.org/10.1007/978-3-030-20641-3_1

co-production by director Franco Rossi and producer Dino de Laurentiis. This Odyssey began with a mysteriously plucked harp string, an intriguing echo from times long forgotten, which struck the right chord inside me. Odysseus, crossing the Mediterranean with his crew, passed islands full of dangers, concealed in mist: from the Sirens' song to the man-eating Cyclops Polyphemus. Adventure, battle, stratagem, ruin and revival: it had it all. As a matter of fact, it all turned on our history with tuna, but I only discovered that much later.

There was plenty more fish-related viewing in this early television era. Of course, there were the adventures of Flipper [99], the madly popular dolphin from the American television series, who clapped his fins, made chittering sounds and let people stroke him. A dolphin was something you saw at most, if you were lucky, on a boat trip at sea, mainly a glimpse of dorsal fins. I would never have foreseen that Flipper would cross my path several decades later as the main player in a vicious power struggle over tuna, involving hundreds of millions of dollars.

But the most important step on my way to tuna addiction was undoubtedly the vision of a tanned figure who—smoking a pipe under a red woollen sailor's cap—made history by being the first to introduce the general public to the world under water on television. Jacques Cousteau (1910–1997) was a French naval officer who devoted his entire working life to the cause of the seas. You could call him the undersea explorer of the twentieth century, who put into practice what his compatriot Jules Verne had described in the nineteenth century. Cousteau was ahead of his time: he understood better than anyone that media in combination with personality is essential in forming the successful packaging of a message. And I was a grateful victim.

Cousteau breakthrough to the general public came in 1956 with the documentary film *Le Monde du Silence* [67], which was shown at Cannes Film Festival. The film about the underwater world created a keen interest that is hard to imagine now. For the first time, a large audience was introduced to the flora and fauna of the seas and oceans which had previously been invisible. Cousteau recorded the film underwater over a year's tour of the Mediterranean, the Persian Gulf, the Red Sea and the Indian Ocean. Diving from his converted minesweeper Calypso, he revealed sharks and whales, coral reefs with their wealth of colours, sponge fishermen in Greece, sea turtles and pufferfish. Never before had the underwater world been brought almost palpably into view.

Cousteau went on to produce further exciting and exotic television series. While the divers were threatened by sharks, poisonous fish and horrifying decompression sickness, Cousteau presided over a positive outcome like a tanned Poseidon with a woollen fisherman's cap. And at the end of such an adventure we always returned to the galley of the Calypso, where the cook prepared the *fruits de mer* and anything else they had come across on their dives. Eating well from the sea was important too.

The sea was still a playground where Cousteau could go about his business unimpeded. In *Le Monde du Silence* we see the captain blow up an area of coral reef with dynamite as a 'scientific experiment' (to count the fish, our marine guide informs us in a neutral tone). After the explosion, a puffer-fish of proportions you wouldn't often see these days is hauled ashore only to suffocate, deflating like a punctured tyre and gasping for breath. Deflating pufferfish still made good entertainment in the 1960s.

So it went on for a while. The big wreckfish Jojo is teased by the divers, but at least the creature doesn't disappear into a bouillabaisse. A baby sperm whale is not as lucky. Accidentally rammed by the Calypso, the creature is harpooned by the crew, while smoking their Gauloises. The carcass serves as food for a shoal of famished sharks, which in turn are lugged aboard with hooks one by one and killed with axes. Not in the name of scientific research this time. 'All sailors in the world hate sharks,' the commentary explains.

Jacques Cousteau knew what his audience wanted to see. The sea meant dangerous monsters, blood and adventure. It sold, and the Calypso's maintenance had to be paid for somehow. Nevertheless, it was Cousteau who was one of the first who made efforts towards conserving the marine environment. As early as 1960 he successfully protested against French plans to dump radioactive waste off the coast of Nice.

Like Peter Benchley, who as the author of the epic shark thriller *Jaws* had some making up to do, Cousteau became an increasingly outspoken defender of the sea. Although he was always on guard against being pigeonholed as an activist, his popular documentaries were intended to impart love and respect for the marine environment to the public. 'You conserve what you love,' Cousteau remarked.

Towards the end of his life, after more than 50 years of research and sea documentaries, the French oceanographer exhibited increasing pessimism regarding the survival chances of the natural phenomena he had been revealing all his life. Perhaps it would be better if, instead of the fish, humanity largely disappeared, the marine explorer concluded bitterly. By then, dynamite under a coral reef, a harpooned baby whale or even beating sharks to death would all have long been unacceptable to the public. But respect for the sea remained limited to a show of good behaviour and had not yet entered the stage of the complex and tricky sustainable governance of fisheries and the marine environment.

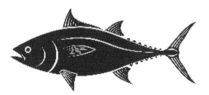

There was no tuna in Cousteau's film. In retrospect that seems odd, as the French sea explorer was well aware of the *almadraba*, the traditional tuna trap which had been operating on the coasts of southern Spain, Sicily and North Africa for three millennia. And the bloody spectacle of this breath-taking fishing method would have been right up Cousteau's street. But remarkably enough the tuna didn't make it into the final version of his documentary. In his book *The Silent World* [65, 66], about the Calypso's travels, Cousteau does describe a visit to an *almadraba* at Sidi Daoud in Tunisia, a tuna port not far from what was once Carthage. It's no coincidence that traditional tuna fishing took place near the remains of the city that was totally destroyed in antiquity's greatest wars. Here the Phoenicians had founded three millennia back the capital of their trading empire and brought large-scale tuna fishing from the east to the west of the Mediterranean.

Cousteau was deeply impressed by what he saw at Sidi Daoud. The *almadraba* is one of the 'most horrible and grand' marine spectacles to be seen, he writes. Having set out with hundreds of Tunisian fishermen on their large flat-bottomed boats, the *raïs* (captain), signals that the *matanza* (tuna killing) can begin. 'A barbarian roar broke from the fishermen and they chanted an old Sicilian song, traditional to the *matanza*.'

This was the sign for Cousteau to descend with his camera into the 'death chamber', the innermost of three chambers into which bluefin tuna are herded before they are hauled up by the fishermen. Cousteau is surrounded by 60 bluefin tuna and hundreds of smaller bonitos. The large tunas are still peaceful swimming past them at close quarters in a school without paying us much attention, then disappearing to explore their prison. They reappear, opening and closing their mouths so that the water flows over their gills, never losing sight of the net. Their metallic bodies are like perfect rockets, at most a little stiff in their movements.

> We found ourselves almost assuming the role of the doomed animals. . . . The noble fish, weighing up to four hundred pounds apiece, swam round and round counter-clockwise, according to their habit. In contrast to their might, the net wall looked like a spider web that would rend before their charge, but they did not challenge it. Above the surface, the Arabs were shrinking the walls of the *corpo*, and the rising floor came into view.
>
> We pondered how it would feel to be trapped with the other animals and have to live their tragedy. . .

As the space around them shrinks, Cousteau is tempted to pull out his diving knife and cut a hole in the net 'for a mass break to freedom'. Instead he continues filming.

> The death chamber was reduced to a third of its size. The atmosphere grew excited, frantic. The herd swam restlessly faster, but still in formation. As they passed us, the expression of fright in their eyes was almost human.
>
> My final dive came just before the boatmen tied off the *corpo* to begin the killing. Never have I beheld a sight like the death cell in the last moments. In a space the size of a large living-room, tunas and bonitos drove madly in all directions. The tuna's right-eyed honeymoon instinct was at last destroyed. . .
>
> . . . With what seemed like the momentum of express trains, the tuna drove at me, head-on, obliquely and crosswise.

Paralysed with fear, Cousteau hears the roll of film in his camera run out. When he surfaces among the compacted fish bodies, there is not a scratch on him. Even while fighting for their lives, the enormous tunas have succeeded in dodging him by a couple of centimetres, the current in their wake massaging his body.

Above them the *raïs* removes his hat in a gesture of respect for those about to die. It is a signal that the slaughter can begin. With their long hooks the fishermen haul the enormous tuna aboard, a job requiring five or six men. The bluefin tuna now lies shaking 'like a gross mechanical toy' on deck. The fishermen wash off the blood; the battle is done.

Anyone who has seen bluefin tuna in action understands Cousteau's fascination. With the volume and weight of several hundreds of kilos, tuna is among the largest and fastest predatory fish in the oceans, an impressive swimming machine which travels thousands of kilometres apparently without any effort on its migratory routes. A fish that combines the appearance and speed of a torpedo with the stamina of a diesel engine and the elegance of a racehorse.

Although the magnitude of today's tuna catches and trade is unprecedented, our social and political entanglement with tuna has been part of our civilisation from its earliest origins. You might say that tuna was already globalised long before the word was invented. Thousands of years ago, tuna was the first fish in human history to be captured, processed and sold on a large industrial scale. Traditional bluefin tuna fishing is thus the oldest form of fishing industry still active and a cultural and industrial heritage that is honoured in practice. We have eaten this species for millennia, raw, later salted, later still from jars or cans, and now raw once again. The fish sauce derived from tuna was among the most expensive culinary indulgences that ancient Mediterranean civilisation could afford.

Tuna was an engine of large-scale trade, wealth and war. The Phoenicians constructed their tuna salting pits near all their trading posts. We still find them today. The Carthaginians fought the three Punic Wars around southern Spain, Sicily and the north coast of Africa, where the core of their tuna industry was based. From the dark forests of Germania to the deserts of Syria and Persia, Roman soldiers set off to war eating a piece of salted bluefin tuna. A bottle of tuna sauce was available to heal their wounds. The fermented bluefin tuna went along on the Spanish naval fleets in the sixteenth century. Throughout our history, war was always closely related to tuna as healthy, conserved proteins and a source of great wealth and power. But however deeply and radically our history has been determined by tuna, the fish only became truly famous in the 1970s. Tuna became an indispensable ingredient of sashimi and sushi, Japan's big contribution to global cuisine, and it was the bluefin tuna, the giant among the tuna species which grew to be the big favourite in this food culture. Raw, thinly sliced or served on a piece of rice, bluefin was eaten to the furthest corners of the earth. The fish became the focus of an industry worth fortunes, complete with wars, scams and piracy, espionage and politics.

At the start of this century, for the first time in our history, the question was seriously raised as to whether the *almadraba* and bluefin tuna would survive much longer. The threat to the species was a sudden wake-up call for many people with a fascination for the sea and fishing. The fish might be regarded as the icon of large-scale changes and overfishing of the seas and oceans, the consequences of which we are ill-equipped to assess, let alone contain. Was Cousteau right in his sombre vision of an ecological race to the bottom, in which humanity had destroyed the oceans? Has his documentary *Le Monde du Silence* taken on a new, less poetic meaning when it comes to the sea? Is there a disaster of unprecedented proportions unfolding in silence, under water, largely invisible to the human eye? And was bluefin tuna the unintentional aquatic equivalent of the tiger, threatened with extinction? Could anything else be done to prevent large predatory fish, particularly bluefin tuna, from dying out en masse? And what about all that other tuna species in our seas?

This book is a quest for answers to these questions. Our tour begins in prehistory, following the traces of tuna from antiquity to medieval times, the rise of the canning industry and the global sushi craze. It is a quest that leads from a world of abundance to one where our tuna populations are fished out of the seas at an unprecedented rate, and to research into initiatives aimed at managing a global sustainable fisheries industry which also ensures a safe future for tuna. The stakes are high and so are the environmental, industrial, political and social interests involved in this issue. Interests that make tuna fisheries a wicked problem, including hybrid tuna wars that are hardly noticed by the public, but that affect deeply society as a whole. If we succeed in saving the fishery of a cosmopolitan migrant fish such as tuna for future generations, using new, inventive techniques and forms of governance to work towards a sustainability, then we must be able to do the same with other fish and animals.

I like to invite to this journey into the world of tuna in particular those who have heard that something is the matter with tuna and sustainability, but who lost track in the labyrinth of the species, its fisheries and its politics and governance. If you have the vague notion that there might be something wrong with the tuna on your plate, I am happy to be your guide on finding your way back on the tuna track. More informed tuna geeks are welcome too. In our journey I will tell some tuna facts that you might have missed, but that are worth to know about (Fig. 1.1).

Fig. 1.1 The skipjack tuna (by courtesy of NOAA Fisheries)

Part I

The Dawn of Giant Tuna

The Cave

2

> *The Mediterranean is . . . 1000 things at once. Not just one landscape but innumerable landscapes. Not just one sea, but a succession of seas. Not just one civilisation but many civilisations superimposed on one another. The Mediterranean is an ancient crossroads. For millennia, everything converged on it, disturbing and enriching its history.*
>
> Fernand Braudel [51]

Civilisation begins with fish; fish in the Mediterranean. Its arrival was eagerly anticipated in the caves of the steep rock faces. The giant tuna, after all, always returned to the Mediterranean on its regular route past their beaches, so it was a question of waiting patiently in the right place. They must have been taught this from an early age, just as they were taught how to make fire and how to sharpen the edge of a stone to make a hand axe. Much about the Neanderthals remains a mystery, but we know that they were partial to bluefin tuna. They knew how to cut large fillets from the spine and roast it over the fire, perhaps with a sprig of rosemary as a first step in culinary evolution. Tuna, in any case, was a welcome change from the menu of shellfish, wild boar or ibex, the local mountain goat. As the days grew warmer, it was time to go in search of the big fish (Fig. 2.1).

So it must have started here, the fuss over tuna. In fact, that is quite surprising. Neanderthals are not the first candidates to spring to mind when it comes to catching and eating giant tuna. Eating tuna is different from plucking a berry from a bush or opening a shell and gulping down the contents. Tuna requires knowledge, organisation and skill. Catching this great fish involved passing on experience, demanding at least a rudimentary language and some degree of social order. Tuna entails governance: without governance there is no tuna.

The long journey of tuna begins here, in Gorham's Cave at the southern tip of Gibraltar. Like the Neanderthals, we look south from high up at the back of this

© Springer Nature Switzerland AG 2019
S. Adolf, *Tuna Wars*, https://doi.org/10.1007/978-3-030-20641-3_2

Fig. 2.1 Gorham's Cave, last refuge of the tuna eating Neanderthals (S. Finlayson, by courtesy of Gibraltar Museum)

enormous cave. The view has changed since their day. They looked out over dune-like steppes which, a couple of kilometres further on, became the beach of the narrow strait dividing the continents. Over the last 50,000 years the sea level has risen substantially, a process that could well continue much further if climate predictions are to be believed. The seawater now laps at the foot of the cave entrance. Around us lie more caves and caverns once inhabited by Neanderthals, now withdrawn from sight, deep under the rising sea level. The view of Morocco remains the same, just as nothing has changed in the migration of the large shoals of giant tuna which pass by every spring towards the rising sun.

So that's how they saw it, Nana and perhaps little Flint too. They look at us in their own modest exhibition room at the Gibraltar Museum: Nana with her arms crossed, hands over her shoulders, a broad smile and an expression that says she's well and truly got the measure of us. That *Homo sapiens* wasn't to be trusted: swimming over from Africa, hunting down all the wildlife, beating the men's brains in and going after you too, given half a chance. But what can you do about it? No point in losing your temper. Flint clearly sees it differently: curiosity and fear battle for priority in his eyes, as so often in children when a stranger suddenly turns up to visit. His arms are flung around Nana for protection.

It is hard not to be moved by Nana and Flint, who have been on show at the Gibraltar Museum since 2016. Both Neanderthals—a woman and a little boy, aged approximately 30 and 5—have been lovingly and respectfully reproduced in life-size models, based on their skulls and other remains, by Dutch twin brothers aptly named

Adrie and Alphons Kennis (Kennis means 'knowledge' in Dutch). You could call it payment of a debt of honour, if you were so inclined. It is little known that Neanderthals were first found on Gibraltar. The skull of a prehistoric humanoid was discovered here in 1848. He was named Homo Calpicus, after Calpe, the Phoenician name for Gibraltar. Calpe or kalph means 'hollow' or 'cave' in Phoenician: the mountain with the caves. The discovery of the 'humanoid' remains received little attention. Eight years later, a couple of thousand kilometres further north, in the German Neander Valley beside the River Düssel, researchers came across a partial skull, radial bone and some other remains which led to considerably more commotion. The 'primitive human' was to derive his name from here.

From the outset there was shame and unease around the Neanderthal and his relationship to man. No one had bargained on this distant family member, who made his entrance clumsily and unannounced. The Catholic Church was compelled to consider whether Neanderthals might also have been descended from Adam and Eve. Then there was further doubt as to whether or not primitive man had a soul. Had there been contact with this species and, if so, in what form? More enlightened minds concluded that this was a failed version of ourselves. To colonial Europe of this period, when superiority was elevated above any suspicion, this was an unwelcome development. The Neanderthal even presented a serious problem for Darwin, or vice versa: the species had substantial brains, judging by their sizable skulls, which did not square with the tiny brains that the founder of the theory of evolution had in mind for the ancestors of man.

In the end it was best for everyone to view the Neanderthal as the backward cousin of the human race, to be forgotten as quickly as possible. Just as the Lombroso type stood for criminal tendencies, the Neanderthal, with his low forehead, monobrow and ape-like appearance, came to stand for the proverbial blockhead. It was also a neat explanation for their losing out to superior man, who was stronger, cleverer and even looked considerably more attractive. Dispersed between small communities across the European continent, things presumably went downhill for the Neanderthal, until the critical biological threshold was reached below which the species became unsustainable. The last of the Neanderthals perished 30,000 years ago, probably here in Gibraltar, where favourable living conditions made it their final reserve and sanctuary. Like the dodo, the Tasmanian tiger and the bubal hartebeest, the harelip sucker and the Adriatic sturgeon, they had irrevocably lost the battle against Homo sapiens (Fig. 2.2).

But we are not there yet, as we stand at the entrance to Gorham's Cave 50,000 years ago. The Neanderthals left their caves, descended and walked six or seven kilometres to the beaches which were their destination. Here, along the deep channel which divides the continents, they gathered food all year round: shellfish, of course, and the odd young monk seal if they were lucky. They knew the big fish were coming when they saw the orcas appear. Like the Neanderthals, these fearsome great black sea creatures were after the bluefin tuna, the difference being that the orcas hunted them from the water. You could see them from far off, with their tall, upright black back fins protruding above the sea. Sometimes their mouths glistened, full of teeth like white daggers. Orcas were monsters. They sometimes leapt out of the

water, falling back with a splash to offer a brief glimpse of their true scale. Then the Neanderthal knew he was in the right place and it was just a matter of time.

When the giant tuna finally came swimming by, everything began to stir. It was as if a strong wind blew up over the sea, without a breath of actual breeze. The water began to boil, fins appeared on the surface. Dolphins tumbled out of the water, alternating with the large fins of the orcas rising as they gasped for air, before smoothly diving down again, after the bluefin.

Catching bluefin tuna wasn't exactly straightforward. The orcas, with their intelligence and collective hunting technique, were among the few creatures capable of catching the lightning-fast tuna, measuring several metres in length. They lay in ambush in the narrow channel, where all the tuna swam, as if into a trap. Once the tuna were surrounded, the orcas could feast on them. Now and again a fish would escape from the circle; orcas may be quick, but tuna are quicker. So it was down to luck: there was always a bluefin that panicked and made a mistake. Instead of heading for open sea, the giant fish would shoot like a silver spear at full speed onto the beach, remaining helplessly stranded a couple of metres up, bulbous tail powerlessly thrashing in the sand. That was the moment for the Neanderthals to strike and pull the beast further up the beach. They had to be quick about it: once the orcas noticed, they would try to shoot up the beach on an incoming wave and swipe the tuna out from under their noses with the retreating surf.

The stranded bluefin tuna was as big as a Neanderthal, if not bigger, but the impressive beast must have been an easy prey: not quick like the red deer or aggressive like the bear, and nothing at all like the rhino, which would charge with thundering steps, turning hunter into prey. The giant fish did not bite or peck, nor did it attack; it simply lay gasping on the dry sand, large eyes reflecting panic. The tuna died quickly once out of the water. At most they would knock it on the head with their axe and job done.

The giant fish was then cut into pieces and dragged back to the caves. By the fire the meat was cut into smaller portions and handed out. The fattiest pieces of the belly had to be used first, as they did not keep for long. The rest could not wait much longer. That was the drawback of this easy prize: it spoiled quickly. But the red flesh was tasty and more resembled that of land animals than other fish, a nutritious meal that fitted nicely into the protein-rich Neanderthal diet. It was a good start to the warmer days. They would never leave here. Life around the great rock was not so bad after all.

Archaeologist Clive Finlayson knows all about this [230]. Today the director of the Gibraltar Museum has invited me on an expedition into the prehistory of tuna. Together with a team—his wife and right hand Geraldine, Richard Jennings of Liverpool John Moores University, Spanish archaeologist José María Gutiérrez López and a group of students—we have descended an endless stairway to Gorham's Cave. This is the military territory at the southernmost point of the small peninsula of the British Crown colony and only accessible with special permission. The place still smells of wartime romance with a touch of James Bond. At the top of the access road where we park the minibus there is still a fully furnished field hospital dating back to World War II boxed into the rock. It would hardly be surprising to find yourself standing at the entrance of a secret submarine base hacked out of the cliffs. Directly opposite is the fictitious spot where the U-96 submarine from the novel and film *Das Boot* sinks into the icy depths of the Strait of Gibraltar. Our descent consists of a stairway stretching hundreds of metres down to sea level. Already sweating under our safety helmets, we move downwards into prehistory. Below, in 1907, Captain A. Gorham of the 2nd Battalion Royal Munster Fusiliers discovered the cave which brought his name to international fame. It is an impressively high crevice, like a cathedral in the rock, with an opening around 40 m high offering an entrance to an 80 m deep chimney leading upwards. The cave's unique secret is located deep in the chimney. Over millennia the harsh easterly Levante wind which howls through the strait here has created sand dunes which slowly filled up the cave layer by layer, thoroughly covering the traces of the Neanderthal. Every time a fire was kindled, a meal cooked and eaten, the traces disappeared under a new layer of sand. This created an angled, rising archive, which, year after year, century after century, neatly preserved the history of the Neanderthal. 'Gorham's Cave is a time machine,' says Clive Finlayson.

It is no coincidence that our history of bluefin tuna begins here, because tuna is a special fish with a rich human history. The two sides come together in Gibraltar. This is the spot where a narrow strait between two continents has always been one of the richest areas of underwater life. Whales were hunted here in the bay right into the 1920s, with three catches in a day not unusual. On the other side of the bay, in the little village of Getares, the dilapidated remains of the whaling station can still be seen, with its sloping landing where the creatures were pulled ashore. You can read the abundance of sea life in the names here. Getares—like Ceuta, the Spanish enclave visible in clear weather on the other side—is derived from the Greek *kethos*, which means 'sea monsters'. It refers to whales and tuna, which swam through the strait here en masse.

Gibraltar also has a particularly rich human past. Few places in the world are such intersections of historical heritage, myth and legend, culture and war, along with a colourful mix of peoples between the continents. Every surface here conceals yet another story. Take 'The Rock', where Finlayson lives, along with his 30,000 compatriots. It sounds massive and hard, and for anyone who drives along the coast from Malaga that's how it looks too. A limestone wall rises steeply 423 m into the air, like an impenetrable fortress. In fact nothing could be further from the truth: the Rock is a Swiss cheese. This 'Pillar of Hercules', which forms the entrance to the Mediterranean Sea, is perforated with caves and passageways. This is mainly down to the British, who, along with the Dutch, plundered Cádiz and set it on fire and, enthused by this success, took Gibraltar from the Spaniards in 1704. Under the Treaty of Utrecht of 1713, which put an end to the War of the Spanish Succession, Spain lost its rock. Gibraltar became a British Crown colony.

That is still noticeable here now. The atmosphere of the Spanish Costa del Sol evaporates on the border. Bobbies patrol the streets, where the greasy smell of bacon and eggs is clearly apparent. A Mediterranean Brighton. The cable car offers a broad view of the Strait of Gibraltar and the Bay of Algeciras. Unwary tourists are regularly robbed of their bags and cameras by the wild Barbary macaque, a rather shabby ape for whom this forms the only more or less wild habitat in Europe. These monkeys have grown to be a real nuisance. They plunder the rubbish dumps and refuse sacks, and from time to time descend to the city to terrorise the population. But shooting apes is frowned upon; the people of Gibraltar love the Rock's fellow inhabitants. Monkeys attract tourists. Moreover legend has it that when the last ape disappears the Rock will return to Spanish hands, and in order to avoid that risk the people of Gibraltar put up with the inconvenience. Rather monkeys than Spaniards ruling the Rock.

From the start the people of Gibraltar entrenched themselves to keep the Spaniards out. At the end of the eighteenth century they chipped out the 'upper galleries' of the rocks, creating a system of tunnels to house their cannons during the repeated siege of Gibraltar by Spain. And they were successful: the Spaniards did not win their rock back. The Gibraltar Museum, run by Finlayson, his wife Geraldine and son Stewart, tells a story packed with war, occupations and changing rulers who have left behind their traces. The cellar of the building contains a bathhouse dating back to the time when the Muslims ruled here, just a millennium ago. The current inhabitants are a mix of Spaniards and Berbers, Sephardi Jews, Brits and Hindus. Those who have attended university in Britain speak cultivated English, but if you address people in Spanish you receive an answer in the heavy, sometimes incomprehensible accent of the province of Cádiz.

Once down the long staircase to sea level, having passed through a narrow passageway over slippery rocks, we walk upwards into the cave. A walkway protects the sand which has blown in here. Up in the cave, in the layers nearest the surface, a team of archaeologists is carefully digging away the sand over an area of a few square metres, working with the patience of saints. An electricity generator hums away, the bluish glow of a laptop screen shines from a small table. So this is the sand archive of Gorham's Cave, with an estimated depth of 18 m. Deposition is not steady through time, Finlayson explains, some levels depositing more sediment than others. The layers cover the period from 127,000 to 2500 years ago, containing a wealth of animal and human remains and paraphernalia, the majority of which dates back to the time when Neanderthals took up camp here.

The layer currently under excavation takes us back 45,000 to 50,000 years. Finlayson describes the scene we see before us, illuminated by the work lights. In the sand lies a black ring marking the perimeter of the little campfire. The fire stones still lie on the ground. The Neanderthals had been working on a hand axe, as shown by the splinters of cut stone. Today the menu consists of roasted goat cheek, mussels and pine nuts. There must have been a robust mixture of smells in the cave: fossilised excrement reveals that after the meal, while the Neanderthals lay asleep in the cave or had left, hyenas came along in search of edible leftovers.

The sand in the cave contains remains from tools, small rodents' teeth, shells, sea urchin spines, vultures and eagles. Black feathers served in all likelihood as decorations and headdresses, the eagle's claw as jewellery. The Neanderthal's menu included rabbit, venison and bear, as well as young monk seals, dolphins, fish, pigeons and goats. The Neanderthals would eat them separately or mixed together, as ingredients of a prehistoric paella, often with clear markings from sharp stones with which the flesh had been scraped from the bones. The prehistoric meal's most conspicuous ingredient was bluefin tuna. The clearly recognisable vertebra of one of the largest and fastest fish was found next to the wood fires.

Initially it was a puzzling finding. You can tackle a deer with a spear, stone bears to death, lure a rhino into a hole, but how did the Neanderthal come by his bluefin tuna? How had this prehistoric man managed to get one of the largest and fastest fish out of the ocean and drag it up to his cave? To catch this fish you need nets, harpoons. Perhaps even boats.

If a boat or fishing net were suddenly to turn up in the excavation, that would have been too wonderful for words, Finlayson laughs. A Neanderthal tuna boat would be world news. Finlayson has always been an advocate of the hypothesis that Neanderthals were much more intelligent than many considered possible, but even he would have been rather surprised by a tuna boat. Wood from the boats and fibres from the nets would in any case long have rotted away, if Neanderthals ever did use

Fig. 2.3 The birth of Art. Is it a hashtag? Or tuna? (S. Finlayson, by courtesy of Gibraltar Museum)

them. The most obvious explanation is in fact already quite intelligent: Neanderthals patiently awaited the arrival of the bluefin tuna. 'Of course we don't know precisely what happened,' says Clive Finlayson. 'But everyone here on the coast knows the stories of how the tuna shot up the beach to escape the orcas. That used to be a regular occurrence on the shores here.'

In Finlayson's view there is no doubt: the shoals of tuna which skirted Gibraltar in prehistoric times must have been enormous, so the number of tuna shooting up the beach would have been commensurately higher. 'In the periods when Neanderthals lived here the climate fluctuated quite a bit. Sometimes icebergs would even have floated past. That mixture of cold and warm water caused an enormous explosion in plankton. The abundance of life and food here in the sea must have been spectacular.'

For Finlayson the discovery of the remains of tuna by the fire was an important step in the re-evaluation of Neanderthals. The varied menu of meat and fish is characteristic of an intelligent creature, one capable of using tools, probably speaking, making fire and using weapons and knives. They ate roast bluefin tuna, using toothpicks after the meal. 'The tuna proves that Neanderthals were far less stupid than was always assumed,' the museum director remarks with satisfaction. A number of other conclusions could be drawn: the Neanderthals ate well and that helped the species survive so long here in Gibraltar. Even those on the point of extinction could have kept going a long time yet with tuna (Fig. 2.3).

Behind the spot where the excavation team is at work, a metal stairway leads to the highest and deepest part of the cave. There, accessible via a metal footbridge, lie more treasures from the past. The greatest trophy was discovered in 2012 and is now protected by a steel door. On a stone ledge 13 lines have been scored painstakingly into the rock to create an image [208]. Anyone who sees the engraving automatically thinks it's a hashtag. The experts agree on a number of points: it dates back almost 40,000 years. Humans had not yet reached Gibraltar, so its authors must have been Neanderthals. The engraving serves no functional purpose, as would have been the case with unintentional scratching of the rock while slaughtering prey. It must therefore be a symbol, placed there by an individual capable of abstract thought.

A hashtag with a message, perhaps part of a ritual, perhaps a territorial marking. 'I call it art,' says Finlayson. He believes the hashtag in any case constitutes definitive proof that the Neanderthal was fundamentally not inferior to man in cognitive intelligence or capacity for thought. What the message behind the primitive abstraction meant, we will never know. Was it simply a symbol or a name? Or perhaps a schematic representation of an animal?

Yes, an animal. Because I immediately see a tuna fish. Head on the right, two dorsal fins on top, two belly fins and even the characteristic tail fin can be discerned in the prehistoric artwork if you look at it in the right way. Off course I know: this is just the compulsive interpretation of an obsessive tuna-lover.

But was it? Just before finishing this book, a group of archeologists published about an landmark discovery in the Blombos cave, situated at the southern coast of South Africa [133].

In the cave the scientist found what is believed to be the oldest abstract painting of early *Homo Sapiens* origin. Blombos has some remarkable resemblances with Gorham's cave in Gibraltar. A sand dune covers different layers that go back in time. The African time machine is explored from the end of the Middle Stone Age about 100,000 to 72,000 years ago. The engraving, found on the 73,000 years old level in the cave, was produced by carving deliberately a ground silcrete flake with a red ochre crayon. It consists of 'six straight sub-parallel lines crossed obliquely by three slightly curved lines'. It was immediately nicknamed as the Blombos hashtag.

Another hashtack, this time even older and created by the human competition of Neanderthal, before they decided to join them in Europe.

Of course it can be coincidental. But the inhabitants of both caves seem to have some things in common. They had both fish on the menu, judging the fishbones that where found. Among them tuna [241].

In the case of the Blombos cave it probably was the Southern bluefin tuna that torpedoed itself onshore escaping the orca's attacks. The ocean here is still an important crossroad on the tuna migratory track for the bluefin.

And consider this: if you suddenly felt the urge to engrave a piece of art for future generations to share one of the most overwhelming emotions you had in your lifetime, why not an abstract representation of that impressive fish that would pop-up out of the blue ocean onto the beach just to serve as the best meal you ever experienced?

The most important conclusion, in any case, remains: the Neanderthals were anything but stupid. Just like their early cousins in South Africa, they were able to eat tuna.

The sacred site of the tuna hashtag in Gorham's Cave continued to attract people over millennia, research shows. Thousands of years after the Neanderthals there were new residents, humans now, who left a bright red drawing on the walls. After this, an archaeological silence lasting millennia. Then all of a sudden a layer of ceramic shards surfaced, dated around 800 BCE, which can be traced back to the first Phoenicians to land on these coasts. For them the back of the cave was a ritual location, with a view of the surf backlit by the sun and full moon. Perhaps they too came across the mysterious Neanderthal hashtag, which contributed to the sacred character of the cave. 'We suspect they prayed to Tanit here,' says Finlayson. Tanit was the great mother goddess of Carthage, wife of the sacrificial god Ba`al. She ruled the sun, stars and moon and represented fertility on land and sea. Now largely forgotten, once the most popular goddess on the western side of the Mediterranean Sea.

By then the Neanderthals had long disappeared from the earth's surface. New inhabitants of the rock, like their predecessors, were attracted to this magical spot between the continents, with its abundance and mild climate. And like their prehistoric cousin, they came for the tuna.

Tuna and caves also remained closely connected after the Neanderthals. If we drive back from Gibraltar along the road to Cádiz, we pass the Sierra de la Plata, the 'Silver Mountain'. It is a modest mountain ridge that rises at the end of the beach of Zahara de los Atunes, 'Zahara of the tuna', 40 km west of Gibraltar as the crow flies. Here too, the rocks are like Swiss cheese, and again tuna has become part of the surroundings.

If I wanted to know more about bluefin tuna I had to visit a cave here, I was told by Luisa Isabel Álvarez de Toledo, 21st duchess of Medina Sidonia, descendant of one of the oldest noble families in Spain. No other aristocratic family was so closely connected with bluefin tuna. So a visit to the duchess was a must for anyone who wanted to find out more. A few months before her death in 2008 I paid her a visit at the ducal family palace in Sanlúcar de Barrameda. There the duchess explained to me where I could find this tuna cave: on the eastern side of a beach extending several kilometres from Barbate to Zahara de los Atunes.

The stretch of beach overlooked by the tuna cave is called Playa de los Alemanes, the 'beach of the Germans'. After the end of World War II many Nazis sought a safe haven here on this forgotten, inaccessible edge of Europe. They built their ample villas with red bougainvillea facing the magnificent view of the strait, giving them

Fig. 2.4   The Orca Cave, Atlanterra. Prehistoric watching post to see the tuna come by (S. Adolf)

names such as 'Mi último refugio', 'my last refuge'. The settlement was christened Atlanterra after the mythical Atlantis, which according to Plato must rest somewhere around here on the seabed. From the 1970s tourists also discovered the refuge with its view of Africa. The last plots up against the mountainside have now been sold by project developers to new owners. They build their design villas between the caves and on top of remains of structures dating back to antiquity.

Here in the surroundings of the Sierra de Plata, amateur-archaeologists have mapped out a couple of hundred of these kinds of caves, complete with prehistoric remains. Our cave is no more than a modest hole with a cross section of a couple of metres in the cliff, hollowed out by millennia of strong wind and rain. The wide opening offers a view over dozens of kilometres of coast and sea. The cave has no official name. Local researchers refer to it concisely as 'the orca cave'. Here, if you time it right, you have a view of passing orcas. A prominent indentation at the entrance of the cave looks like an orca's dorsal fin (Fig. 2.4).

The key to the cave gate was lost. Months of phone calls to the relevant offices of the Andalusian regional government remained fruitless. One office referred me to the next. The official who everyone was eventually sure must have the key turned out to be on sick leave, indefinitely. On closer inspection the padlock turned out to be rusted into place and it was fairly simple to climb over the fence. I was not the only one to have discovered this: a certain Dany and Pablo had inscribed their names on the back wall of the cave. Below that, the traces of the original cave inhabitants were clearly visible: sketches and signs, painted in a warm shade of terracotta produced by crushing up iron residues and diluting them with lard to make a red paste. After thousands of years the paint could no longer be wiped away and had become an indelible part of the rock.

The cave of Atlanterra must have been a sacred, ceremonial place where sacrifices were made and the gods worshipped. From an artistic perspective the rock drawings of the Orca Cave are not particularly impressive. No colossal images of wildlife, as in the caves of Altamira in the north of Spain. Instead, these were modest attempts to express something that was manifestly greater than everyday life, which the makers believed should be immortalised. There is an animal figure that looks like a deer, abstract signs—a cross, an arrow—and there are series of dots on the rock face, neatly ordered in a rectangle, like the sun loungers we can see lined up below on the beach as we look down from our eagle's nest. In some drawings the dots have red borders. Were they keeping a kind of calendar here by adding a dot for every full moon? Was it a tally of the surrounding tribes?

The duchess of Medina Sidonia stuck to the subject of giant tuna. She believes that one of the first instances of record keeping for the tuna catch was maintained here in the cave, she told me during our meeting at her family palace. Each dot represented a bluefin catch. The lines were nets. It reflected an early fishing method. 'These original coastal inhabitants left behind their method of catching tuna in the caves,' the Duchess declared with certainty. As is customary, her tone left no room for doubt or contradiction. Certainly not when it came to tuna, the fish which had brought her forefathers wealth and fame reaching far beyond the borders of Andalusia.

Not everyone took the tuna theory of an eccentric duchess seriously, but no one wanted a fight with the headstrong resident of the ducal palace in Sanlúcar. Many of her theories, expounded in self-published books, have been politely ignored by specialists. Nevertheless, this self-made historian commanded respect for the way in which she held the sceptre of her family archive, dating back centuries. It is the largest private historical archive in Spain, containing centuries of bluefin tuna history recorded in old manuscripts, and constitutes the first serious account of fishing in the world that has been meticulously maintained. Since the records relate to bluefin nets in fixed places on the Andalusian coast, the archive is also of immeasurable scientific value. Thanks to the persistent efforts of the duchess, the recording of centuries of bluefin tuna fishing has been saved from destruction and is still available to the outside world.

The orca cave looks out over the bay where the tuna came swimming in of old. Profiting from the three-dimensional pattern of in- and outgoing currents between the Atlantic Ocean and the Mediterranean Sea—currents in the middle of the strait, along the coasts, deep down and on the surface—the tuna allow themselves to be swept inwards. Preferably at high water, so that they can dive down into the depths if necessary. You have to be an oceanographer to understand the tuna's route. Orcas are naturally good at that. The family group of 30 orcas to be found here are experts in the swimming patterns of the fish and lie in wait in spring. You can see from the orcas appearing at the surface where the tuna is located in the water. Right under the cave there is a gathering point where they wait for the tuna. From this natural watch tower the coastal inhabitants had an excellent view of the sea. And for that reason many here share the belief that this cave and its wall carvings tell a story of orca and tuna.

The deer here in the orca cave was drawn 20,000 years ago. The more abstract symbols such as arrows, lines and dots date back 5000 or 6000 years. All that time the bluefin tuna determined the view. As they do today. In the coastal water below are the buoys which keep the traditional *almadraba* nets of Zahara de los Atunes afloat. In the twenty-first century the hunt for giant tuna still takes place in the location where it once began.

The Purple Folk's Tuna

3

The ships hastened westwards through the thin Mediterranean morning mist, sailing with a following wind, the water foaming, waves breaking over their prows. The 50 oarsmen in the slender *penteconter* kept up the momentum: bearded men, dressed in simple cotton tunics, stirred up by the thud of the ship's drum. In the hold it smelt of sweat and bitumen, the coal tar pitch with which they made their wooden ships watertight. With the Levante wind in the mainsail of this war and cargo ship, specially designed for long distances, they easily achieved a speed of nine knots (17 km per hour). Dolphins crossed their prow, playing with the big, fast ship they encountered en route. The sailors pressed on at night, navigating by the Pole Star which they discovered worked as a handy beacon in the Mediterranean sky. That meant gaining time on the way westwards, and time was money, especially if you wanted to be the first to land in Tartessus, the mysterious kingdom with its silver mines on the edge of the known world. There in the west, the Occident, was where the sun set between the Pillars of Hercules, and darkness started [52]. This was the land of Ereb, the Phoenician Princess Europa, mother of King Minos of Crete, abducted by Zeus in the guise of a white bull. In Europe's waters lived the *Leviathan* or *Thanin*, the seamonster from the deep. And actually this giant tuna was one of the reasons that they were heading for the West (Fig. 3.1).

The ships of the Phoenicians were famous, and sometimes feared. Elongated boats, swift, manoeuvrable and propelled forward by sails and oarsmen. Prow and stern featured horse's heads in honour of Poseidon, god of sea and earthquakes,

© Springer Nature Switzerland AG 2019
S. Adolf, *Tuna Wars*, https://doi.org/10.1007/978-3-030-20641-3_3

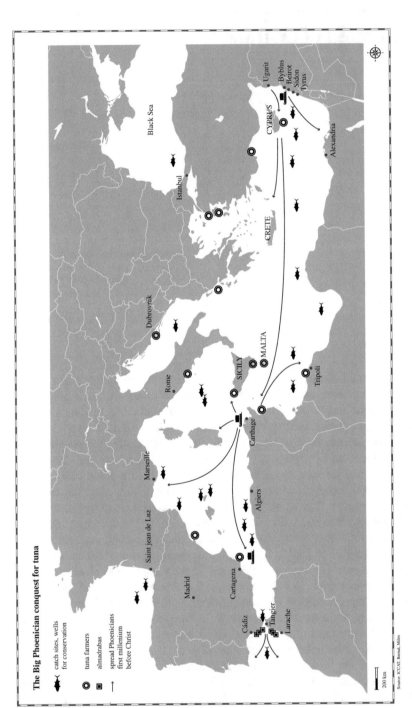

Fig. 3.1 The Big Phoenician conquest for tuna. From their base merchant cities in the Levant at the Eastern side of the Mediterranean, the Phoenicians spread out at the end of the second millennium BCE to the West to establish a network of new trade settlements. Their large-scale tuna fisheries and salting pits to conserve tuna travelled with them. Today ancient tuna salting pits are excavated next to the archaeological sites of Phoenician origin. At the same spots we find today's traditional almadraba's, large scale purse seine fisheries, modern tuna factories and tuna farms (infographic Ramses Reijerman)

protector of horses. The god with the trident, the weapon used to kill the captured tuna and haul it aboard. On the prow the feared ram was mounted on a long pole, with which opponents were driven aground. Courage was needed to undertake the journey to the Far West. The sea was full of dangers: pirates lay in wait and depending on the season, violent and unpredictable weather might prevail. New, unknown shores with unexpected shallows and sharp reefs could put a sudden end to the journey. They needed to be able to forage. Initially, the local inhabitants would not always have offered the newcomers from the east a friendly welcome, although they would soon come round, perhaps with the occasional brawl on the way: trade was not served by sabre-rattling or other troublemaking.

Odysseus preceded the Phoenicians on the route west on his return from the Trojan War. A storm had driven him far west beyond his home island of Ithaca to the mysterious land of the lotus-eaters. Legend had it that they too had once fled Troy, when the city was definitively destroyed in 1184 BCE and had settled on the coasts of the Levant.

Unlike Odysseus, the Phoenicians' goal was not to locate the island of the lotus-eaters or slay the man-eating cyclops Polyphemus. These were well-planned discovery expeditions to found new trade colonies with new wealth. These journeys were clearly different from plundering missions carried out by pirates, who raided local settlements and sold the inhabitants as slaves. This was about establishing lasting settlements to support their trade network, which was rapidly expanding westwards. Outright violence was applied if necessary when the local population put up a fight against the intruders. But in the Wild West of the Mediterranean Sea there was not all that much organised resistance to be expected from the local population and their trifling kingdoms. The territories generally did not yet form part of the surrounding empires with which the Phoenicians would normally have come into conflict. Some Greeks maybe, but no Egyptians, and in particular no Assyrians [168], the powerful neighbours in the Levant, who had to be paid off with forced levies of silver, iron ore and other goods. In exchange, the Phoenicians were permitted to enjoy the coastal cities in relative freedom. The Phoenicians themselves were in fact only moderately interested in imperial subjugation and conquest. Trade: that was their focus. Gradually the Phoenicians extended settlements westwards along the coasts of North Africa. Ipqy and Sabrata in Libya, Motya and Panormos in Sicily, Tharros in Sardinia, Rusadir in Algeria, Ebusos on Ibiza, Malaka in Andalusia, Tingis in Morocco, Gades on the Iberian coast opposite, and of course Carthage, near what is now Tunis, the Phoenician colony which was to grow into the metropolis of the powerful western trade empire and which would acquire its own status as a major autonomous power through its position as a 'Punic' satellite. Little is known about this time in which the trade posts were being established, the data is lost in the Mediterranean mist. The originally leading city of Tyre (in what is now Lebanon) was mentioned by the Egyptian pharaohs as early as the fourteenth century BCE. The Roman writers are certain that Cádiz, Utica and Lixus were founded two centuries later, but excavations confirming this antiquity remain elusive and it is still doubtful whether anything will ever be found of the beginning of these ancient cities.

We find ourselves aboard a Phoenician ship, probably somewhere around the start of the eighth century BCE. Our team of oarsmen come from Tyre. They are adventurers, entrepreneurs and go-getters. Their god is Melqart, also known as Baʻl Ṣūr, Lord of Tyre. They are familiar with the myth that their god already travelled to the far west in the guise of the adventurous Hercules, to mysterious locations such as the Garden of the Hesperides, in search of Atlas and the island of Erytheia. There, near the submerged city of Atlantis, two powerful mountains marked a strait which offered an entrance to a new, unknown ocean, where the seawater rose and fell each day. In spring the water's surface around these pillars of Melqart changed into a boiling current as the great tuna entered together in large shoals, often chased by the monstrous orcas. A couple of months later the fish returned to disappear into the enormous expanse of water behind the gates. That was the story. Now they were effectively chasing after their tuna, on their way to the exit of the Mediterranean Sea. Tartessus was the name of the largely unknown, barbarian kingdom beyond the strait, their desired destination. All the conditions for interesting trade were there for the picking. There were rich silver mines, salt could be obtained in the coves and there was a surplus of tuna. Speed was of the essence if they were to stay ahead of competition from their sister city of Sidon and be the first to establish their trade posts here. That was all the more true when it came to their arch rivals, the Greeks, who already had their greedy eyes on the western side of the Mediterranean. The early bird catches the worm. There was the prospect of a tidy fortune in the Wild West.

The Greeks regularly told stories of the kingdom of Tartessus, the mysterious country on the western edge of the ocean with its wealth of gold and silver. The Old Testament notes that Solomon, the last king of Israel, received a visit from a handful of trade ships every 3 years from 'Tarsis', loaded with gold, silver, ivory, monkeys and peacocks. Theologians quibble over whether Tarsis and Tartessus are the same place, but the historians agree that something must have existed between the Atlantic coast and the rivers now known as Guadiana and Guadalquivir, around the south-west side of present-day Andalusia. Silver and gold were supposedly picked up in Tartessus, which is certainly possible as the province of Huelva had had silver mines dating back to ancient times. However, there is surprisingly little left of the kingdom and its supposed wealth. Spanish academics have searched industriously for the capital city of Tarta, using aerial photos and satellite navigation. Tarta, which means cake in Spanish, may have been situated at the mouth of the Guadalquivir, where Sanlúcar de Barrameda is now located. Remains of a civilisation have been found: some smaller settlements, beautifully worked gold jewellery, small figurines and inscriptions in a kind of cuneiform script most resembling Berber. The only king whose existence is more or less accepted as historical fact is Arganthonio, 'man of silver', who must have lived sometime between the seventh and sixth centuries BCE. He is probably a mythical conflation of a number of local rulers. Arganthonio lived to be 150 years old, writes the Roman general and writer Pliny [196]. The Greek Herodotus offers a more modest 120 years. Arganthonio's death was the end of Tartessus. After the sixth century BCE the kingdom disappeared from the chronicles without a trace.

Considerably more is known about the Phoenicians. They were famous for being men and women of the sea, foreigners, wanderers: a trading people who felt very much at home on the water. They were often called 'Peleshet' or Philistines in the Bible and are supposed to have given their name to Palestine. Their Greek rivals called these inhabitants of the coasts of the Levante Phoinikers or Phoenicians, after the much sought-after purple dye for which they had a monopoly on production and trade. The Greeks were not keen on the Phoenicians. Perhaps that was why they picked a name which was directly reminiscent of the rotting process of the sea snail from which the exquisite purple colour was obtained. The harbour districts of the Phoenician cities must have reeked. The purple made them famous in the entire eastern Levant. The Phoenicians themselves preferred to call themselves the people of Canaan, that much was clear. They had settled in their city states of Ugarit, Byblos, Sidon and Tyre on the coasts which now belong to Syria, Lebanon and Israel. Despite mutual hostilities, the style of their ceramic work shows that they originated from the Greek people [1]. Sometime around 1200 BCE the surrounding major powers (early Greece, the Egypt of the pharaohs and Mesopotamia) were struck by a mysterious disaster (something biblical like a plague or flood) and suddenly suffered a dramatic loss of influence. The Phoenician trading cities profited from this vacuum and flourished. Trade and technical innovation were the basis of their power. They perfected the Greeks' boat construction. Their boats were slender, the innovation consisting in their being propelled forward by a double line of oarsmen, making the ships a good deal more seaworthy and faster. They discovered tar as a means of making ships watertight, and salting as a means of preserving fish. Formations of large boulders as anchor points are an indication that the Phoenicians had developed the fixed nets which were later perfected in the *almadraba* for catching the bluefin tuna [36].

Early on, the Phoenicians had to deal with an image problem that never disappeared. You need only look at films and television coverage. The Romans have *Spartacus*, *I, Claudius*, *Rome* and *Gladiator*. A film about the Greek Alexander the Great can always snag an actor of the calibre of Richard Burton or Collin Farrell. The Greek heroic epic about the Spartans taking up arms against the supremacy of the Persian invading army at Thermopylae (480 BCE) was turned into the muscle-bound Hollywood film *300*. But the Phoenicians? They have always been somewhat neglected. At best the scene dancing around the golden calf in Cecil B. DeMille's biblical epic *The Ten Commandments* (1956) might be seen as the imagining of a piece of Phoenician history that has made it into a film background, even then with negative connotations, depicting the fall of the Canaanites when faced with temptation.

So the Phoenicians are not cool. That's the way it often goes with peoples who have been conquered and disappeared. They liked trade and profit, according to the Greek Homer (approx. 800–750 BCE), and that was not a compliment coming from a Greek. As the author of the *Odyssey* Homer must have felt some affiliation with the seafaring people, but he showed little appreciation for their role as one of the founders of Western civilisation. Nevertheless the Phoenician cuneiform was the first alphabet to be broadly disseminated and formed the basis of the Greek script

[168]. Of course what also did not help was the line-up of gods maintained by the Phoenicians. Although the Romans smoothly took over a few of these gods and the Berber tribes in North Africa also saw the charm of them, to the Jews, Christians and later Muslims the Phoenician gods were tantamount to personifications of the devil. The sacrificial god Baʿal in particular, also known as Melqart in one of his guises, or Moloch in the Bible, was generally depicted as a terrifying man with a bull's head. In the Old Testament we find him called Ba-al-Zebub, the Lord of the Flies—none other than Beelzebub, Satan himself. Bible exegetes suspect that the Golden Calf of the Ten Commandments is in fact Baʿal, so there must have been something seductive about this sacrificial god of the Phoenicians. Not only did Moses' followers dance around his golden image; the Romans, too, had the greatest difficulty in knocking his worship on the head. Perhaps that was the reason for the terrifying stories—greedily spread by the Greeks—of Baʿal Hammon, who as Lord of Furnaces propagated purification by fire. It is true: Baʿal Hammon was not an easy god. He was not satisfied with a bit of chicken or tuna as a sacrifice. Baʿal demanded people, babies to be precise. As the story goes, the new-born offspring of prominent citizens would be thrown alive onto the fire to the sound of festive lute music to appease Baʿal Hammon. Carthaginian excavations of urns holding remains attributed to young children constitute strong evidence that such sacrifices did indeed take place, although we cannot tell whether they were living or dead children who were thrown into Baʿal's furnace. In any case, the image was striking enough to survive the millennia. Baʿal lived on in the Roman god Saturn, who in his Greek instantiation (Cronus) devoured his own children. The god of the Christians from the Old Testament, that of the Jews and the Muslims, in fact also preferred to leave the door open to child sacrifice, instructing Abraham/Ibrahim to sacrifice his child as proof of loyalty. As the prospective child murderer is on the point of cutting his son's throat he is rewarded by his god for his good behaviour: an angel descends and tells the unhappy Abraham that he can put a ram in the place of his son. Saved by the bell, so to speak.

Appreciation for the Phoenicians came only much later with the French historian Fernand Braudel (1902–1985). Braudel regarded the Mediterranean as the great breeding ground of culture, civilisation and trade; perhaps the most beautiful gift of history, in the Frenchman's view, having spent most of his life describing his Méditerranée with lyrical pen [46–51]. A unique universe of different peoples from Europe, Africa and Asia had arisen around this inland sea. The cross-pollination produced mixed cultures of exceptional quality. Romans in North Africa, Greeks in Sardinia, Celts on the French coast, Phoenicians and Berbers in Spain and North Africa, Turks in Yugoslavia: a crossroads of peoples, plants, animals, fish, religions, ideas and trade. Not that it could be described as a multicultural idyll: war was more the rule than the exception. The Greeks hated the Persians (more than the other way around), the Phoenicians the Greeks (and vice versa), the Romans wanted to destroy Carthage, the Syrians and Palestinians were petrified of the crusaders from the North, the Spaniards hated the Turks, and when there was no group left to hate, there were always the Jews. Nevertheless the lively trade brought exchange and interaction which became essential characteristics of Mediterranean culture. Imported products became so thoroughly assimilated that they were soon seen as 'typically Mediterranean'. The tomato was introduced by Spaniards from America, as were maize, cactuses and peppers; Arabs imported citrus trees from the Far East; cypresses came from Persia and eucalyptus from Australia. Whatever came with the trade caravans from the Sahara, whatever was contained in the cargo holds of the ships returning from the other side of the Ocean, Mediterranean culture made them its own.

In this mixed culture the bluefin tuna played a prominent role as a cosmopolitan fish that successfully overcame barriers and distances. Wherever the bluefin tuna surfaced, it conquered the market and the cuisine. In ancient times tuna fishing and production had already spread across the Mediterranean Sea to become a multinational, cross-border enterprise. On the one hand this was because of the salting technique, which represented an important innovation for prolonging the shelf life of fish [36, 55, 76]. Salted bluefin tuna appeared on markets from the north of Europe to deepest Africa and far into Asia. It also had to do with biological characteristics which made the bluefin tuna an odd man out in the Mediterranean. Compared with the Atlantic Ocean, where larger fish such as cod moved in enormous shoals, larger species of fish were far less abundant in the Mediterranean. If any fish were prevalent in huge numbers it was the smaller species, with large shoals of big fish generally absent. Apart from tuna. Many tens of thousands a day could pass the Mediterranean fishing areas during their migration. The tuna expert among the Greek natural scientists was the geographer and philosopher Strabo (64/63 BCE–25 CE). Born in what is now Northern Turkey, not far from the Black Sea, he had presumably seen with his own eyes that the young tuna came from that region. Once they were bigger, he noted, the fish swam back via the narrow strait at Byzantium, later Constantinople and Istanbul, towards the Mediterranean Sea [229]. There, in the Golden Horn Bay, they swam by in such closely packed shoals that the fishermen could scoop the still small bluefin tuna out of the sea with their bare hands. Later Strabo again described the tuna, this time the shoals of mature fish swimming into the western Mediterranean from the Atlantic Ocean. Now that the tuna were more than three metres long and weighed hundreds of kilos, scooping them out by hand was out of the question. Shoals of giant tuna entered the Mediterranean en masse like a herd of bison in the Wild West. That soon became a challenge to the fishermen. If they succeeded in catching such a shoal, the profit was enormous. The only thing tuna fishing could result in was a large-scale catch with thousands of kilos of fish. A bloody slaughter. Tuna fishing was war on the water. No wonder the important sea battle at Salamis (480 BCE), described by the Greek author Aeschylus in his tragedy *The Persians*, was compared with the catch of a shoal of tuna. 'The seawater was concealed under a floating carpet of bleeding bodies. The Greeks slaughtered the Persians, as if they were fishing for tuna. They struck at their bodies with clubs and pieces of wreckage, Aeschylus writes, having reputedly taken part in the battle himself [238]. Not unimportant detail was the fact that the Persian forces consisted of an important contingent of the Phoenician fleet, that was crushed by the Greek, in the same way like they liked to catch their tuna.

The two great innovations which made tuna fishing possible—clever net constructions to catch the fish and a salting technique to conserve the enormous quantity of fish afterwards—both came from the Phoenicians [36]. They combined their fishing techniques with another important innovation: Braudel regards the Phoenicians as one of the first peoples capable of organising geopolitical trade powers [46] as a world-spanning economy, a kind of global value chain, an approach which would only later be taken over by the Romans. This capability puts us strongly in mind of the later Italian city states such as Genoa and Venice, as well as the later major colonial powers such as Portugal, the Netherlands and Great Britain, powerful states supported by strong fleets, which succeeded in organising expansive economic powers. The state offered a guarantee of sufficient stability for trade and the economy to flourish and the trade routes were kept safe. Far more than classic empires with their military dominance, the power of the Phoenicians was organised around that of a multinational business and capitalism *avant la lettre*, Braudel effectively states. The Phoenicians built their society on family businesses. Their leaders and kings came from the trading elites. Tyrian purple was the linchpin of exclusive trade in a broad range of luxury products, from ceramics to furniture, metal-forging, woodwork and jewellery. They exchanged it for metals, ivory, gemstones and salt. Products from the sea were always closely bound with this maritime trade power, particularly the purple dye of their sea snail *murex* of course. It was only recently discovered that the Phoenicians also sold salt fish on a large scale in their amphorae [55, 84]. Tuna was an important core ingredient for this, especially bluefin tuna, the fish measuring several metres in length, which they caught with their ingenious nets. The Phoenicians understood the art of the catch and of salting tuna, which went considerably further than artisanal curing: they were the first people to process bluefin on an industrial scale. It was a large-scale source of protein for establishing their empire. In modern terms, fishing and trade in bluefin tuna could only sustainably flourish if supported by a cross-border partnership organisation bringing together state and enterprise. Tuna therefore became an indispensable component of their success.

'Where the Phoenicians were, there was tuna.' For Spanish archaeologist Darío Bernal there is no doubt about it. We drink coffee at a roadside restaurant near Barca de Vejer, a settlement at a crossroads on the provincial road linking the southern Spanish city of Cádiz with the rest of the coast. It does not amount to much more than a couple of restaurants, an old petrol station and a bus stop. A little further on is Barbate, the base port for what remains today of traditional bluefin tuna fishing. But the attentive passer-by sees more: the holes in the rocks give away the beam constructions for the houses and forts which stood here millennia ago. Towering above them is the 'white town' of Vejer, thousands of years old, on the steep ridge of the hill. Down below there must once have been a settlement beside the inland sea, with a ferry or bridge connecting the river banks. The name even might suggests a village of some importance: Barca was the name of the illustrious Phoenician family clan who had crossed over from Carthage and taken control of the Spanish south coast.

A select little club of tuna experts among Spain's historians and archaeologists can quibble over many of the details, but they agree on one thing: the tradition of a large-scale, industrial bluefin tuna catch is directly inherited from the Phoenicians. The river delta here, where flocks of storks forage in the tidal marsh, is a good example of this. The Phoenicians obtained the salt they used for conserving fish from the dry pans. The coast here is littered with remains of pits where the tuna was salted and ceramic kilns for the production of the amphorae in which the valuable fish was transported [198].

The Phoenician Tuna Economy

Far more than we knew until recently, the bluefin tuna was an important engine behind the classic prototypical trade economy which allowed us to discover and connect the world, from the old continent around the Mediterranean Sea to the New World of the American continent, Africa and Asia. The Phoenicians followed the great, spectacular fish westwards and constructed salting pits around their trading posts, that later where continued and amplified in the Roman era. The expansion of larger scale amphora producing furnaces in the area around Gadir (Cádiz) from the sixth century B.C.E. indicates that from the sixth BC indicates a large scale production of salted tuna. Together with the industrial scale fisheries, salt-production and large and complex production factories and shipyards they formed a basis for the first 'global value chain' of fish conserves. The amphorae containing bluefin tuna were exported on a large scale to the consumer markets in the Western world and what is now the Middle East. It was a whole millennium later when the Vikings and later the Hanseatic cities exploited their stockfish or salt cod on a comparable industrial scale. The fishing empire of the Phoenicians with their commercial family enterprises contributed to our modern world in the form of money, trade, entrepreneurial spirit and curiosity. That ambition to look beyond the borders and develop a new reality was certainly also driven by a fish which itself provided a prime example, crossing the oceans without borders and growing to become an icon of the world's seas.

The paradigm is that the Phoenicians went westwards after the bluefin tuna, bringing with them their fishing techniques to the coasts on both sides of the Strait of Gibraltar. 'They had a culture and lifestyle on a substantially higher level than what originally existed on the Iberian peninsula,' Bernal tells me. He is dressed in the uniform of an archaeologist on the way to work: the annual summer excavation in Baelo Claudia, the archaeological site of the tuna town half an hour's drive away. 'The problem of the role of the Phoenicians is hard scientific proof,' Bernal sighs. 'We don't have that. There are hardly any written Phoenician sources. There is no iconography from the ninth and eighth centuries and of course the organic material from the fish has long rotted away.' But there are plenty of other indications. The superior Phoenician boats for one. Their sophisticated fishing technique. The knowledge of fish salting. And the converse: the complete lack of any trace of large-scale tuna fishing or salting factories before their arrival.

There's more too. What is continually left unsaid by historians and writers who have immersed themselves in the misty past of the Phoenicians, is the remarkable similarity between the places where the Phoenician colonists travelling west founded their trade posts and the evident remains of the installations for conserving the tuna, features which have survived the millennia. On the coasts of Carthage, Tangier and Cádiz, the original salting pits still exist. As far as Lixus, the most important outpost on the Atlantic coast founded by the Phoenicians, not far from what is now Larache in Morocco, they can still be admired today. Most date back to a time when the Romans had taken control, but the origin of the tuna business was directly related to the founders of the cities. The tuna has long escaped the attention of historians, who prefer to focus on trade in purple dye, luxury goods and exploitation of silver and other metals. The tuna salting industry was generally completely disregarded. We can put that down to a blind spot in historical consciousness, because tuna fishing and preservation actually expanded to a large-scale, important sector of the Phoenician economy. Take the bronze replica of a common coin from Gades or Cádiz, dating back to the second century BCE. On the head side we see the god Melqart, adorned with a lion's head in the style of Hercules. Between the Phoenician inscriptions on the tail side are two clearly recognisable fish: bluefin tuna. By depicting tuna on their coins the Phoenicians signalled the crucial importance of the fish. Hercules in turn was the divine hero who watched over the tuna. Everywhere he went and carried out his work, the bluefin tuna was close by.

Only from the third century BCE does tuna begin to appear on those coins, archaeologist Bernal tells me. Before that there was not much small change. But what was only recently discovered from the early days of the Phoenician colonists are the heavy ballast stones which held the large fishing nets in place: it is quite likely that we are dealing with the *almadraba* of ancient times. In the oldest pits of the salting factories of Cádiz, the neatly cut remains of bluefin bones, from loose vertebrae to complete backbones, have been found along with other types of fish. On the outside, indentations can still be discerned where the flesh was cut away with knives. They date back to the fifth century BCE. Further evidence comes from the amphorae in which the salted tuna was traded by the Phoenicians. The ancient packaging is an important indicator of growing trade in the product from the sixth century to mass export around the second century BCE [36]. Trade and consumption spread over what was then the Western world in a manner remarkably similar to the later explosion of sushi tuna: from an exclusive luxury product for the elite classes, to a mass market product which took off on a large scale as a useful, healthy source of protein. 'The amphorae of salted tuna from Cádiz are found all over the Mediterranean, as far as Greece. We can recognise them from the shards,' says Bernal.

These are important findings for the history of tuna. We know that the Phoenicians set up the first industrial tuna businesses with cross-border production which covered the complete chain from fishing to processing to trade and transport. The bluefin catch took place in their innovative, complex fixed nets, from which the *almadrabas* were developed. The processing by salting in the tuna pits took place on an industrial scale in special buildings. They transported the end product overseas in amphorae in their sophisticated cargo boats via regular ports and sold them in the final consumer market through a well developed trade and distribution network. In this way the Phoenicians were the founders of global tuna value chains which have survived centuries of antiquity and conquered the modern world. It was nothing less than a revolution. All conditions for modern global tuna chains were brought together by the Phoenicians: innovative large-scale fishing methods, organised industrial preservation and efficient bulk transport. And above all: a spirit of trade without borders.

Tuna City of Lixus

How the Phoenicians continued their journey beyond the Pillars of Hercules remains the subject of discussion. Some people believe that they reached France via the Atlantic Ocean and even landed in Britain, where they found the rich tin mines which supplied the material for bronze alloy. They may even

(continued)

have gone as far as the Azores. Some believe the Phoenicians even reached America. There is little evidence for the latter claim, but an expedition by seafarer Hanno in the sixth or fifth century BCE is suspected to have made it to somewhere around Cameroon. There Hanno killed three female apes, probably chimpanzees, whose skins remained preserved for centuries in a temple in Carthage [168]. The northern parts of the Atlantic coasts of Africa were in any case used to found settlements. The Phoenicians populated the coast as far as Salé and Essaouira in Morocco. In Lixus, on the northern side of what is now the Moroccan town of Larache, lie the ruins of what must once have been an impressive Phoenician city, still largely unexcavated, near a hill. If you didn't know it, you could easily drive on past the archaeological excavations by the river Loukkos, which bends sharply a little further on to cover the final couple of kilometres to the ocean. A small signpost announces that Lixus is located here, a city founded by the Phoenicians, later to become the southwestern border of the Roman Empire after the fall of Carthage.

El Mokhtar el Hannach, an elderly, frail man, acts as a tourist guide here. There are not that many visitors to this UNESCO world heritage site. '*Bonjour, bongiorno, buenos dias*, good morning,' he routinely greets his visitors. Guiding is in El Hannach's blood. From 1925 his father Tajib el Hannach worked as a guide for the Spaniards. His son Jallal is to continue the family tradition. 'God willing, in any case,' says El Hannach. This friendly, modest man is the only living mortal who can honestly claim to have been a resident of Lixus. His family lived at the bottom of the hill, where a house had been built under Spanish colonial rule so that his father could keep watch and protect the archaeological excavations from plunderers.

'You're looking for tuna?' Smoothly and dexterously avoiding the stones, El Mokthar el Hannach races ahead of us to the right along the hill. There they lie in neat formation, the square pits where the fish were salted, amongst weeds and shrubs, but still amazingly intact: concrete from antiquity, overgrown with orange lichen. Remains of a factory building with installations for cleaning the fish. And still more pits. More than 150, El Hannach tells us. This was the city's factory district, built by the Phoenicians and later further expanded under the Romans. Bluefin tuna, bonito and mackerel were preserved here. The pans for salt extraction lie a little further on by the meandering river in the valley. They are still in use. 'The scale of tuna preservation, you don't see anything like that these days,' El Hannach sighs. From the 1940s to the 1960s the Spanish Crespo family still had two *almadrabas* for the tuna here, but Moroccan independence soon put an end to that.

El Hannach sprints up the hill and points out the places where piles of stone mark what must once have been a building. This was where the Phoenicians had their olive oil factory, and over there lie the remains of the amphitheatre. The area offers a vast panorama of the river delta, beach and sea. 'This

(continued)

entrance was for slaves,' says El Hannach, pointing to the wall enclosing the arena. 'And the one next to it was for lions.' Bordering on the theatre at knee height lie the remains of the wall of what must once have been the temple of Poseidon. Higher still is the temple of Ba'al Melqart. Here we find the marble sacrificial blocks, completely intact. 'Children were sacrificed here,' says El Hannah with visible revulsion. 'Islam did well to put an end to those practices.'

With rapid steps he runs down between the weeds, shrubs and goat droppings with which ancient Lixus is covered. The guide is not particularly satisfied with the state of his tuna city. The house in which he grew up was pulled down in the early 1990s, because it was deemed to stand in the way of archaeological research, yet only a fraction of the city has been excavated even now. Of all the buildings it is Lixus's tuna factories that have best stood the test of time.

Three Classic Tuna Wars

4

Tuna was money, big money. Money was power. Tuna also had a direct military strategic importance. Salted tuna had a long shelf life, making it a practical food on board ships and for the army on land. So it's not surprising that tuna amphorae were found where armed conflicts had taken place. Those conflicts were to result in the greatest wars of ancient times.

Barca, *Baraq* in Phoenician and Hebrew, means lightning. A force of nature that can strike swiftly. There is no doubt that the Barca family belonged to the Carthaginian elite class. They almost certainly obtained part of their fortune from fishing and trade in bluefin tuna. But that is not the reason Hannibal Barca, the most famous descendant of the line, is the only Phoenician to have remained in the collective memory of history. In the second half of the first millennium BCE, Carthage had grown to be the most important metropolis in the west of the Mediterranean. The Phoenicians had established their hegemony in the area, after their *penteconters* had vanquished a Greek colonial fleet in the Sea of Sardinia around 540 BCE. This impeded their Greek arch rivals from establishing large settlements in the south of Italy and Sicily. With their colony of Massalia (Marseille) and a number of posts on Sicily, the Greeks already had too much of a presence in the western markets where the Phoenicians had set up their trade networks. That had to be stopped. Carthage emerged victorious from the sea battle with the Greeks, but it was also the starting shot of a continuous battle between Phoenicians and Greeks on Sicily. In 264 BCE the chaotic accumulation of skirmishes on the peninsula led to the intervention of the region's expanding major power, the Republic of Rome. The Romans were growing nervous about the rumblings in their back garden. Carthage threatened to win supremacy over Sicily.

The battle between Rome and Carthage, the two major military and economic powers, would go down in history as the three Punic Wars which held the western Mediterranean Sea in their grip for almost 120 years [168]. They were undeniably the world wars of antiquity. Never had such enormous and horrific conflicts taken place on land or sea, involving so many different peoples from around the Mediterranean for such a long time. Hundreds of thousands of fighters died in the violence. To give an idea

© Springer Nature Switzerland AG 2019

S. Adolf, *Tuna Wars*, https://doi.org/10.1007/978-3-030-20641-3_4

of the magnitude, the Greek historian Polybius counted almost 600 warships in the 256 BCE sea battle at Eknomos in western Sicily. War historians see it as one of the greatest sea battles in history. It was a war between two city states of epic proportions. Here, world dominion was determined for the coming centuries, for the economy and for communities in Europe. The manufacturing of countless products, along with trade routes and raw materials, was at stake, and with it supremacy over lucrative tuna fishing which could be found everywhere in the disputed territories. The loser threatened to be erased from the map on all fronts, including that of tuna. This made the Punic Wars also the first major 'tuna wars' in world history. One of the issues at stake was supremacy over the bluefin tuna industry.

With their superior fleet and naval power, Carthage as an operational base and Cádiz as a trade metropolis, the Phoenicians had the upper hand to start with, consolidated by the monopoly on the tuna catch on the western coasts of the Mediterranean and the Atlantic. The first war of 264 to 241 BCE began with an attack by a Carthaginian fleet on the pirate base of Messina in Sicily. What appeared to begin as an incident to teach a few sea wolves a lesson grew into a war after Rome decided to tackle the military presence of the Carthaginians, using it as an excuse to conquer all of Sicily. The result was a sea war in which Rome landed on the African coast and attempted to take Carthage. They failed, but Carthage lost its colonies along with the salting industry on Sicily and Sardinia. No more Sicilian tuna for Carthage.

Carthage had a naval power Rome could not hope to emulate. The Romans seemed to lose out due to their outmoded ships and lack of seafaring skills. However, the war took a new turn when the Romans realised the importance of the fleet as a part of their military power and decided to create their own navy. The approach was as practical as it was inventive: a few captured Phoenician ships were imitated in every detail and at surprising speed. The Romans were quick to achieve greater success with the copied fleet. The first Punic War ended with their victory. The Phoenicians might have lost, but General Hamilcar Barca emerged from the battle as a legendary figure. As the general of western Sicily he had led the Punic troops in a successful battle against the Romans. As a war hero he became a negotiator around the peace treaty with Rome. After the war, in 237, he decided to leave Carthage for Cádiz, along with his nine-year-old son Hannibal. The southern Spanish coast could offer a lucrative alternative in compensation for the losses in trade and tuna in Sicily and Sardinia [1].

The journey, stretching hundreds of kilometres, must have been hard for little Hannibal. On the way they would undoubtedly have eaten salted tuna and probably treated their wounds and blisters with *garum*, the fermented fish sauce to which healing powers were attributed. But it was worth the effort: once they had landed on the Spanish south coast the Barca family successfully founded a colonial trade empire which quickly gained the status of a separate regional power, supported by the city states of Carthage and Cádiz. Hamilcar entered into alliances with the local Iberian inhabitants on the southern coasts of the Spanish peninsula. He gradually built a land army of more than 50,000 soldiers. He founded new trade colonies far along the Spanish east coast near to what is now Alicante. His successor and son-in-law Hasdrubal established Qart Hadasht, 'New Carthage' or Cartagena, the Spanish city where the remains of the military city wall, built by the Punic colonists, can still be visited. Cartagena served as a bridgehead with Carthage on the other side. In addition to mines for silver, lead and zinc, the city had an important salt extraction area for tuna. There were six tuna factories, four kilns for amphorae and six tuna ports. The silver trade may have remained the backbone of the Barcas' fortune, but the rapidly expanding production and trade in preserved tuna soon gained in importance.

The humiliation of Carthage after the first Punic War had been serious. Not only did the country lose its settlements on Sicily and Sardinia; the peace treaty also stipulated a war tribute of 66 t of silver, which the Carthaginians had to pay within 10 years, with the clear intention of holding the economy of Carthage hostage. As with the Treaty of Versailles after World War I for Germany, in Carthage the seed was planted for a second war against Rome. Legend has it that before departing for Cádiz Hamilcar made a sacrifice to the god Ba'al at the temple in Carthage. With his hand on the sacrificial animal, Hannibal was made to swear revenge against Rome [168]. His family's wealth, his father's military vision and the strategic base in the south of the Iberian peninsula all pointed to Hannibal Barca as angel of revenge. Even his name committed him: in Phoenician Hanni-ba'al means 'he who has Ba'al's blessing'.

The story of Hannibal in the Second Punic War (218–201 BCE) has survived the millennia and continues to appeal to the imagination due to its daring and inventiveness, not to mention the elephants in his army, 40 or 60—academics dispute the exact number—with which he crossed the high passes over the Alps. At the outbreak of the Second Punic War Hannibal had succeeded his father and brother-in-law as the leader of the military force formed under the rule of the Barcas in what is now southern Spain. The conflict with the Romans started with Hannibal, on the Spanish east coast in the border region between the Roman sphere of influence in the north and the southern part of the Spanish coast ruled by the Barca clan. Hannibal used Cartagena as his operational base. He launched his attack in 219 by conquering the Iberian city of Saguntum (now Sagunto in Valencia), which was allied with Rome. The taking of the city led to diplomatic protests from Rome to Carthage. The issue became more serious when Hannibal subsequently ordered his army, elephants and all, to cross the Ebro river. This was the beginning of one of the most ambitious invasions in military history. The aim was to cross the Pyrenees, then the Alps, and then onwards towards Rome with those elephants that had survived the cold. The city state was under direct attack, and on land, where the Romans traditionally held military sway. The Roman army sustained heavy losses practically on its own doorstep. The Roman general Scipio struck back by driving the people of Cartagena from their city on the Iberian peninsula and besieging Carthage in North Africa with the help of the local Berber armies. That proved successful. Once again Rome defeated Carthage. The city state was condemned to pay immense tributes, the military fleet was banned and Carthage was no longer permitted to conduct its own politics without Rome's permission.

The victory was complete, leaving no room for misunderstanding as to who was now the boss in the region. Nevertheless, this did not reduce the Romans' deep mistrust. The trade city seemed to have displayed remarkable resilience. Despite the immense war tributes, the Carthaginians had proven themselves capable of maintaining their trade positions and generating prosperity. After the Second Punic War, in contrast with Rome, Carthage had no expenses financing a costly army and navy. That made a considerable difference. In 20 years Carthage had succeeded in paying off its swingeing war tribute. Once liberated from that burden, the economy really took off. The recovery was paralleled by explosive growth in tuna trade and consumption, as the fish had changed from a luxury product into something everyone could afford. At the entrance to the famous rounded military port, now useless, a big new trade port was established. Carthage rose from the ashes due to its traditional strength: a powerful empire which continued to dominate key economic trade routes. Their prosperity was so great that in order to appease their conquerors they took the initiative in making large gifts of grain to support the Romans in their wars to the east. The gesture might have come in handy, but it also fed into Rome's envy, mistrust and greed. The Phoenician trade empire had to be destroyed, its wealth confiscated. Carthage became synonymous with treachery and danger. 'By the way, I am still of the opinion that Carthage must be destroyed,' the elderly Roman senator Marcus Porcius Cato (234–149) invariably concluded his speeches, regardless of the subject of discussion. As a statesman and former soldier who had earned his laurels in Spain, Cato had witnessed with his own eyes Carthage's success in expanding to become a prosperous trade community even after losing the Second Punic War. Luxury, that was something Cato could not stand in any case, let alone the luxury that prevailed again in defeated Carthage.

With his harsh, forthright and often reactionary ideas, Cato may have been a misfit and a buffoon in the Senate. As far as is known there was no direct motive behind his constant hammering on about the spectre of a Carthage risen from the ashes and reappearing at the gates of Rome. But his persistent polemics eventually bore fruit.

The Third Punic War (149–146 BCE) was really no more than an ordinary massacre. Citing the excuse that Carthage had been looking for a fight with their former Berber allies, Rome started a siege of the city. After 3 years of isolation Carthage was occupied by the Romans under the leadership of General Scipio Africanus in a manner quite unprecedented in history. The streets of the trade metropolis turned into a gruesome hell of murdering and plundering Roman troops. Tens of thousands of citizens were slaughtered, the remaining 50,000 carted off as slaves. The city was broken up and basically razed to the ground. Carthage was no more [168]. Even the tuna factories vanished.

The destruction was an act of revenge which got out of hand under the influence of the grim hate propaganda of a Senate ex-consul. The Romans deliberately made an example of Carthage, showcasing the fate of those who resisted their power. But they also clearly feared the resilience and opposition of an empire that might have been militarily disabled, but as a trading power still had access to an unprecedented economic engine in the Mediterranean Sea. An example had to be set. By literally breaking up the city the Romans attempted to wipe from the face of the earth the memory of Carthage, the city which had come within a hair's breadth of beating them and continued to sparkle even in defeat. The works of art in their temples, their silver and jewellery and the content of their richly stocked warehouses were carefully loaded up and taken on board the ships [168]. From then on it was Rome that would rule Mare Nostrum, dominating the trade routes and the fishing areas and their salting industry for the bluefin tuna.

Gadir, the Punic settlement on the other side of the water, in time choose the winning side and thus escaped the commotion. After the fall of Carthage the city grew to be an unconditional, loyal ally of the Romans. The merchants could be seen in the capital city of the new empire, as could the bankers, who succeeded in acquiring a key position through loans to the Roman financial elite. In gratitude for their loyalty and credit, Julius Caesar even awarded Gadir the official predicate of *civitas federata*, a city allied with Rome.

Thunnus Nostrum

5

The tuna is exceptionally large. I have heard of one weighing more than 800 pound. . .

Gaius Plinius Secundus (23–79)

Abundant and wonderful is the catch of the fishermen when the army of tuna sets out in spring.

Oppian, Halieutica, second century BCE

Quintus Pupius Urbicus could afford to be satisfied. Having been named governor of the town of Baelo Claudia in the province of Baetica, Hispania Ulterior, he was a source of pride to his parents. They had a marble altar stone specially chiselled for their Urbicus so that everyone could read it: 'For Quintus Pupius Urbicus, beloved son of the family of Galeria, co-governor of Baelo Claudia. From your father, Quintus Pupius Genetivus and your mother, Innia Eleuthera.'

There it stood, chiselled for eternity in the little white marble altar, easily legible for anyone who rode into the city through the eastern gate at the forum. We find ourselves now in the second century CE in Baelo Claudia, city of Ba'al and the Roman Emperor Claudius. With a few tens of thousands of inhabitants this was a hub of industrious activity on the southern edge of the Iberian peninsula. It was the boarding point for the ferry to Tingis, Tangier, which is visible on the other side of the glistening water. Moreover, it possessed a substantial industrial district for salting fish, particularly tuna. Bluefin tuna was handed from boats onto the wooden beach jetties in the bay where the fish was caught. Once salted, the fish was transported in amphorae onto the boats to be taken to the farthest corners of the empire and probably far beyond.

More than 2000 years on, Baelo Claudia provides a window on the flourishing of the tuna industry under Roman rule. We know relatively little about co-governor Quintus Pupius Urbicus. What did he think of the little altar his parents had erected? Did they make coarse jokes about it at the forum or among the fishwives on the market a little further on? It is not inconceivable. In any case, the marble slab

© Springer Nature Switzerland AG 2019
S. Adolf, *Tuna Wars*, https://doi.org/10.1007/978-3-030-20641-3_5

remained in place, even centuries after Quintus Pupius Urbicus had disappeared and the walls around the forum had collapsed. The co-governor had enough reason to be satisfied in life in any case. The tuna that was prepared in the big factory complexes which housed his salting pits was famous to the farthest corners of the Roman Empire. Tuna from Belon, as Baelo Claudia was called for short, salted and packaged in amphorae, was exported to ports all over Mare Nostrum. The brand guaranteed the highest quality. *Garum*, the fermented fish sauce prepared from tuna blood and fish remains, found its way to Roman kitchens and medicine cabinets for astronomical sums. The tuna had brought wealth and prosperity to Baelo Claudia and the governor received his share of respect and authority. When Quintus Pupius ascended the steps of the temples of the Capitol, it was to show his gratitude to the gods.

The archaeological excavations of Baelo Claudia (Fig. 5.1) offer a good insight into how a Roman city revolved around tuna. Although lacking the monumental glory of cities such as Itálica near Seville, Emérita near Merida, Corduba (Cordoba) or Tarraco (Tarragona), this industrial fishing town has remained well preserved. The magnificent bay of Bolonia, as the little hamlet populated by fishermen and surfers is known today, remained practically untouched for centuries, complete with yellow-white sand dunes and cauliflower-shaped Mediterranean pines rising bright green against the clear blue of the sky. This view has survived the millennia, as have the nets of the *almadrabas* which can still be found along the coast here in the tuna season. From the three temples dedicated to Minerva, Jupiter and Juno the governor had a good view over the lower-lying town, the market square with the covered shop stalls. To his right lay the large amphitheatre, and below it the thermal bath houses. Lower down, towards the beach, lies the forum. Behind it was the basilica, where the bust of Emperor Trajan looked down between the pillars on the meetings of the tribunal. Out of sight behind the city wall were the fish factories with their tuna pits,

Fig. 5.1 Baelo Claudia, the smelly tuna pits of the industrial quarter (S. Adolf)

against the panoramic backdrop of the bay where the nets had been set out. In the courtyards of the spacious factory buildings stood tables for removing the tuna flesh, processing it and preparing the salting sauces. Of course you could buy the end product, the salted tuna, here, fresh from the pit as it were, from the shop of Decumanus Maximus in the market or the sales counter of the Barcino fish factories outside the city wall. It made a nice present if you were in transit and happened to have to wait in Baelo Claudia for the ferry to Tingis at times when the fierce Levante wind made the crossing impossible. Baelo Claudia was a provincial town, but not without ambitions. That was clear from its entire structure, planned according to classical Roman principles, with an efficient administration under direct jurisdiction of the empire. The spirit of Emperor Trajan (52–98), the first Iberian Caesar of Rome, born in Itálica, kept watch to ensure that justice was upheld. Gades, as the Romans called Cádiz, was a day's journey on horseback. Plenty of money was earned there by the bankers and merchants with their supplies and loans to the Roman elite. Baelo Claudia was where the fish were caught. The town was at the start of the tuna value chain. Out of more than 100 settlements with fish factories beyond the Pillars of Hercules, Baelo Claudia received special mention in the Greek geographer Strabo's travelogue. 'After that comes Melaria (Tarifa), which has salting factories, and then the city and the river Belon. Belon is the port where people normally embark for Tingis in Mauretania. It is a merchant town and has salting factories for the fish' [229].

Tingis was the capital of the Roman province of Mauretania Tingitana, still visible in the distance directly opposite, to the right of an extensive beach delta with apartment blocks which now forms the beach boulevard of Tangier. At the time a watch tower must have been visible, just next to the salting factory, which was used during the tuna migration to spot the large shoals along the coast in good time. To the right of the bay of Tingis lies the hill which ends in Cape Spartel, the western point at which the African continent abruptly disappears into the vast nothingness of the Atlantic Ocean. There lie the caves named after Hercules. Tuna has swum past here in large shoals every spring since time immemorial, funnelling into the Strait of Gibraltar and past the Pillars of Hercules, into Mare Nostrum. And there lay the *almadrabas* of Baelo Claudia and Tingis ready to catch the shoals of bluefin. With a slight stretch of the imagination, the Romans could have called it *Thunnus nostrum*, our tuna.

Between the city wall and the beach, where the tuna lay fermenting between layers of salt in the pits of the preserving factories, the stench was terrible. There was little point in complaining. Everyone knew where Baelo Claudia's money came from. In the forum or at the Capitol temple it would not normally be so noticeable. The prevailing wind in Baelo came from the east or west, as it does today, so the stench blew away along the beaches. The people most bothered, perhaps, were the prostitutes who hired little rooms beside the fish factory, but many of their clients were fishermen, and they would have been used to it.

Baelo Claudia must have begun as a humble settlement on the coast at the end of the second century BCE, after the fall of Carthage [109]. Garcia Munoz The south of Spain had been definitively placed under Roman rule, but the heritage of the

Phoenician Barca clan was clearly present everywhere. This town is named after none other than the god Ba'al. One of the temples in the city was of course dedicated to the Phoenician god; the Romans were always quite liberal when it came to worshipping different gods. Caesar Augustus (27–14), who was familiar with Spain from the campaigns he had undertaken under Julius Cesar, was the first to see the strategic opportunities of the town. He gave the order to expand Baelo into a real city, complete with a theatre for 3000 men, a forum and basilica, temples, three aqueducts, bath houses and city walls to keep the bands of pirates from Mauretania out [42]. The new rulers took over the tuna catch and processing, probably including personnel. Under the regime of the Emperor Claudius (10 BCE–54 CE, grandson of Augustus), sometime between 50 and 60 CE, the city was destroyed. It is unclear what happened. Speculation often involves mention of an earthquake, although a tsunami seems more likely: the seaquake of 1755, which destroyed much of Lisbon, also wiped out a number of the tuna towns in the southern Spanish region with an immense tidal wave. The silver lining to this cloud was that the destruction took place under the reign of Claudius, maybe the greatest bluefin tuna emperor. The man who was dismissed by his opponents as a stuttering, slobbering idiot proved to possess the intelligence, strategic insight and organisational capabilities required to get the bluefin tuna industry back on its feet. Perhaps he saw the importance of the salted tuna as a source of protein with a long shelf life, to feed the armies. Perhaps it was his preference for ambitious building projects. Or perhaps Claudius just liked tuna. In any case, like his grandfather, the emperor recognised the strategic importance of the tuna industry and had the city restored, including the preserving factories outside the city gates.

It was the beginning of a new period of economic prosperity for Baelo Claudia, which proudly bore the name of Claudius from then on out of gratitude to the tuna emperor. Tuna fishing and processing had meanwhile been further organised to become a complete industrial sector, with several factory buildings where large quantities of fish could be preserved together in neatly symmetrical built-in salting pits. The tuna fisheries that started under Phoenican rule now truly expanded on a industrial large scale, with real production volumes of conserved fish, archaeologist Dario Bernal explains [36]. Bernal Garum. 'Just take the tuna pits, the *piletas* here in Baelo Claudia. They are two by two and two metres deep. That's a volume of eight cubic metres, five of which would be fish. Then we're soon talking about several thousand kilos per pit.' A factory would have 10–15 pits, providing considerable production and trade volumes.

'The first and second centuries BCE were the absolute pinnacle of the tuna industry in Baelo Claudia,' says Bernal. Like the earlier Phoenician period, this can be observed from the many amphorae found, with their clearly recognisable markings. The volume of salt production in the area is another indication. As in Baelo Claudia, tuna became a massive fishing industry around the whole Mediterranean, with a concentration of production sites around the entrance of Mare Nostrum. Tuna had decisively developed from a luxury product to a commodity for large-scale consumption, at a price that made the food affordable to a broad range of city residents. There was also a clear military interest in tuna production. Salting gave the fish a long shelf life and it was easy to transport, ideal for reserve stock or on journeys and campaigns from Britain to Tingitania. Wherever war was waged, we find amphorae with salted tuna from Cádiz. 'It was the canned tuna of ancient times,' says Bernal. Really the only difference was that the packaging was not quite as easy to shove into a kit bag as today's cans, as an amphora contained at least 15–20 kg of salted fish.

Delving into the physical remains of tuna has its limits in archaeology, Bernal explains. The salting process might have allowed the tuna to keep for longer, but the preservation technique was clearly not intended to bridge a period of 2000 years. Little remains of the organic material from the fish other than backbones, loose vertebrae and skulls. So there was considerable excitement among archaeologists when, in recent excavations, the remains of tuna really were found in a layer of salt on the bottom of the salting pits [39]. 'The organic material rots away, but not if the skin is still in place. Then it remains preserved in the salt. That was what we found on the bottom of a pit: tuna fins, skin and even some remaining meat.' They were authentic Roman *tuna loins,* cut from the fish they had just caught. The tuna head was sometimes cut off in the boats so that the fish could bleed out. The rest of the filleting took place on the beach or the paved areas of the factories. The technique was easily recognisable by the cuts found on the vertebrae. The way in which the head and tail had been separated and the loins cut loose from the body seems to differ little from the techniques still used today in the nearby ports where the *almadraba* tuna are brought ashore [36]. The tuna was salted in the pits which were dug out and lined with *opus signinum,* the Roman hydraulic concrete. Seeing it after two millennia, the quality and finish of the material still inspire admiration. The numerous salting pits can be seen in many locations around the western Mediterranean Sea and on the Atlantic coasts. Based on the amphorae found, neatly marked with the brand and specification of contents, we can conclude that the Romans started up their salting factories and traded tuna products wherever there was a coastal settlement of any significance. After their victory over Egypt, gaining them access to the Red Sea, trade missions brought the amphorae of Roman tuna as far as the Bay of Bengal.

Tuna, the Classic Cuisine

'I recommend that visitors to the holy city of Byzantium in the high season try tuna steak; the flesh is tender and has an excellent flavour,' Archestratus, Greek poet of Sicily, foody and fish lover of the classical poets, sang the praises of the grilled slices of giant tuna as far back as the fourth century BCE. The tuna of Sicily might have been good, but that of the Bosporus really was a must, claimed the author of culinary poetry specialising in fish. Many centuries later the historian Aelian (170–235 CE) mentioned tuna as the fish by which you could recognise the elite on the island of Rhodes. A fatty tuna was far and away the best and finest fish available.

(continued)

The name 'tuna' comes from the Greek *thunnos*, derived from the verb *thuo*, which means 'to dart' or 'to rush'. Clearly even in ancient times the tuna's speed did not go unnoticed.

The Greeks loved tuna. Poets dedicated their verse to the fish, scientists described them in their reference works. The tuna was caught and boiled, roasted or salted. Its flavour was widely praised. Tuna was an addiction for those who could afford it, initially mainly the elite. Satirical poet Hipponax of Ephesus of the sixth century BCE wrote a famous poem about a rich man who burned through his entire inheritance pigging out on tuna. What makes the tuna irresistibly delicious, according to Hipponax's description, is a sauce made from cheese, honey and garlic. The Ancient Greeks, like the Japanese and the Spanish today, showed a preference for the fatty belly cuts of the tuna, and, like the Japanese, they considered the head a delicacy.

Tuna had status. It was the only fish used as a sacrifice to Poseidon, god of the sea, horses and earthquakes, the one you had to keep happy to avoid disaster. What could be a better sacrifice than this giant fish with its dark red, iron smelling blood?

The Greek traveller Pausanias (110–180 CE) notes that the inhabitants of Corfu had erected an enormous bronze bull in honour of Poseidon, commemorating a bull escaping and leading the inhabitants to the beach, where a large shoal of tuna was just swimming by. At festivals dedicated to Poseidon the first tuna of the season was given to the god of the sea.

Greek vases picturing fishing boys and the impressive Roman floor mosaics remind us of the wealth of fish in antiquity. Images remain of moray eels, sole and stone bass, sardines, anchovies, gilthead bream, conger eel, angler and John Dory, scorpionfish, ray, sea lamprey and mackerel, as well as octopus and squid, whelks, mussels and oysters, lobsters, crabs, prawns and langoustines. In the relatively monotonous cuisine of the Ancient Greeks, who mainly stuck to cereal products and a bit of fruit, fish was the highlight. At the end of the second or beginning of the third century Athenaeus was the authority on culinary culture, fish and literature. His *Deipnosophistae* [224] is seen as the classical reference work on the 'dinner philosophers'. And according to Athenaeus the true gourmand was to be found on the market among the fish sellers: 'It is no wonder, dear friends, that of all dishes we call delicacies, fish is the only one which truly appeals, due to its excellent nutritional quality and because people love it.'

The Greeks ate tuna in various dishes, according to culinary archaeologist Daniel Levine of the University of Arkansas [158]. Grilled tuna loin was a simple but highly valued dish, and of course there was dried and salted tuna, introduced by the Phoenicians and perfected by the Romans.

Archestratus, the best known culinary writer of the ancient world and prominent humorous dinner philosopher in the *Hedypatheia* poem, emerges

(continued)

as a great fan of tuna caught in the Bosporus area. The central cut of this tuna from Byzantium is broadly praised as a delicacy, as is the tail, which is generally seen as the least edible part of the fish. The tail was the food of the gods, Archestratus wrote, although he warned that this was only true of the tails of the female tuna. If in doubt as to the gender it was best not to eat it.

Classic Greek-Style Tuna

Cut the tuna into slices and grill evenly.
Avoid overcooking.
Sprinkle with a pinch of salt and a few drops of olive oil.
Serve warm.
Dip into spicy brine if desired.
NB: Do not sprinkle with vinegar, as this can spoil the flavour.

(Dutch food writer Johannes van Dam advised: fry the tuna briefly on both sides, allow to cool and fry again on both sides. This prevents the tuna from overcooking and drying out in the middle. Use MSC-certified tuna.)

Bonito à la Archestratus

You do not really need a recipe, writes Archestratus: the clumsiest oaf couldn't mess up tuna if he tried. (Personally I do not agree with my fellow tuna-lover: tuna is easily spoiled by frying it too long, resulting in a dry and leathery steak, in particular those tuna-species that have less fat or are badly defrozen).

Wrap the tuna in a fig leaf with a little oregano.
(No cheese or other additions, Archestratus warns.)
Tie up the fig leaf with string.
Lay the leaf with its contents in hot ash.
NB: Do not allow it to burn!

Those who do not have Byzantine tuna immediately to hand need not despair, writes Archestratus. Tuna caught elsewhere can also be tasty.

Grilled tuna à la Romana

Except for the dried fermented *mojama*, salted tuna no longer forms part of world cuisine and is therefore rather difficult to obtain. Nevertheless to give an idea of a simple but tasty Roman dish, this is a grilled tuna recipe from the culinary reference work of the Dutch food historian Patrick Faas, *Around the Roman Table* [90].

Make a vinaigrette with the following ingredients:

3 tablespoons of strong vinegar
2 tablespoons of *garum*, or vinegar with anchovy paste or fish sauce
5 tablespoons of olive oil
4 finely chopped shallots
1 teaspoon of pepper
1 tablespoon of freshly chopped lovage
25 g of fresh mint

Remove the skin and bone from the tuna and cut the fish into slices. Brush these with oil, pepper and salt, and grill them on one side on a hot barbecue. Turn the slices over and brush the grilled side with the vinaigrette. Repeat and grill until the flesh inside is pink. Serve with the rest of the sauce. This Roman tuna recipe also works best with MSC-certified tuna.

Until recently it was assumed that in antiquity no complex net constructions were used to catch the bluefin tuna. Historians and archaeologists, in so far as they took an interest at all, assumed that the tuna was caught using no more than simple fishing methods, such as hooks, harpoons or spears. But if we listen to Oppian, the Greco-Roman author of a didactic poem about fishing from the second century CE, we gain a radically different picture. 'The construction of the system of nets resembles the layout of a city. (...) There are entrances, gates and access routes. The tuna follow one another into the nets, squashed together like the mass of a stream of strangers. There are young, mature and older tuna. A multitude pushes through, deep into the nets, and the current only stops when the fishermen consider it enough or no more fish will fit' [182].

Further descriptions appear from Oppian's time which point to large-scale, complex, organised fishing methods. Inscriptions found at the Roman settlement of Parium near the Hellespont speak of fishermen who had a lease for use of crucial lookout towers from which the passing shoals of tuna were tracked. The catch is described in detail: there were as many as 70 fishermen involved, each with a fixed job, a very similar fishing crew to the teams who now man the *almadrabas*.

In his description of the animal world the Greco-Roman writer Aelian (175–235 CE) discusses the Sicilian tuna catch in detail. While certain descriptions in his *De Natura Animalum* are impressive nonsense, like the story of a beaver that performs self-castration to distract hunters, that of tuna fishing is surprisingly accurate [3].

'When spring begins to sparkle and the sea breeze gently blows, when the clear air smiles down on the lapping waves, the watchman, who combines a secret gift with a sharp eye for spotting fish, warns of the direction the tuna will come swimming from. The fishermen spread their nets on the beach and receive instructions shouted down to them from the watchman, like a general giving orders, how to lay the nets in the sea, closer to or further from the beach. He also gives an indication of how many fish there are in the shoal. His guess is usually close.'

Boats, each with 12 strong oarsmen are needed to position the nets, Aelian writes. The nets are of a considerable length and are set out vertically in the water, with cork floats at the top and lead weights down below. The fishermen embark on the water in a closed formation of rowing boats, the tuna is closed in and has nowhere left to go. 'The oarsmen then move the catch in their net towards the coast. As a poet might say, like a conquered city with fish as residents.' After the catch they take a moment to thank Poseidon and to ask him to bless the next catch, and especially to keep the nets in one piece, as sometimes the shoal of tuna conceals unpleasant surprises. 'The fishermen ask that no swordfish or dolphin find its way into the shoal of tuna in the nets,' Aelian recounts. 'The noble swordfish often cuts open the nets, allowing the entire catch to escape. The dolphin is just as much an enemy of the nets, because of its skill in biting its way out through the mesh.' Substitute 'orca' for 'dolphin' and the picture is complete. Even more important: here too we find large, complex vertical trap nets to enclose the tuna. The description seems to relate to a large net that was towed to the beach as soon as the fish had swum in, like a 'conquered city' being taken captive.

'Two fishing techniques were applied in Baelo,' says archaeologist Muñoz of the Baelo Claudia archaeological site. One involved throwing out large nets with floats, which were then pulled in from the beach. Another made use of a fixed system of floating nets in the sea, driving the tuna into a trap chamber where they were slaughtered. The latter, technically more complicated fishing technique was probably most often used, Muñoz believes. It is the technique still used today just around the corner, off the beaches of Zahara de los Atunes. New discoveries around the Mediterranean Sea, the Strait of Gibraltar and the Atlantic coasts seem to confirm the theory: in the entire region of Mare Nostrum tuna was caught using a technique which is known as the *almadraba* in Spain, *tonnare* in Italy and *madrague* in France.

Salsamentum and Mojama

The Romans referred to salted fish using the term *salsamentum*, derived from salt (*sal*) and in principle covering all kinds of salted products, but since it generally related to fish this came to be the primary meaning. Depending on the season, the part of the fish (fat belly, dorsal loins, central cut steaks) and the amount of salt used, different names were used for salsamentum. The misconception of rotting fish might be down to the fact that fermented fish products had long enjoyed rather lukewarm interest. But in recent years we have seen a true revival in research into the salted and fermented fish of ancient times, which again enjoys the attention of food historians and archaeologists. In fact there was never any question of rotting: the technique involves a chemical process for preserving the vulnerable proteins, so that they keep even in the often high temperatures around the Mediterranean. It is not as filthy as it sounds: the preservation takes place by means of a chemical reaction involving the activity of certain enzymes found in the fish's intestines. The salting products of the *cetariae*, as the salting factories were known, were the classical mixture of techniques we find recurring in salted herring, canned tuna and fish sauce.

The salting involved layering tuna flesh, alternated with sea salt, according to a fixed pattern. The oval or rectangular pits were sealed with another thick, heavy layer of salt. The mixture remained in the pit for a number of months. We do not know how the fermented tuna that emerged looked. The final texture would have depended on its fat content. It probably somewhat resembled the brown pieces of fermented tuna or *mojama*, which are still sold in southern Spain. *Mojama* is a tasty tapa. The Spaniards cut it into thin slices, drizzle a little olive oil on it and serve it, sometimes accompanied by roasted almonds. Depending on the quality, the flavour is mild and not directly reminiscent of fish. If you did not know better, you could easily confuse it with a thin slice of raw salted ham. It seems reasonable to assume that the culinary form of salted tuna must have borne some relation to this.

Garum

The tuna sauce of the gods, more expensive than the heavenliest perfume and more powerful than the strongest wine: *garum* is the legendary heritage of the bluefin tuna of classical times. This liquid fish extract was viewed as the gold of the ocean, for which high prices were paid throughout the Roman Empire. The ingredients of this highly exclusive delicacy, however, were not as grand as one might assume based on the value of the fluid. Pliny (23–79 CE) was one of the first to publish a recipe of sorts in his reference work on nature, the *Naturalis Historia* [196]. It was a sauce of spines, fish remains and salt, the brief description claimed. 'Liquid from rotting material,' the writer recounted

(continued)

dryly. That was true, in so far as the preparation of the *garum* involved mixing tuna offal—intestines and other organs, gills, heads and blood—with salt water and storing it in the smaller pits of the fish factories. Often smaller fish such as anchovies, sardines or horse mackerel were added to the mixture, and the fish could be dried first, to obtain the right bacteria for the fermentation process. To mitigate the pungent odour, the sauce was seasoned with herbs, honey or wine. In medieval times, Greek texts, probably based on lost manuscripts, re-emerged, detailing the recipe for *garum*. The blood, intestines and remains of the backbone of the tuna were salted along with smaller fish. In some cases oysters, sea anemones and urchins were added to the mass. The mixture was regularly churned during the fermentation process. After a couple of months' brewing in the sun, the liquid was pressed out of the fermented mass and seasoned with mature wine, oregano or other herbs. The result was a clear, light-coloured liquid with a strong flavour. What was left was a sort of thick, brown fish paste, known as *allec*. The *allec* was also used in the kitchen, and presumably most resembled what we know today as Indonesian *trassi*. The neck of the amphorae into which the *garum* was poured were marked in black ink with the brand of tuna extract and other product information. For example: 'Premium quality *garum* with mackerel from the preserving factory of Aulus Umbricius Scaurus, matured for 2 years in the pit, lightly spicy in flavour.' Or, for the really wealthy, the costly *garum haimation* prepared from the blood of the tuna.

The exquisite fish extract was something that initially only the rich could afford, but gradually cheaper variants became available to a broader public, as seems to be the fate of all periods of bluefin tuna. Just as millennia later with sushi, the consumption of tuna products became democratised, probably as a result of its popularity, although the advantages of scale in the mass production of *garum* will also have made it ever cheaper. As with sushi, *garum* also benefited from excellent marketing: an exclusive image, portraying it as magical and sexy.

Garum was not only cool and delicious, but also healthy, according to the greatest physician of ancient times after Hippocrates, Claudius Galenus (130–210 CE). At the top of his list of favourite garum varieties is that of the bluefin from the coasts of Cádiz. *Garum* served as an aperitif to whet the appetite, as a pungent condiment, and as a remedy for burns, gastric ulcers and blood poisoning. *Garum* lowered fever and disinfected open wounds. Those troubled by haemorrhoids were advised to take a sip. Was *garum* the inspiration for the magic potion of the Gallic strip-cartoon hero Asterix? For the real bluefin tuna addict this seems highly plausible, especially given that in one of the episodes [115] go in search of fresh fish as one of the indispensable ingredients. When the fish needed for the potion is not available, eventually overfished, they cross the ocean and end up at the US East coast. These are,

(continued)

maybe not coincidentally, waters where the Atlantic bluefin can still be found today.

Despite its fame *garum* has disappeared from the culinary stage. The fish sauce, renowned for centuries as one of the top products of Roman cuisine, went down with the empire. That was to a large extent thanks to the fanaticism of the Christians who came to power. The joys of creative cuisine were suspicious in the eyes of this rapidly rising cult. *Garum's* reputation for enhancing potency—tradition had it that *garum* could also enhance sexual performance—made it no less reprehensible in their eyes. This was illustrated by the recommendations of the Egyptian monk Pachomius (287–347), inventor of the Christian monastic life, who knew precisely what was and was not good for other people. *Garum* clearly belonged to the latter category. Too strong in flavour and therefore too stimulating to the senses. A sauce for Ba'al, Poseidon and Hercules. According to Pachomius, the Christian god certainly did not like that. *Garum* came to be officially banned from Christian cuisine. But as with other pleasures that were stimulating to the senses, the Christians were unable to completely stamp it out. In his reference work on smallpox and measles the famous Persian doctor and alchemist Rhazes (865–925) continues to recommend the sauce as a remedy, which means that Muslims had meanwhile become aware of its existence. Centuries later *garum* gained a kind of cult status among European heretics and the early enlightened souls of the sixteenth century. The French writer and doctor François Rabelais (1483–1553)—author of the stories of Pantagruel and Gargantua which were also forbidden by the Church—mentions *garum* as a sauce which crops up here and there in European cuisine and was also used for medicinal purposes. 'If you are always busy with your nose in a book, there is no better means of awakening your lost appetite,' he writes of a mixture of *garum*, oil and vinegar.

Garum succeeded in maintaining its position for longer outside the Christian sphere of influence. The sauce found its way via India to the Far East. It still forms an essential component of oriental cuisine today in the form of *nuoc mam* or *nam pla* from Vietnam and Thailand, or *patis* from the Philippines and Chinese oyster sauce. A brown, salty, pungent fish sauce with the same function as condiments in Western cuisine or soy sauce in that of the Chinese and Indonesians.

In Europe *garum* still runs into problems, although no longer of a religious nature. The tuna restaurant El Campero in Barbate once thought it would be nice to serve its clients *garum*, but on reflection owner and chef José Melero Sánchez decided against it, fearing food safety inspections. 'Leaving fish out for a while in the sun, then collecting the liquid which runs out—if they see that you're in trouble,' says Sánchez. In laboratory conditions *garum* production in fact turned out to be a cinch, archaeologist Dário Bernal comments

(continued)

from experience. Mixed teams of archaeologists and food experts from the Universities of Cádiz and Seville got to work with various mixtures of fish offal, flesh and salt to bring *garum* back to life. As sources for the recipes they used a number of classical manuscripts, including the famous *Geoponica* which appeared in Constantinople in the tenth century and in turn drew on records from the Romans and even Mago of Carthage, the universally respected first great writer on agriculture and fishing. After 2 or 3 weeks the mixtures had already produced enough liquid bearing some resemblance to the oily substance released in the production of raw cured ham. This fluid, *muria*, proved fit for consumption. Pressing the mixture released the real *garum*, which was less oily and more liquid [110].

Those who would like to try making *garum* themselves will have the best chance with the recipe for an oriental fish sauce by Thai chef Kasma Loha-unchit [138]. Use fresh sea fish, preferably of a variety which has little value in direct consumption and is noted in the Monterey Bay Aquarium Seafood Watch, the UK Marine Conservation Society's 'Good Fish Guide' or the Dutch Good Fish Foundation 'VISwijzer' Seafood Guide as a species which can be eaten without any objections. Rinse the fish, dry it and mix with salt in the weight ratio one part salt to two to three parts fish. Place the fish in large ceramic pots, with further layers of salt on the bottom and top to cover the mixture. Seal with a woven bamboo mat, weighted down with stones so that the fish does not float on the liquid released during the process. Place the pots in a warm, sunny spot for 8 or 9 months so that the fish decomposes faster. Replace the mat from time to time. The more it is exposed to the sun, the better the quality of the mixture and the red-brown colour of the sauce. Once the sauce is sufficiently preserved, allow it to run out of the containers, filter and allow to mature properly over a couple of weeks in a well ventilated, sunny spot to eliminate overly strong fishy flavours. Bon appétit.

Of course oriental fish sauce can simply be bought in the shops. Kasma Loha-unchit warns that there is a good deal of messing around with large-scale production of fish sauces. Enzymes and acids are added to accelerate the preservation process, the mixture is laced with caramel, flavour enhancers and colourings. Her advice: pick a fish sauce with a clear, light colour, similar to sherry, and without residues at the bottom of the bottle.

Baelo Claudia's decline began at the end of the second century. From the north the Goths and Vandals advanced into the Iberian Peninsula, reaching North Africa via the south coast. Archaeological remains show that the town was again severely hit by a disaster. Was it a big fire, an earthquake or another tsunami? Was it plundering pirates and rebellious Moors? History does not specify, but like everywhere else around the Mediterranean, the decline went hand in hand with the fall of the western Roman Empire. The demise of the existing order soon brought trouble for the tuna industry. Based on the remains of amphorae found, from the third century onwards there appears to have been a dramatic drop in tuna production. Lack of written sources means we cannot be certain but it is also possible that the enormous expansion in fishing had struck a drastic blow to the existing tuna population by overfishing, which would have added to the troubles.

Were the Romans the first to be faced with the industrial overfishing which has threatened the bluefin tuna? A combination of factors, however, appears more probable for the sudden downfall of the tuna empire in the western Mediterranean. The complicated organisation behind tuna fishing, which demanded a good deal of investment, hiring workers and the deployment of slaves from the opposite shores and safe transport of the amphorae around the Mediterranean. All this contributed to making the bluefin tuna a catch which could only thrive in an environment with a high level of economic and social organisation. The fall of the existing order [111] caused the monetary economy and the markets to collapse as did the naval control of the trade routes. Pirates became a menace to sea transport. Without guarantees for the funding of complex fishing methods and secure trade routes, the tuna fisheries could no longer flourish. Corruption did not make matters easier. The Roman Empire fell, and so did the tuna industry.

It was a pattern that would repeat itself later. Sustainable management of large-scale border-crossing tuna fishing is a matter of high-level policy and governance. When the social and economic machine faltered for one reason or another, large-scale tuna fishing, along with salting, trading, transporting and financing—the value chain as a whole—was soon a thing of the past. From early in our history the tuna production chain had been elaborate, with complex, large-scale fishing methods, the need for factories to process the fish, salt extraction and trade, as well as well equipped boats for transport over long distances. This was the unravelling of civilisation, decline of order and breakdown of international markets: large-scale fishing for tuna, a migratory fish which crosses thousands of kilometres of ocean, could only flourish sustainably as an industry in a developed, economically organised society.

In 330 the capital of the western Roman Empire moved to Constantinople. The western edge of Mare Nostrum began to fray; Rome's military and economic power was ending. Baelo Claudia fell into ruin. No money and no manpower were available anymore to make the *almadrabas* work. Fish salting stagnated, export stalled. People moved away. The fishing town's factories kept going for a long time. The now roofless halls with their pits were still used for hundreds of years for salting tuna, archaeological research indicates. But that bore no relation to the glory days of more than five centuries of tuna fishing. The coasts fell prey to petty local wars with invading hordes, rebels and adventurers. Amongst the clash of weapons tuna fishing continued on a modest scale for a handful of fishermen who built their huts on the remains of the town.

Tuna Renaissance

6

Ir a por atún y a ver al Duque.
(Go after the tuna and to see the Duke) Old Spanish proverb

For centuries this was the last part of Europe seen by seafarers as they set sail in their galleons and frigates from Spain to the New World in search of wealth. If they were lucky and survived the journey and the stay in America, it was also the first European city reached on the way back, before their ships set off upstream along the great Guadalquivir river towards Seville to deliver the captured gold and silver. Sanlúcar de Barrameda is situated on the eastern bank at the mouth of the broad river which runs into the Atlantic Ocean. The streets of the old city have a salty sea smell with hints of the special *Manzanilla* sherry made in countless bodegas. Sanlúcar's extensive beach is famous for horse races and for its restaurants next to the port with their shellfish, lobster and renowned *gambas* from Sanlúcar. On the other side of the river lies the vast dune and marsh area of Coto Doñana, a nature reserve known for lynx and migratory birds which forage here on their way from Europe to North Africa and back (Fig. 6.1).

Every spring at Pentecost Sanlúcar's beach is filled with pilgrims of the brotherhoods of the Virgin of El Rocío, the Holy Maiden of the Morning Dew. Landing craft carry the pilgrims with their ox carts, horses and four-wheel drives, carriages and covered wagons over the river. Singing, dancing and drinking, they worship the Virgin in the sandy village on the edge of the nature reserve, where a large cathedral has been built in her honour. More than a million Spaniards come together like this in a ritual which seems more heathen than Catholic mass. Like the tuna catch these are ancient rites, much older than most people realise. For instance, the Holy Virgin of El Rocío exhibits a number of similarities with the Egyptian goddess Isis, who was worshipped in this region in ancient times. She was imported by the Phoenicians from the east and also enjoyed great popularity under the Romans. That was no less true in southern Spain: excavations of Baelo Claudia

© Springer Nature Switzerland AG 2019
S. Adolf, *Tuna Wars*, https://doi.org/10.1007/978-3-030-20641-3_6

Fig. 6.1 Cleaning and salting tuna in one of the factories of the Dukes of Medina Sidonia (Joris Hoefnagels, 1575 by courtesy of The National Library of Israel)

have revealed a complete temple dedicated to Isis which awaits restoration. Only in the sixth century did the Christians succeed in ending worship of Isis once and for all. But the goddess turned out to be stronger. Her fertility cult and sacred mother worship appears to have successfully slipped into the Catholic Church via the backdoor with the Holy Virgin of El Rocío and still lives on today.

The countless palaces and churches of Sanlúcar exude the faded glory of an important harbour city. We are visiting the palace of Medina Sidonia, which overlooks the old town. After all, if you want to visit the duchess of tuna, you won't find her on the beaches of Zahara de los Atunes; I was required to present myself at the palace for an appointment. This is where the Medina Sidonia family, one of Spain's oldest and most aristocratic families, has resided since 1297. The building is bordered by the lower-lying city wall. Directly below that is the daily market, the place to go for lobster, prawns and Venus clams, shark, gilthead bream and stone bass. And tuna of course. Wherever the dukes of Medina Sidonia are, the giant tuna is never far off.

Inside, Luisa Isabel Álvarez de Toledo Maura, 21st Duchess of Medina Sidonia, 17th Marchioness of Villafranca del Bierzo, 18th Marchioness of Los Vélez, awaits. Three times Grandee of Spain. Álvarez de Toledo has the right to a dozen other titles, some Italian principalities and noble decorations of the more frivolous sort. But perhaps most importantly: the blood of Guzmán *el Bueno*, Guzmán the Good, hero of Tarifa, conqueror of Gibraltar and founder of a medieval tuna empire which has withstood the centuries, runs through her veins.

The duchess apologises: she is a little hoarse today, a bit of a sore throat. Whispering and regularly interrupted by bouts of dry coughing, she invites me into her office, a large room under an arched ceiling, filled with bookshelves, computers and a seat overlooking a courtyard. On the other side of the room her friend Liliane Dahlmann, also president of the trust that manages the palace and family archive, sits behind a screen. 'We are working on digitising the archive,' Álvarez de Toledo explains in a whisper, reaching out of habit for a pack of Marlboro Lights on the table. The duchess is not the type to be held back by a bit of a cough. She has no interest in denying herself a little smoke, it's too late for that anyway.

Seventy-one, she is now. A fragile, small figure, clad in faded jeans, a pair of sturdy workman's shoes underneath and a pink sweater over a white blouse. A tanned and wrinkled face with sharp features. Dark eyes with the calculating gaze of a conqueror ready to attack. Luisa Isabel Álvarez de Toledo has never been the accommodating type. 'The Red Duchess', she was called, due to her left-wing sympathies. She provided financial support to destitute orphans from Sanlúcar, and had houses built on her land for the labourers. The duchess always preferred to call a spade a spade. Opponents were soon dubbed a 'bunch of assholes' or 'imbeciles'. All that never stood in the way of a fine sense of etiquette. A gardener who forgot to remove his cap when entering the noble office could expect an earful. A certain level of respect was owed to the descendants of Guzmán *el Bueno*, or Alfonso Pérez de Guzmán (1256–1309). Indeed, in 1294 the king of Castile rewarded Guzmán for his immense services and proven heroic courage by granting him the rights to tuna fishing on the Spanish south coast.

Like her illustrious forefather, the duchess lived a combative life. Álvarez de Toledo was born in 1936, the year the Spanish Civil War broke out. She married young and had three children whom she never raised herself. She divorced her husband in a time when that was absolutely not done in heavily repressive Catholic Spain, certainly not in the circles in which she moved. She got into trouble with the dictator Franco. In 1966, when an American B-52 bomber armed with atom bombs crashed near the Spanish town of Palomares, she campaigned for the local population who were threatened with radioactive plutonium poisoning. Franco twice had her imprisoned for her big mouth. At the start of the 1970s she began a 6-year voluntary exile in France. Did she inherit that rebellious streak from her distant ancestor? The duchess takes a moment to consider this. 'I don't know if it's something genetic or whether it comes from my upbringing, but I like to draw my own conclusions,' she says with a faint smile.

In life the duchess was the custodian of the largest private historical archive in Spain. A library of unique documents tells the story of the rise of one of Spain's oldest and most important aristocratic families, their victories, their defeats. Their role in Spain as a world power. And their fishing and trade in tuna. An important pillar of the influence and power of the dukes of Medina Sidonia was bluefin tuna. They were rightfully the 'tuna dukes'. When Álvarez de Toledo returned from exile in 1975, the year the dictator Franco died, she found the palace in Sanlúcar a decaying ruin, its archive mouldering away. It was only by a hair's breadth that the building escaped demolition behind her back to make way for new apartment blocks. Most of the palace dated back to the fourteenth century, so the duchess had it registered as a monument and invested in a café and a hotel to generate income for restoration and maintenance. Emulating her uncle, a famous politician and writer, she trained herself through self-study in the family archive, becoming an amateur historian. 'I have placed the archive and the palace in trust. Two universities and the regional government of Andalusia now participate in this project,' she says proudly.

It led to an enormous fight with her children, who saw the family palace with its valuable documents removed from their inheritance.

The bluefin tuna left plenty of traces behind in the documents archived here. For centuries logbooks painstakingly recorded the daily catch of the various *almadrabas* along the coast. It is a goldmine of statistical material for today's scientists interested in tuna. It also gives an insight into the enormous magnitude and productivity of the tuna empire. Take for instance the *almadraba* of Zahara de los Atunes during the fishing season of 19 April to 12 June 1554: 52,663 tuna caught. The *almadraba* of Conil de la Frontera of 27 April to 28 June 1564: 49,409 tuna. In 1554, at the start of the season, the price of a tuna was 9.5 reals, but due to the abundant supply fell to six reals. Ten years later, in 1564, the price fluctuations may even have been greater: from 15 to 16 reals the tuna price plummeted to three reals under the influence of the surplus catches. The records contain details ranging from the salt purchases for conserving the tuna to the wages of the army of fishermen who had to be mobilised to man the *almadraba*. The crew of the boats, boatswain and fishermen who brought in the tuna, as well as oarsmen, ox drivers, butchers and filleters, the warehouse supervisor and the youngest servants, everything was carefully recorded.

During the course of the sixteenth century bluefin tuna was bigger business than ever. An average seasonal catch on the south coast could land up to 80,000 tuna. At a price of eight reals per tuna that meant an annual turnover of 640,000 reals, or more than 45,000 ducats. That was roughly equivalent to the annual salary of 600 trained workers, a small fortune in a time when large-scale enterprise had yet to be invented. 'The tuna was our family business right into the nineteenth century,' the duchess whispers with the distant pride of someone stating an undeniable fact. She has written a book about tuna, self-published of course, with reproductions of the drawings, accounts and other documents in the family archive [6, 201].

Bluefin tuna fishing had once again become possible on a large scale on the coasts of Cádiz in the late medieval period. After almost 1000 years of chaos and insecurity there was sufficient organisational and administrative power to get this capital-intensive, complex industry and its international value chains off the ground. The times of the Phoenicians and Romans were revived and may well have been surpassed in catch and trade volumes. There was still fighting at sea, but once salted in the pit, there were enough secure trade routes to export the fish. From the surrounding towns on the south coast the tuna was transported by boat to the markets of Alicante, Valencia and Barcelona, or further towards Italy. The export market, in contrast to classical times, now also stretched in the direction of the Low Countries to the north. Again, the rise of the large-scale tuna industry was made possible by a strong navy, which ensured sufficient security for the trade routes on the one hand, while the Spanish war fleet on the other hand had a great need for protein with a long shelf life for their maritime expeditions and wars, a need fulfilled by the salted tuna.

The tuna catch still required enormous organisation. Hundreds of boys and men were needed to place the nets in the water, enclose the fish and bring them onto land. At the start of every fishing season the dukes of Medina Sidonia sent their drummers through the streets of Tarifa, Vejer, Conil, Chiclana and other little towns and villages along the coast to recruit teams for the *almadrabas*. If people did not come willingly then force could be used on the local population of serfs. In 1540 there was so much tuna that all the strong men and boys from Vejer simply received orders in the name of the duke to help bring in the fish. The mid sixteenth century was a period of abundance: the tuna was fished out of the sea in hundreds of thousands in the *almadraba* nets. Moorish slaves and gypsies were all welcome on the beach to help with the catch. Sometimes even the prisons were emptied to mobilise sufficient manpower. They whiled away the hours waiting for the tuna by playing dice and drinking. A fixed garrison of ladies of pleasure were deployed to meet the demand for distraction. The value of the fish was clear from the recurring problem of robbery. Like later workers in diamond mines, the fishermen, oarsmen and sailors who worked on the boats had to agree by contract to comply with inspections at any time of the day. The punishment for discovery of stolen tuna sent a clear message: 100 lashes of the whip and public humiliation with a crier, accompanied by a drummer, who proclaimed the thief on all the beaches.

Miguel de Cervantes (1547–1616), Spain's greatest writer, author of *Don Quixote*, described the *almadraba* in 'The illustrious scullery maid', one of 12 short stories which he published in 1613 [57]. *Don Quixote* was by then a great success and the public wanted further spectacle. Their demands were rewarded with a description of the tuna catch. 'O kitchen-walloping rogues, fat and shining with grease; feigned cripples; cutpurses of Zocodober and of the Plaza of Madrid; sanctimonious patterers of prayers; Seville porters; bullies of the Hampa, and all the countless host comprised under the denomination of rogues! Never presume to call yourself by that name if you have not gone through two courses, at least, in the academy of the tuna fisheries.' That academy was the beach of Zahara de los Atunes during the *almadraba* fishing. The freeloaders and adventurers from Andalusia and the rest of Spain camped out on the beaches. Between fishing sessions they taught one another the craft of deception.

'There it is that you may see converging as it were in one grand focus, toil and idleness, filth and spruceness, sharp set hunger and lavish plenty, vice without disguise, incessant gambling, brawls and quarrels every hour in the day, murders every now and then, ribaldry and obscenity, singing, dancing, laughing, swearing, cheating, and thieving without end. There many a man of quality seeks for his truant son, nor seeks in vain; and the youth feels as acutely the pain of being torn from that life of licence as though he were going to meet his death.'

This was an excellent place for those fleeing justice: no one asked awkward questions, and the names given in the contracts were generally made up. You were guaranteed to be left in peace, because the soldiers of the kings of Castile had little authority on the beaches of the Duke of Medina Sidonia. Cheap labour outweighed a criminal record. Corporate social responsibility was yet to be invented. Tuna was money, lots of money, and the fish would not wait around to be brought in. Officially gambling and prostitution were banned, but the captain of the *almadraba* was happy to turn a blind eye. As long as it didn't come to stealing tuna or cutting into the catch from the nets, it was all the same to him. Cervantes did have some experience when it came to adventurers and thieves in the south of Spain. In 1587, in the absence of success with his plays, he was hired in Seville to levy food supplies for the Armada—the feared, invincible fleet with which Philip II aimed to teach England and the Low Countries a lesson. Salted and fermented tuna was probably one of the foodstuffs with which the tens of thousands of troops of this fleet were sent into battle.

By then Cervantes himself had had the opportunity to enjoy salted tuna in the service of the Spanish navy. In 1571, still a young man, he stepped on board a Spanish warship to sail against the Turkish sultan. In the sea battle of Lepanto in the Gulf of Patras near the Greek coast, under command of John of Austria, the Spaniards achieved a resounding victory. Cervantes was wounded but was person-ally rewarded by the Spanish admiral for the courage he showed. In 1573 he was among the troops who took Tunisia. But it all went wrong in 1575. On the way back, somewhere between Italy and Spain, his boat was hijacked by Moorish pirates. The letter of recommendation he was carrying from John of Austria, with which he hoped to gain a good job under King Philip II, now got him into serious trouble. After all, the Moorish hijackers reasoned logically enough, a man recommended by the hero of Lepanto must be important. So they decided they would free him only in exchange for a high ransom. For 5 years the unfortunate writer was imprisoned in Algiers along with around 25,000 other Christian slaves. A lucky coincidence—ransom money intended for a Spanish noble turned out to be insufficient, but just enough for Cervantes—he was bailed out and thus spared an inglorious end as a galley slave.

Cervantes was therefore closely familiar with tuna fishing on the beaches of Andalusia. The Moorish hijackers who had imprisoned him were the same people who regularly attacked the bluefin tuna fishermen, scooped them up and robbed them of tuna and all. The coastal night watchmen standing guard in the towers commissioned by Philip II could do nothing to prevent it, we read in 'The illustrious scullery maid'. 'It has sometimes happened that scouts, sentinels, rogues, overseers, boats, nets, and all the posse comitatus of the place have begun the night in Spain and have seen the dawn in Tetuan.' Tetuan is on the opposite shore, a stop-off on the way to Algiers, and had a lively market of Christian slaves. During his imprisonment there as a ransom slave, Cervantes heard first-hand the stories of the day labourers who had travelled to the coast to fish for tuna but ended up as slaves in Algiers.

The palace of the dukes of Medina Sidonia in Zahara de los Atunes, mentioned in Cervantes' writing, is still standing. The large walled building by the beach dating back to the fifteenth century can best be compared with an industrial complex avant la lettre: a curious combination of a fish factory, a palace and a fortress. Cervantes describes how the dukes inspected the catch in the courtyard, while the labourers rested a little further on in the sun. From their watch towers they looked out on Africa, Tangier and the Atlantic Ocean. Beyond them lay the Canary Islands, the new world and even larger oceanic regions, where one day the tuna industry would be developed.

This ducal fish palace is now in a pitiful state. The walls have crumbled and the enormous courtyard, where once the tuna were hung up in racks and cleaned, serves as an open-air nightclub. The harsh Levante wind has free rein between the clumps of wild daisies which grow along the walls. Of the watch towers Cervantes mentions, only the westernmost remains standing. The only thing left to remember the palace's former glory is the white church of the Virgen del Carmen, protector of fishermen, in the large hall where the fish were once salted.

In spite of her apparent fragility, the duchess of Medina Sidonia ignites into exasperated rage when the building is mentioned. 'It's a scandal,' she says furiously. 'The palace is a ruin, a total abandoned mess.' The local and regional governments do nothing to protect the cultural heritage relating to tuna, she notes angrily. The salting pits remain intact; she has often protested that something should be done with them, but nothing happens, there is no money to renovate this industrial palace.

The Great Tuna Conqueror

7

The tuna empire of the modern age was born in the fragmented chaos that was late medieval Spain. After the fall of the western Roman Empire the northern barbarian tribes of the Vandals and the Visigoths took over the Iberian peninsula, leading to centuries of obscure conflict between kings with unpronounceable names such as Amalaric, Ataulf, Gesalec and Liuvigild. They barely put up a fight in the spring of 711 when the Berber troops led by Tariq landed from what is now Morocco. It was the beginning of the Islamic conquest of the Iberian peninsula. The tuna fishermen must have been the first to see the invasion coming over the strait. It was a walkover: before winter that same year the Moorish troops had broken through to Toledo, capital of the empire under the Visigoths, right in the heart of the Iberian peninsula. Large parts of Spain and Portugal subsequently remained under Muslim rule for centuries. The area around the Strait of Gibraltar was of great military importance as a bridge between the Muslims in Spain and Muslim rulers in the Moroccan hinterland. Tuna fishing again found itself at the centre of an area of geostrategic value. There is no doubt that the tradition of tuna fishing continued under the Muslims: the name *almadraba*—the place where the blows are struck—is of Moorish origin. Nonetheless the great Muslim historians and explorers of the western Mediterranean region—Ibn Khaldun, Ibn Battuta and later Leo Africanus—turned out not to be particularly interested in fisheries. Muhammad al-Idrisi, the Moorish poet and famous cartographer who drew the world map for the Catholic King Roger II of Sicily in the twelfth century, is the only one who writes about tuna. But then, he was born in Ceuta, right next to the main *almadraba* fisheries area. The Sicilian king, himself ruler of an island where the Phoenicians had left behind an important tuna industry, was probably equally interested in the fish. According to al-Idrisi, a large fish called tuna was effectively caught near Ceuta. The fishermen hunted the tuna with a kind of harpoon, according to the Muslim scholar. Large-scale fishing of any importance seemed not to exist, or was lost in the course of time.

This obviously changed with the Christian conquest of the Moorish regions on the peninsula—known as the *Reconquista*. King Alfonso X, nicknamed Alfonso the

© Springer Nature Switzerland AG 2019

S. Adolf, *Tuna Wars*, https://doi.org/10.1007/978-3-030-20641-3_7

Wise, decreed in 1268 that the knights of Santiago be rewarded for their services in battle against the Moors with a licence to the '*almadrabas* of the tuna, the right to the seaport, fishing and salting'. It went hand in hand with a new period of stability around the peninsula which enabled the resurrection of large-scale bluefin tuna fishing and trade. As one of the greatest conquerors in the *Reconquista*, the famous Guzmán *el Bueno* would eventually obtain the most extensive rights to fish for bluefin tuna.

Guzmán, along with El Cid, belongs to the most important Spanish heroes to emerge in the late medieval period. Both evolved into icons of a Holy War to drive the Muslims out of the Iberian peninsula. But as always, behind both myths lay a considerably less romantic reality of money and power. Take El Cid, who was praised in epic poems and immortalised on the silver screen by Charlton Heston in a 1960s film as the Catholic hero who liberated the peninsula from its villainous invaders. Historians currently believe that El Cid is a hotchpotch of several figures, who were more robber knights and mercenaries than idealists. Whoever offered the most could avail himself of the services of El Cid, be he Christian or Muslim.

Alonso Pérez de Guzmán, in contrast to El Cid, actually did exist. The founder of the tuna empire, according to official, carefully rewritten, history, was born in León, a proper Castilian environment, as the illegitimate son of a noble family. According to tradition (where history begins to find itself on shakier ground) we also know that he had six children and that his eldest son was kidnapped by an army of Moors. He died, less controversially, in Gaucín near Malaga. In fact it was not tuna with which Guzmán el Bueno made history, but his conquest and defence of the strategically important port city of Tarifa. Situated on the narrowest point of the strait between Spain and Morocco, Tarifa was a military and trade crossroads, both between Al-Andalus and the sultanate, in what is Morocco today, and between the Mediterranean Sea and the Atlantic Ocean. And around Tarifa, of course, lay the most important beaches for bluefin tuna fishing. 'Guzmán el Bueno knew that,' the duchess of Medina Sidonia told me when I visited her at her palace [6]. Alvarez In her view there was no doubt: in addition to Tarifa as a strategic military base, Guzmán wanted to conquer the tuna trade. He understood that large-scale tuna fishing was a key to power. In the summer of 1292 Guzmán succeeded without much difficulty in taking Tarifa from the Muslims under orders from King Sancho IV. Two years later the Moors tried again to get their hands on the city. But Guzmán had withdrawn behind the city fortifications and refused to give up. Legend has it that the Muslims, weary from their resistance, had Guzmán's imprisoned eldest son dragged before the city wall with the message that his throat would be cut if Tarifa did not surrender immediately. Of course Guzmán would not be blackmailed. He pulled out his own knife and threw it down from the ramparts at the feet of his besiegers. 'Here is a knife for you to cut my son's throat with. Rather my son's glorious death than a cowardly life,' the hero of Tarifa is supposed to have cried on this occasion. It became a favorite, mythical slogan in Spanish collective memory that popped up in several other occasions in the centuries to come. We find the motto repeated in the family crest of Medina Sidonia on the duchess's wall in the palace of Sanlúcar. 'Praeferre patriam liberis parentem decet.' A good parent prefers a free fatherland. The story is most likely fabricated. If we are to believe the duchess, Guzmán did not care in the slightest about a free fatherland. He believed in fisheries. In her view the family motto should have read, 'Praeferre Thunnus Thynnus parentem decet.' A good parent prefers tuna.

In Spain at the end of the medieval period the dividing lines between religions and loyalties in the population were never so sharply drawn as later myth would have us believe. The population of Tarifa at the end of the thirteenth century consisted of the Christian descendants of the Berbers from Morocco, Islamised Goths and Vandals, progeny of the Phoenicians, a mishmash of slaves and the usual colony of Jews. The conflicts which were fought out resembled a more or less permanent state of gang war in which different factions and warlords attacked one another in quickly alternating combinations. In the same way Guzmán was anything but a pure representative of the Christian warriors. 'He himself was a Muslim,' said the duchess of Medina Sidonia about her distant ancestor. She was convinced that Guzmán had been a rich Moorish knight at the court of the sultan in Féz, a fortune seeker who crossed the Strait of Gibraltar as the clan of Hannibal Barca had done. 'In the documents he is described as someone who comes 'from the other side', from Morocco. The rest is an accumulation of tall stories,' she said. Tarifa was therefore defended by Christians under the leadership of a Muslim with a serious business interest in conquering the tuna monopoly on the southern coast of Spain.

Guzmán won, became a legend, was awarded the fiercely sought after licences for catching bluefin tuna and started a noble family line that would turn into the most prestigious ducal title of Medina Sidonia. In 1520 the dukes of Medina Sidonia were even elevated to 'Grandes de España', the highest league of the Spanish nobility. But among friends the duke of Medina Sidonia was known as the duke of the tuna. His enemies spoke scornfully of the 'tuna of the dukes'. By the time of his death in 1309—murdered by the Muslims according to the official version, victim of a conspiracy said the duchess—he had the exclusive right to the *almadrabas* along the entire Atlantic south coast, from Gibraltar to the border with Portugal. The king also made him lord of Sanlúcar de Barrameda and he had gained property in Conil, Chiclana, Santi Petri, Vejer, Barbate and Zahara de los Atunes. The *almadrabas* were closely managed by his wife, a descendant of a rich Jewish family from Andalusia. The first modern European empire of bluefin tuna was born from a marriage between a Muslim and a Jew in the territory of Christian Reconquista. It empowered a dynasty which became one of the richest and most important noble families in all of Europe.

The Tuna Dukes

8

Supported by their tuna monopoly and its profits, the Guzmán dynasty succeeded in strengthening their powerbase through useful alliances in the wars and the endless internal power struggle of the kingdoms on the Spanish peninsula during the fourteenth and fifteenth centuries. Hendrik II of Castile rewarded Juan Alonso Pérez de Guzmán (1342–1396), fourth lord of Sanlúcar, with the prestigious title of count of Niebla, which covered the majority of the Atlantic coastal province of Huelva, including a number of important tuna fisheries. From 1445 Juan Alonso Pérez de Guzmán y Orozco (1410–1468), sixth lord of Sanlúcar and third Count of Niebla was also permitted to call himself the first duke of Medina Sidonia. The Guzmán family thus obtained the only duchy that existed in Spain at that point. From that time on, the dukes of Medina Sidonia ruled like viceroys in the area from Gibraltar to the border with Portugal. The fact that the noble line was now named after Medina Sidonia was more than a remarkable coincidence. The small town, a few dozen kilometres inland from the coasts where tuna is fished, is more than 2500 years old. Beneath the northern side of the old city centre lie the oldest remains to which the city owes its name. Medina Sidonia—the meaning is literally 'city of Sidon'—founded by the Phoenicians, who came from Sidon. They were the men in fast boats who had brought the tuna trade to the west.

The duke and his heirs now formally had a monopoly on all tuna fishing from the border with Portugal to the coast of Granada. The bluefin tuna that came from the Atlantic to the Mediterranean to breed almost inevitably passed the nets of their *almadrabas*. If records are to be believed, at its peak the tuna empire comprised more than a yearly catch of 100,000 fish. After a millennium of decline the *almadraba* had been restored to industrial proportions. Centuries of tuna trade once again allowed wealth and power to accumulate. The peak of the tuna empire was without a doubt under Alonso de Guzmán (1550–1615), seventh duke of Medina Sidonia. Around the beginning of the seventeenth century Alonso kept a court worthy of a king. The duke owned vast lands with 90,000 vassals. His annual income is estimated by historians to have been more than 160,000 ducats at the start of the 1580s. Comparisons are a slippery slope, but based on a labourer's annual salary of around

© Springer Nature Switzerland AG 2019
S. Adolf, *Tuna Wars*, https://doi.org/10.1007/978-3-030-20641-3_8

75 ducats, Guzmán's income was equivalent to that of more than 2100 labourers. Converted to an annual salary of 30,000 euros the duke's income would therefore have been more than 63 million euros. In terms of income from assets: with an annual interest rate of 5% the ducal income would be the equivalent of assets worth a cool 1.3 billion euros today. If *Forbes*' list had existed in the sixteenth century, the duke of Medina Sidonia would have been at the top. Alonso seems to have preferred a low-profile existence among the army of noble layabouts populating Spain in this period. Instead of the royal court in Madrid and its imperial hassles, he liked to be on the beach near his tuna fisheries. But wealth brought obligations. The Spanish King Philip II had grand plans for the wealthy tuna duke from the south. Whether it was because of the tuna, his knowledge of the sea and coastal defence, or the contribution to the war chest is not entirely clear, but the king pronounced him commander-in-chief of the greatest war fleet of the early modern age: the 'invincible' Armada. The seventh tuna duke of Medina Sidonia would thus go down in history as the admiral of the greatest naval force designed to safeguard Spain's position as a world power.

Fishing Methods

The archives of the dukes of Medina Sidonia indicate that the modern tuna era produced a number of techniques for catching tuna [6]. Building on ancient gear, different systems of long nets were deployed, enabling the capture of a large shoal. The *almadraba de tiro* was initially used in the first centuries of the new tuna era. Nets were towed into the water from the beach by smacks in order to enclose and haul in the tuna. The stretches of coast consisting of shallow sandbanks and beaches were particularly suitable for this, as the tuna swam past in large shoals not far from the beaches. According to the last duchess of Medina Sidonia this was also the classic form of tuna fishing in antiquity. Until the twentieth century this form of *almadraba* was still found in North Africa and South America. In actual fact the *almadraba de tiro* is the original form of a circular net, an unmechanised prototype of the *purse seine* which at the end of the twentieth century grew to be the most important large-scale tuna fishing technique in the oceans.

The *almadraba de tiro* mobilised a small army of fishermen. As many as 300 men with a strict hierarchy and division of labour were deployed to pull the fish ashore. That was largely a matter of strength, but getting the tuna into the nets required more than brute force. It was a process in which everything hinged on timing, knowledge of the sea and of tuna. A team of 80 to 120 fishermen had to set out the nets so that the shoal of tuna was slowly closed in, involving startling the fish by beating the blades of their oars on the water. Once they were enclosed, the nets were pulled in at both ends by the beach team.

There was a crucial role for the lookouts in the boats and from the towers on the beach to communicate the position of the tuna. Once the nets were hauled

---

(continued)

in and the colossal struggling fish no longer had anywhere to go, the tuna was pulled out of the water with lances and hooks and if necessary clubbed to death. The fish was hoisted onto ox carts, transported to the racks where they were cleaned, cut into pieces by specialised filleters, and subsequently salted.

Probably under pressure from Italian financiers who were looking to cut costs, halfway through the eighteenth century the 'Sicilian' *almadraba* was introduced to the coasts of the Strait of Gibraltar. This method, originating in Italy and Portugal, worked with a fixed system of long nets. Basically it was the same system that according to some archaeologists was used in antiquity by the Phoenicians and Romans, and had survived in other areas of the ancient fishing grounds. These vertical nets were held in place with floats and heavy anchors in the shallow coastal waters. On its journey the tuna would encounter the wall of the *almadraba* nets, swerve towards the open sea and be led into a number of fixed chambers from which they could not escape. Once they had swum into the central chamber, the entrance was closed off. A horizontal net was then raised up from the bottom, eventually bringing the tuna to the surface, where it could be pulled out of the water. The Sicilian method meant a considerable saving on staff, so it was no wonder that the fishermen initially attempted to boycott this form of *almadraba*. In the end economics won. The fixed-net *almadraba* is still used off the Spanish south coast, in the waters of Morocco, Algeria and Tunisia, as well as Libya, Portugal, Italy and Canada.

Among fish archaeologists and tuna experts there is some dispute as to whether the 'Sicilian' *almadraba* was really used in ancient times alongside, or perhaps instead of, the *almadraba de tiro* [39]. Descriptions by the classical writers seem to indicate that the Sicilian method was in fact used even in ancient times. The formations of heavy weights found in the sea in the places where the tuna fishing took place also point to this conclusion. The classical descriptions, however, also mention scouts standing watch, which may be an indication that the circular nets might also have been set along the shoreline.

In Spain, King Philip II, undisputed ruler of the sixteenth century empire over which he reigned from his monastery palace El Escorial, is known as 'el Prudente' (the Prudent) The rest of the world, however, is clearly less favourably disposed towards Philip [187]. Avoidant, unenlightened, fanatical and belligerent, more Catholic than the Inquisition. In short, the Philip of the Black Legend, a scourge in 1587. The battle against the resistance in the northern Low Countries of the empire had brought Spain to the verge of bankruptcy. Persistent attacks on Spanish ships by the Dutch and the English brought substantial financial and geostrategic damage. The pirate Sir Francis Drake, one of the most infamous plunderers of the Spanish fleet, had attacked Cádiz and Queen Elisabeth of England had had the Catholic Mary Queen of Scots executed. The Low Countries were in revolt. This demanded action. Philip came up with an ambitious military operation which was to humiliate his opponents and throw them off balance. A massive invading fleet of unprecedented scale would attack England and then help the troops, led by the duke of Parma, to cross the Channel from the Netherlands. England and the Low Countries could be controlled in this way. And who better to lead this monster fleet than the duke of Medina Sidonia? He knew the sea, the tuna trade benefitted from stable shipping routes for export, and, perhaps most important of all, in this way the immensely wealthy duke could also make a direct financial contribution in exchange for this great honour.

The 'invincible' Armada was a terrifying war machine [162]. On 20 May 1588, 130 warships, half of them made from the greatest converted trade ships of the Spanish, Portuguese and Italian fleet, manned by 30,000 crew members, set out from the port of Lisbon. Initially the stubborn marquis of Santa Cruz was to be the commander-in-chief of the Spanish fleet, but 3 months before departure he had unexpectedly given up the ghost. The seventh duke of Medina Sidonia became the new admiral. In Spain opinions still differ today as to the quality of the appointment. Besides some coast guarding and his annual fishing campaigns at the tuna palace in Zahara de los Atunes, the tuna duke had little to no maritime experience. Was this fishing nobleman not appointed solely on account of his fortune, his high birth and family relationships with the king? Or did Philip II require a docile, prudent type for his plans? From the 'top secret' documents sent to the admiral in sealed envelopes and only discovered years later, it appears that Philip, contrary to popular belief, was not planning a military occupation of England. Instead the whole operation was intended to be a punitive expedition to frighten the English Queen Elizabeth. He really wanted to force peace on his terms: the departure of the English from the Netherlands and the religious freedom of the Catholics in England. The law-abiding, diplomatic duke of Medina Sidonia was the man for the job.

The duke himself, however, had very different views on the matter. As we can read in a letter to the king he begged to have someone else appointed in his place. 'I kiss Your Majesty's feet and hands for having chosen me for a task of such magnitude. I wish I possessed the talent and strength to carry it out. My Lord, my health prevents me from boarding a ship because my scant experience at sea has taught me that I soon become seasick and am also plagued by rheumatism' [140]. Furthermore, despite the flourishing trade in tuna, there were money problems. The duke noted that he had mortgaged his palace for 900,000 ducats. His majesty needed to know that he therefore did not have a cent to put into the Armada. The nobleman had even had to borrow money for his last trip to Madrid. The king was kindly requested to look for another helmsman. 'It is not right to accept the position of commander-in-chief of such a great war machine, such an important enterprise, with no experience of sea or war. I would be a blind man (. . .) and would have to be led by others without knowing who was good and who bad, who wanted to betray me or throw me into the ravine. I beg you, in all humility, not to burden me with a business for which I cannot accept accountability, because I do not understand it, because I do not have the health for it, nor the necessary money . . .' It is a long letter, perhaps a little plaintive in tone, although exquisitely penned. Philip II, however, was unrelenting and sent the duke personally to Lisbon with no further delay to equip the fleet for the operation.

Initially things went smoothly. Working with precision, as with an *almadraba*, Medina Sidonia had the fleet rigged in no time. The navy which left the mouth of the Tagus on 29 May 1588 was terrifying in its magnitude. At the head of the fleet sailed the galleon San Martín, the flagship carrying the duke. A cannon fired from the admiral's ship announced the departure of the war machine.

Was the duke really troubled by seasickness as he told the king? Then there was little hope for him. No one could remember such horrendous ocean weather as in the spring and summer of 1588. As it passed Cape Finisterre the Armada met with a lengthy summer storm in the Bay of Biscay which already left the fleet badly damaged and delayed. At the end of July the Armada finally entered the Channel in its feared crescent formation. No one on either side would deny that it was an impressive sight. Nevertheless, in just 4 days the English fleet under the leadership of Charles Howard—and in particular his vice-admiral Francis Drake—wrought serious damage on the Spanish monster fleet. The smaller, more agile English ships with their more modern artillery succeeded in trumping the cumbersome Spanish colossi with clever manoeuvres.

While the Spanish force had so little result at sea, the land invasion was also scuppered [186]. The duke of Parma was not even considering embarking for

England from the Low Countries with his enormous army, for the simple reason that he only had flatboats for the crossing, river boats which were far from seaworthy and would be defenceless prey in an attack by the Geuzen (the Dutch sea thug opponents of the Spanish). The duke of Medina Sidonia by contrast depended on Parma coming to his aid at sea against the English attacks and sent a series of increasingly desperate letters ashore. The duke of Parma's aid never came.

The situation only worsened. When the Armada moored at Calais, the English launched 'Antwerp Hellburners' at the fleet. These ships, burning and loaded with explosives, were categorised as secret weapons and sowed the same panic as torpedoes in World War II, or drones nowadays. Dispelled from its moorings the fleet became a puppet of wind and tide. Part of the fleet threatened to land on the sandbanks of Zeeland. The Armada set off on a chaotic flight towards the cold northern seas of Scotland and Ireland. Many ships went down there in storms off the coast. Since messages took a long time to get through, at home in Madrid, King Philip II lived for weeks in the belief that the operation was a success, his invading fleet terrorising the English and the Dutch. But in the autumn and winter of 1588 little that remained of the Armada returning to the ports on the Spanish north coast could be called invincible. The crews were sick with typhoid fever and exhaustion and a number of ships were in such a bad state that they sank in sight of port. The total number of Spanish ships lost has never been established, but it is estimated that half the fleet perished and 10,000 men never returned. The admiral made it home. He was one of the first to arrive in Santander in the San Martín, the colossal, gilded flagship. The Duke of Medina Sidonia left as a broken man who did not even wait until his ship had been sailed into the port, but was hastily set ashore in a dinghy.

From the letters and reports the tuna duke emerges as a man who was sceptical from the start about the chances of the entire operation succeeding and continually warned the king of practically insoluble problems, but who, despite all setbacks, tried to make the best of it. He had proven himself a loyal and courageous executor of an impossible task. According to British historian Geoffrey Parker the failed adventure with the Armada was a decisive turning point in the Eighty Years' War between Spain and The Netherlands. Philip II, *el Prudente*, never recovered from the blow [186]. The Spanish king treated the demise of the Armada as a test from God, laying all the blame on the duke-commander. The dynasty of Medina Sidonia was reminded of the defeat of the Armada for generations to come. Even Cervantes mocked the tuna duke in a comment on Spanish nobility by Don Quixote to Sancho Panza. He scornfully calls the duke 'the ever victorious and never vanquished Timonel of Carcajona, prince of New Biscay' [58].

The duke hastily withdrew to the palace of Sanlúcar de Barrameda. From there he requested that the king never again require him to serve in a sea war. That request was granted, but, as a reminder, even along his own coast the duke now often had to deal with the English and Dutch fleet, the new rulers of the waves. As captain general of the southern coast guard at the end of June 1596, just as he was inspecting the local tuna factory on the beach of Conil, he received word that an enormous Anglo-Dutch fleet had appeared off the coast. From a distance the duke had to watch, powerless, as the port of Cádiz was invaded and the city thoroughly plundered. In

1606 the Dutch made mincemeat of a squadron under leadership of the duke near Gibraltar and, a year later in a bold strike by Admiral Jacob Van Heemskerck, the Dutch fleet destroyed the entire Spanish fleet in the Bay of Cádiz. The thundering of the cannons must have been audible as far as the tuna palace of Sanlúcar.

Alonso de Guzmán, seventh duke of Medina Sidonia died in 1615. He had led the most successful tuna multinational ever to have existed. It had brought him wealth and fame, but also heavy humiliations. Jokes would be made about his naval defeat for centuries to come.

The loss of the Armada also signalled the end of the *almadraba's* glory days. From the end of the sixteenth century statistics reveal a clear crisis in the tuna market. It is not clear if the two events were related. 'We don't know the cause of the demise of the tuna,' the duchess of Medina Sidonia told me. Not only did the tuna catch dramatically decrease, but the demand for tuna also plummeted. International purchasers lost interest, prices collapsed. The *almadraba*, expensive to set up and maintain, inevitably ran into crisis as a result. It was certainly related to the continuing wars, which complicated trade on several fronts. The financial pressure from the Spanish treasury, which had to keep the war machine going, did not help either. In order to supplement the constant shortages the eye of the state had fallen upon the salt used as a preservative by the fishing industry. As early as 1562, Philip II expropriated the salt pans in Andalusia. Only part of the production was reserved for the tuna fisheries. The resulting shortages made salt expensive. The duke of Medina Sidonia was forced by the exchequer to hand over an annual sum of 300,000 ducats in order to lay claim to a guaranteed share of the salt produced. The wars with England and the Netherlands also placed Spain in commercial isolation. The routes to large trade cities in the north, for which Amsterdam had become the lively centre in the seventeenth century, were blocked to Spanish ships. In turn all trade goods, ships' cargos and possessions destined for or originating from Holland, Zeeland, or one of the other rebellious provinces were seized by Spanish royal decree. The merchants, captains, sailors and others on board were arrested. The bilateral boycott served a strategic political goal: given that an important share of the economic motor of the rebellious Low Countries was driven by trade with Spain, the thinking was that the blockade would place the North in a strangle hold [154]. This politics was conversely applied to cut off Spanish tax revenues from the southern Netherlands, with Antwerp as a port metropolis. The Dutch Republic also gave everyone in their extensive fleet carte blanche to capture and rob the Spanish ships. On the Spanish side the execution of the Spanish trade embargo lay in the hands of commissioners in the coastal regions. Thus the seventh duke of Medina Sidonia, and after his death in 1615 his son, were closely involved in executing the boycott as far as it concerned the Spanish south coast. As a commissioner in service to the king, the duke had little choice but to faithfully enforce the embargo. But as the owner of the monopoly on tuna fishing he saw free trade thwarted. In his role as commander of the southern Spanish coastal guard the CEO of Europe's greatest tuna enterprise was cutting off his nose to spite his face.

The Tuna King of Andalusia

The seventeenth century was a difficult time for the tuna dukes. Under pressure from falling tuna income and heavy taxes on salt, the situation for the tuna fisheries worsened. Gaspar Alonso Pérez de Guzmán, ninth duke of Medina Sidonia and grandson of the unfortunate Alonso de Guzmán who had led the Armada, was still one of the richest men in Spain, but was also deeply in debt. He was looking for a way to hold onto his position of power. By all accounts it appears that in 1641 the duke was involved in a conspiracy against the Spanish King Philip IV [154]. The aim: to found the kingdom of Andalusia. Over the centuries the dukes of Medina Sidonia had reigned as vice-roys in Andalusia. Under the command of the weak Spanish King Philip IV it seemed that the time was ripe for the duke to become king himself. The opportunity came when Portugal declared independence and the Catalans rebelled—also due to the expense of salt. Spain went into institutional crisis. This was not helped by the fact that, in 1628, the pirate Piet Hein from Holland captured the entire Spanish fleet with silver from Peru, looting half a billion euros' worth.

The crisis was a good moment to strike. The plan was for Andalusia to split off as a separate kingdom under leadership of the local Spanish aristocracy. The duke of Medina Sidonia would of course be the new monarch. However, the uprising of the tuna king of Andalusia turned out so messy that the precise identities of the conspirators and their motivations remain a subject of dispute today. Moreover, the conspiracy had all the appearance of a family feud between the various branches of the Guzmán family in Spain, particularly between the duke of Medina Sidonia and his ambitious distant cousin Gaspar de Guzmán, count of Olivares. The latter, who as deputy to the weak Philip IV acted as the de facto ruler of Spain, had ordered his cousin as army commander to defeat the rebels in Portugal. The duke, however, procrastinated so long over mobilising his army that instead rumours began to circulate that he

(continued)

himself was devising a coup in Andalusia, to be supported by the Portuguese rebels. France and the Dutch Republic, always eager to make the life of the Spanish crown difficult, would launch a fleet to help.

The kingdom of Andalusia never got off the ground. The majority of the Andalusian aristocracy in the end turned out not to have any appetite for rebellion, nor were the Andalusian people very interested in the issue. The duke of Medina Sidonia admitted his involvement, but was quick to accuse his co-conspirators of masterminding the plans. The latter had their throats cut, but the duke—still the holder of the most important noble title in Spain—received more clement treatment. He was not to show his face again at court, a few titles were removed from his list and he was fined an exorbitant 200,000 ducats.

Part II

The Modern Tuna Era

The Tuna Saint of Sustainability

9

*Despite their journeys back and forth the tuna … cannot
avoid being eaten by larger fish, and especially by man.*
Brother Martín Sarmiento, 1757 [114, 214]

From the beginning of the eighteenth century a whole new crisis suddenly arose in the tuna industry, with geostrategic factors playing a subsidiary role. There was no question of existing society or its institutions collapsing, nor of trade routes being completely wiped off the map. Nevertheless tuna undeniably lost its importance on the markets and in the game of power. The primary reason: catch volumes suddenly dropped dramatically, leading to reduced trade. Yields fell, causing further problems in financing this labour- and capital-intensive fishing business. There was a downward spiral, with low supply leading to market crisis. For the first time in history the realisation dawned that there was more going on than the usual problems of funding, sufficient market demand volumes and the waning power of a tuna monopoly. Something was missing from the production side of the value chains: the supply of bluefin tuna was dropping. The fishing industry found it increasingly difficult to catch tuna in quantities previously considered normal. The number of places where tuna was fished dropped due to lack of sufficient catch. One way or another, fewer tuna were swimming into the nets. Ingenious tricks were conceived to improve the catch. Models of swordfish fashioned from wood were thrown out into open sea in the hope that the tuna would be startled from its migratory route into the nets. But every inventive attempt failed. Where had the bluefin tuna got to? Were they swimming elsewhere in the sea? Were there fewer of them?.

As early as the fourteenth century the realisation had dawned that perhaps there were certain limits to the catch which should be observed. At that point policy regulations were drawn up stating that tuna fishing could only take place during a certain period in order to guarantee that the tuna could reproduce and sufficient fish remained in the sea. It was a first cautious step in the direction of the concept of MSY

© Springer Nature Switzerland AG 2019
S. Adolf, *Tuna Wars*, https://doi.org/10.1007/978-3-030-20641-3_9

(maximum sustainable yield), developed later to restrict the maximum catch to so that fish stocks were maintained over time. This was accompanied by a ban on fishing for tuna in open sea. But the coast was long, the sea vast and checks on the measures did not amount to much. Even specialists could hardly imagine that there was a limit to the quantity of tuna that could be caught, let alone that it served any purpose to design some kind of fishery management policy. Take the eminent British biologist Thomas Henry Huxley, for example, who declared in 1883, 'Probably all the great sea fisheries are inexhaustible; that is to say, that nothing we do seriously affects the numbers of fish. Any attempt to regulate these fisheries seems consequently, from the nature of the case, to be useless.'

More than a century earlier in Spain, however, people were developing ideas on the need for sustainable fishing. The direct cause was the tuna crisis. It all began with a pink flamingo in a monastery cell in the capital city. Madrid might not be the first place to spring to mind in connection with tuna. The nearest coast is three hours' journey by high-speed train from the capital city, while those wishing to see tuna have to undertake a journey twice as long to Cádiz, or to the north coasts of Galicia or the Basque Country. Nevertheless, there was a pronounced smell of fish in Madrid. Madrid's citizens are among the greatest fish eaters worldwide. After Tsukiji in Tokyo, the wholesale market Mercamadrid has the highest trade volume of any fish market in the world. At every daily market or supermarket the consumer will encounter an extensive range of fresh fish. ICCAT (the International Commission for the Conservation of Atlantic Tuna) has its headquarters in Madrid. Spain still has the largest fishing fleet in Europe and the biggest industry for canning tuna.

During the mid eighteenth century the Spanish capital could also lay claim to the venerable Benedictine monastery of San Martín where the illustrious biologist and monk Martín Sarmiento resided in his cloister cell. The learned Benedictine monk was famous throughout Spain as a botanist, but this sharp-witted natural scientist had broad interests. So no wonder that Pedro de Alcántara Pérez de Guzmán y Pacheco, 14th duke of Medina Sidonia thought of Sarmiento to work out what was going wrong with his tuna.

The flamingo was a gift from the duke to the monk to persuade him to investigate the state of bluefin tuna off the south coast of Spain. The duke had the bird caught in the salt marshes of Cádiz, where these pink stilt walkers forage en masse on their migratory journeys to and from Africa. In January 1757 the learned monk received the bird in his study cell. Could this flamingo perhaps inspire the monk to investigate the migratory patterns of tuna—a fish which, like the flamingo, was known for travelling great distances on its journey to regions with food? So what in heaven's name had happened to the bluefin tuna, which had swum by every spring in living memory, but was now suddenly caught in such depleted numbers in the nets of the *almadrabas*?

The tuna duke was at his wits' end. His tuna empire was on the verge of collapse. Over the course of the eighteenth century the catch had dropped to less than 10% of the quantity caught in its heyday. The records at the palace of Sanlúcar de Barrameda dating back to 1756 register a catch of 6000 tuna: a joke compared to the 130,000 caught by his forefathers in the peak years. Far too little money was coming in; the noble debts were piling up. Meanwhile the market was crying out for tuna. The Spanish economy prospered in the eighteenth century. Money flowed in from the colonies and was quickly spent, the plague was under control and the population was growing rapidly. There were more mouths to feed, and in Spain that meant more fish, more tuna, more *almadrabas*. The causes of the disappointing catches had to be traced and remedied as quickly as possible in order to provide the rising consumer market with a robust supply.

Martín Sarmiento initially protested: he had never seen a tuna in his life. Apart from short research trips into the monastery garden he barely left his cell. At best he was familiar with the salted or fermented pieces of tuna sold on the market in the capital city or in Santiago de Compostela, where his younger brother lived.

But the duke's flamingo unlocked something in him and the puzzle of bluefin tuna roused his interest. Sarmiento had piles of books and records of the Medina Sidonia family's tuna dynasty brought from the palace in Sanlúcar and knuckled down to work. It was the first time that the archives had been systematically studied. In their manuscripts the ducal bookkeepers had noted catches from 1525, making them the first fishing statistics ever recorded in the world. Hundreds of years of figures indicated trends which were also indicative of tuna stocks dating back much further. The fishing equipment, methods and even the places where the dukes fished do not differ fundamentally from those of the Phoenicians and the Romans. Environmental factors around the catch remained practically unchanged all that time.

The decline was remarkable. The books note that from 1673 more and more *almadrabas* were discontinued. The tuna stock dropped. Perhaps that was in part in God's hands, but it was primarily the work of humans, Martín Sarmiento found out. He had studied the biology of bluefin tuna in depth and thus was the first to describe the state of knowledge at the time. Many of his ideas still influence thinking on tuna, which is remarkable considering the fact that the monk had never been at sea. Fortunately the well-stocked library of the monastery of San Martín offered a solution. The classical Greek and Roman scientists had already written about tuna thousands of years before.

Having studied the classical literature, Sarmiento concluded that every year bluefin tuna swam into the Mediterranean from the Atlantic, bred and left. 'The tuna is the vagabond among fish,' he wrote. 'Always on its way, with no fixed abode', a fish 'without a fatherland, . . . a while here, then a while there'—constantly in search of nutrient-rich water and fleeing danger.

Within a couple of months Martín Sarmiento published his findings in a book with a long title: '*About tuna and its migrations and hypotheses about the decline of the almadrabas and the means to recover them*' (Fig. 9.1). The document of 56 handwritten folios would go down in history as the first scientific study on sustainable tuna management worthy of that name. It was also the first time that the idea of sustainability was expressly linked to fishing policy. The conclusions were not particularly hopeful: bluefin tuna might conceivably disappear from the oceans Martín Sarmiento concluded. Bent over his books and puzzling over the tables of centuries of tuna catches, the monk had gradually become convinced that the problem of decreasing catches had nothing to do with weather conditions, as was

Fig. 9.1 Martín Sarmiento,
Tuna Saint (cover page
manuscript 'De Los Atunes',
S. Adolf)

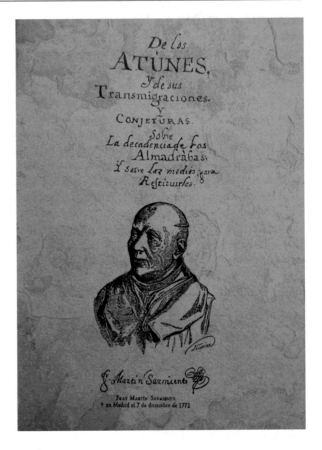

sometimes claimed. Neither was it to do with the Portuguese catching the first shoals
of tuna to come swimming in from the Atlantic in their westernmost *almadrabas*, as
others suspected. Martín Sarmiento became convinced that something was wrong
with all fishing of bluefin tuna in the oceans. This was due to man, his lack of
planning and policy, and his penchant for quick, short-term profits.

This was new. No one had previously suggested that overfishing and exploitation
could be a direct cause of a reduction in the population or drop in the catch. The
monk from Madrid even pointed to the guilty party: fishermen who continued fishing
during the breeding season despite the ban. In his view that was not only biologically
catastrophic but also economically unwise. Maximising profits from the catch in the
short term was a disastrous business model for a fishing economy that took itself
seriously. 'The intent of fishing as much as possible is the worst kind of fishing
business. It will lead to the disappearance of the fish,' he wrote in ornate letters on
the parchment folio. 'Those fishermen were not remotely interested in the fact that
tuna fishing was declining due to their actions, as long as they were able to reap large
short-term profits for the owners of the *almadrabas*, while in doing so they were

behaving terribly destructively towards the business, with fool's gold which raised their own income in the short term, while destroying the source.'

'Greed and gluttony' were in his view the cause of the tuna's disappearance. It was like the tale of the goose with the golden eggs, wrote the monk. If you slaughtered the goose, then the eggs soon ran out.

Brother Martín Sarmiento, the learned monk at the San Martín monastery in Madrid, thus became the first to warn of overfishing for bluefin tuna. He also made a clear connection between biological fish stocks and economic pressure coming from the fishing industry. He considered restricting the catch and other protective measures.

Unfortunately a Saint Martin already existed in the shape of the former bishop of Tours who lived from 316 to 397, protector of weavers and patron of France. Nevertheless, in these times of anxiety about sustainability and the environment, the Catholic Church should seriously consider adding Martín Sarmiento to the pantheon of saints. Saint Martín Sarmiento, protector of the bluefin tuna. Patron of sustainability. At any rate this monk had two miracles to his name. One: he found out that the survival of the tuna could be threatened by overfishing. Two: that was as early as 1757, before anyone else had ever considered the idea.

It wasn't just about the fish, but mainly about the fishermen, the monk believed, thereby for the first time acknowledging the human influence on the marine environment. Sarmiento soon discovered that fishing methods had become considerably more efficient over the centuries. The classic *almadrabas*, whereby the nets were thrown into the water from the beach to pull in the tuna, had gradually been replaced by Sicilian *almadrabas*. The fixed wall of nets, which led the tuna into closed traps, provided a much higher yield for the effort required.

In his manuscript the monk explicitly warned that certain fishing methods could end up being devastating to the survival of a species of fish due to their extreme efficiency. He described them as 'destructive' fishing nets which reach to the seabed, allowing no fish to escape. No fish, no eggs, no reproduction. The fish stocks were thus wiped out by the fishermen 'all together in a single strike', Martín Sarmiento wrote in horror.

Although he does not mention it explicitly—the terminology was only introduced two centuries later—Martín Sarmiento's study is about the lack of sustainable production chains. There were plentiful examples in Spain of how things could go wrong with raw materials, he wrote. A shortage of charcoal and wood had arisen due to insufficient planting of trees. There was too little meat, because the cows and pigs were eaten young. Fish disappeared from the sea and rivers because the fishing ban during the mating season was not respected.

The monk thus revealed himself as a warm advocate of public regulation of the fishing sector. His brother, who as regional minister of fishing was responsible for the Spanish region of Galicia, preceded him. Galicia had to deal with fishermen from Catalonia, who fished for sardines on a large scale, then left for their own region taking profits and all, to spend the money there. The problem was solved by only allowing fishermen from outside the region to fish with a licence. The rest were banned from fishing. Catch quotas and licences: something similar could also be introduced for tuna.

It was a matter of maintaining a critical population of tuna, sufficient to keep the fishing industry going. With that aim the Benedictine monk made a series of recommendations which were supposed to bring the ailing tuna stocks back to health. One was to feed the fish, with seaweed and small crabs which could be scattered in the water. Rather like Strabo and Aristotle, who were convinced that tuna, like pigs, were fond of acorns, Sarmiento proposed planting holm oaks and cork oaks along the coasts, with which the tuna's feed could also be supplemented.

The monk also commented on the economic side of the fishing gear employed. The duke was politely advised to replace the *almadraba de tiro*, with its round nets deployed from the beach, with fixed nets at sea as soon as possible. That would not only save a great deal of manpower and wages, it was also significantly more thrifty in terms of food used by the teams. As things stood at the time they were almost using more fish to feed the fishermen than they were catching in tuna. Sarmiento also felt that they should consider returning to old fishing methods with harpoons and lances. It might be less efficient but was cheap and easy.

Despite all recommendations for savings and efficiency, the bottom line in the monk's study remained the sustainability of the tuna population at sea. If this was endangered then there should be a forced moratorium on tuna fishing, Sarmiento judged.

'If the lack of tuna threatens the survival of the species, the solution must be sought in protection of the young tuna. It would therefore be useful to suspend application of the *almadrabas* in spring for 2–3 years, to enable the fish to produce and fertilise their eggs ... During this suspension they could be provided with extra feed and cared for without being caught.' It was all about reproduction and protecting the young fish so that they could grow to maturity. In particular they needed to prevent people from catching the female tuna for their roe, according to the monk, as, 'From every tuna a million larvae and small tuna grow ...' Right he was.

History makes no mention of the duke actually following Martín Sarmiento's recommendations to throw crabs into the water and plant seaweed and oaks on a large scale. What is certain is that despite the monk's advise the tuna catch from the *almadrabas* remained meagre. Many of the solutions proposed by Martín Sarmiento were probably ineffectual, while others were difficult to execute. It turned out not to be easy to gain control of the entire tuna value chain, formulate policies and enforce measures to protect the juvenile tuna. This does not detract from Martín Sarmiento's great accomplishment: his ideas of fisheries policy and sustainability were far ahead of their time. This was also acknowledged by the duke of Medina Sidonia, who remained friends with the 'green' monk all his life. When Brother Martín Sarmiento died at the age of 77 the duke paid for the funeral and had a bust made of his friend the tuna scholar.

In 1773 the duke of Medina Sidonia was so desperate that he begged the Crown to exchange the tuna monopoly for a privilege which at least generated some money. He wanted to get rid of his fishing licence, which had turned from a privilege into a poisoned chalice. The king refused. So in the end, with great difficulty the house of Medina Sidonia continued to maintain three *almadrabas*: one at their fishing palace in Zahara de los Atunes, one off the coast of Coto Doñana opposite their palace in Sanlúcar and one in Huelva. Catches were meagre; there was no longer any question of real trade. A natural disaster also worked against them. In 1755 the Atlantic Ocean off the southwestern tip of Portugal was again hit by a severe earthquake. According to the chronicles a tsunami claimed the lives of around 1200 victims on this part of the Spanish coast. The tuna town of Conil was virtually wiped off the map, with many boats and nets destroyed. Substantial sums of money must have been tapped for investments in new materials, and bankers were only willing to advance money in an organised market which offered guaranteed returns. In his lifetime the 14th duke of Medina Sidonia mainly lost money on tuna. He died childless in 1777. On his death all his titles and possessions—including the *almadrabas* and Brother Martín Sarmiento's manuscript—passed to his cousin, José Álvarez de Toledo y Gonzaga, 11th marquis of Villafranca, and great great great great grandfather of the latest duchess of Medina Sidonia.

The *almadraba* now definitively fell into decline. Attempts to make the tuna marketable to more sophisticated consumers ended in failure. The better classes in London and Paris turned their noses up at the fish, which was largely seen as food for the common people. The fishing palace in Zahara de los Atunes was no longer the economic motor of times past. Most fishermen lost their knack for the business, which required complex knowledge of tuna. Mistakes were made: the boats went out to sea in stormy weather and *almadrabas* were deployed out of season, when no tuna were swimming by. Inadequate control of illegal fishing and tuna theft further undermined the little trade that was left. The decline of the dukes of Medina Sidonia's tuna empire brought the disappearance of the final form of governance for regulating the catch according to Sarmiento's advice.

In 1816, 4 years after the new Spanish constitution came into effect, the tuna privileges of the dukes of Medina Sidonia officially ended. This also brought an end to a tuna fishing monopoly spanning many centuries. The dukes fished on until the 1870s, now more as a pastime than for serious trade. Only a few private ship owners were interested in taking over the fishing rights. Fishermen had to be hired in from Portugal to keep the fleet up and running.

Gradually new, large-scale marine protein sources overtook tuna. In the North Sea, for example there was a surplus of herring, and from the seventeenth century, with the exploitation of rich fishing grounds on the Atlantic coasts of North America, cod soon conquered the market. This fish abounded and could be relatively cheaply caught. Dried and salted cod was easier to transport than salted tuna. Salmon from Alaska and the Pacific Northwest of the United States conquered the market from halfway through the nineteenth century, first as salt fish and later as a pioneer in a great industrial canning industry.

It was to be another century before bluefin tuna would return to conquer the world market once more.

Luisa Isabel Álvarez de Toledo, 21st duchess of Medina Sidonia died of pneumonia at the age of 71 in the family palace of Sanlúcar de Barrameda on 7 March 2008, 6 months after we met. She remained in all things and to the last the defiant heiress of Guzmán *el Bueno*. Shortly before her death the duchess made an appeal for the legalisation of drugs and euthanasia. She criticised the new digital age and predicted that individual liberties would come under pressure. Spain sometimes seemed even less free than under Franco, she complained. During her lifetime she had one way or another succeeded in making money to save the palace and the family archive of the tuna dukes. On her deathbed Luisa Isabel Álvarez de Toledo married her secretary Liliane Dahlmann, who after her death took on management of the trust in which the centuries of tuna chronicles are now kept. After more than 700 years the house of Medina Sidonia's close personal involvement with bluefin tuna came to a definitive end. What remained preserved were the parchment folios recording the catch of bluefin tuna, the first ever known fishing statistics of inestimable value [108].

The end of the tuna privilege of the house of Medina Sidonia marked the end of centuries of history of tuna fisheries with their origins in the ancient times of the Phoenicians and Romans. It had combined a large-scale fisheries with a processing industry and trade network that created a surprisingly modern, borderless supply chain and laid the foundation for even bigger industrial fisheries. This sophisticated production and trade could only flourish with organised ownership and finance, safe and open trade routes and state governance to guarantee basic stability along the value chain. When the old order started to crumble this inevitably had consequences for large-scale tuna fisheries. In the next chapter I will describe how the dawn of a new industrial and capitalist era completely reinvented the large-scale tuna industry and brought it back in a new global order. The governance of the tuna chain proved capable of successfully conquering the world once more. So successfully that everybody completely forgot the wise warnings of a monk from Madrid.

The Rise of the Can

10

> *Before you finish eating breakfast this morning, you've depended on more than half the world. This is the way our universe is structured. . .we aren't going to have peace on earth until we recognize this basic fact of the interrelated structure of reality.*
>
> Martin Luther King Jr.

After its rise and decline during the late medieval period and the Renaissance, the modern tuna era began in the nineteenth century, silently but persistent as an inseparable consequence of an industrial era that changed the world. Just as the steam engine was at the core of this change, it was the tin can which returned tuna to the global consumer's plate as a popular fish. The concept was not much different from the salted tuna, *garum*, or dried *mojama*. The key was shelf life, an innovative preservation technique for the larder, assuring the consumer of healthy protein which would keep for a long time. The tin can made an underestimated contribution to the world population, enabling a definitive, worldwide breakthrough for tuna as a popular staple. As so often in the history of tuna it was also directly related to war.

It was the French chef Nicolas Appert (1750–1841) who won an award from the French ministry of defence in 1810 for his method of packaging cooked food in airtight containers in order to extend its shelf life [7]. That was a wonderful invention in service of Napoleon's armies. The preservation industry—working first with glass and later with cans—gradually conquered Europe and the United States over the course of the nineteenth century. Among fish it was mainly sardines that had the greatest success in canned form. The first Italian businessmen travelled to the southern Spanish coastal provinces as early as 1879 to set up factories for canning fish [150]. Andalusia was late to fish canning: the fish-rich Spanish regions such as Galicia and Cantabria had already been doing it for 30 years. The Italians introduced the cans for preserving tuna as well, a new method in an industry in which fish had mainly been salted for thousands of years. From then on the tuna was boiled or

© Springer Nature Switzerland AG 2019
S. Adolf, *Tuna Wars*, https://doi.org/10.1007/978-3-030-20641-3_10

grilled, then canned under a layer of oil. Parallel to this development, the United States also discovered tuna in canned form, more thanks to coincidence than a conscious strategy. In 1903, after an unusually bad sardine catch caused by overfishing, a canning factory owner on San Pedro Bay, southern California, decided to fill his supply of empty cans with tuna. He used albacore tuna (*Thunnus alalunga*) for the purpose, also known as 'white tuna', a smaller sibling of the bluefin, immediately recognisable by its very long pectoral fins. In Spain this fish is known as *bonito del norte*. The flesh is a little lighter in colour and less fatty than the bluefin tuna, but in no way inferior in flavour or nutritional value [45, 223]. Basically the fishing was done with labour intensive baitboats using pole and lines with live bait. Only after the Second World War the development of nylon fishnetting and power operated hauling blocks facilitated the so called purse seine fisheries.

The clash of weapons ensured that the newly canned tuna swiftly conquered the market. The big breakthrough came with the Great War, 1914–1918. Tuna proved to be an excellent choice of fish for feeding soldiers on the frontline. Even Spain, which had remained outside the fray, rediscovered tuna as a new market, offering a practical method of preserving nutritious fish in cans for times of uncertainty. The fishermen on the Atlantic coasts of Andalusia saw demand rise. Exports to Italy flourished, and nature also lent a hand. The bluefin fishing malaise of the nineteenth century was now over: by the beginning of the twentieth century the fish population seemed to have made a good recovery. In 1914, the first year of the war, the *almadrabas* caught 90,000 tuna and in 1918 even achieved a catch of 100,000 [114]. Those figures were very similar to the golden years of the tuna empire under the dukes of Medina Sidonia. The gloomy predictions of the monk Martín Sarmiento appeared outdated.

The arrival of canning appeared to lend new impetus to an industry which had entered a serious crisis in the nineteenth century. Everything was going wrong at the time: tuna was imported cheaply from Portugal, the demand for salted fish dropped steadily and catch yields fell. The Italian export market turned out to be a solution, particularly in the uncertain times of the World War I. The number of Spanish *almadrabas* doubled to around 20 and the tuna ship owners emerged from the war as rich men. Growth continued. In the 1920s Spain achieved a record tuna export figure of more than five million kilos, although the *almadraba* fishers only partially benefited. The market was overrun by tuna and the powerful Italian trading houses colluded to command a low purchase price at the start of the tuna season. Trade and distribution were Spain's Achilles' heel. The Italian and French intermediaries achieved the highest margins. The Italians bought the large 5 and 10 kg cans from the Spaniards, repackaged their contents into smaller cans with the labels of their own Italian brands, and sold the same tuna on at a considerably higher price. This was an old Italian trick. Since ancient times Italians have imported olive oil on a large scale from Spain, which is far and away the world's largest producer of olive oil. They then sell the oil on, attractively rebottled, at a substantially higher price. It was no different with tuna.

The bluefin fishermen and owners of the fishing grounds thus lost their grip on the value chain, to the benefit of powerful intermediary trade oligopolies who were able to control the market. This powerlessness to achieve a good price for tuna soon began to break the Spaniards. The situation was not helped by the United States and Japan beginning to set up their own tuna fleets, canning industries and trade channels from the end of World War I in 1918. Supply grew. Tuna once again began to acquire a place as an important international food commodity in the global value chains.

Fishing still demanded large investments, certainly when it came to catching bluefin tuna in *almadrabas*. The enormous nets with their heavy anchors, staff and boats all required substantial investment. The annual levies for fishing rights were a serious expense, even before there was tuna in the nets. In order to keep up the growing market, *almadrabas* were also started up in northern Morocco, now under a Spanish protectorate, in the same spots once used by the Phoenicians and the Romans. Trawlers set off for the Canary Islands to fish for tuna.

The growing power of middlemen became the characteristic pattern of business governance in the new tuna era, setting the tone in a worldwide market. Indeed, tuna went global all along the chain: the tuna fishing regions expanded, new fishing methods were introduced and copied, freezer technologies and transport over long distances opened up the great world seas and oceans to industrial tuna fishing. The processing of the fish took place in factories in the United States, Latin America, Asia and Africa. Once again the world had fallen in love with tuna. As a cheap fish it conquered a broad public in international consumer markets. That was how the global value chains of fishermen, processors, traders and distribution centres came into being, directing the global stream of tuna from the point when the fish was caught to the moment that it finally arrived on the consumer's plate. A limited number of large traders in those value chains were successful in achieving a central position of power on the market. For the fisheries, canneries and retailers, there was no way around them. The fishing industry, with its monopoly-like tuna fishing concessions, was overtaken by traders as the new oligopolists and rulers of the tuna market. They were the ones who set prices. High costs, shrinking margins and uncertain catches: everything pushed the Spanish *almadraba* owners to merge, leading them to grow in scale and enabling them to survive. It was eat or be eaten. The new tuna corporations belonged to the largest food conglomerates that existed in Spain at the time.

For Spain the nineteenth century had not only been a lost century when it came to bluefin tuna. The country was in a near-permanent state of political crisis, with successive military coups and civil wars. At home military rebellions and governments followed rapidly on one another's heels. The last of the colonies, Cuba and the Philippines, had been lost in 1898 in a short and disastrous war with the United States. Spain could no longer maintain the pretence of a global colonial empire. Things did not look much better at the start of the twentieth century. In 1921, under the personal leadership of King Alfonso XIII—last in a long line of incompetent monarchs—the Spanish army suffered a devastating defeat at the hands of Abd el-Krim's Berber army in the Moroccan Rif. Even the protectorate over the North African mountain region seemed too great a task for the Spaniards.

'Today, rather than a nation, Spain is no more than a dust cloud left behind after great peoples have galloped off down the highway of history,' that was how Spanish philosopher José Ortega y Gasset described the feeling of national humiliation. Under those circumstances, colourful figures found an opportunity. In 1923 General Miguel Primo de Rivera, an Andalusian marquis from Jerez, seized power in Spain. Primo de Rivera saw himself as the saviour of the fatherland. He would carry out his ambitious plans with a firm hand to clear up the mess and allow Spain to rise from the ash of its downfall. Critics, however, saw the general as a naive scatterbrain with authoritarian tendencies. More of an armchair politician than an incisive decision maker. But this general from the province of Cádiz nonetheless placed the tuna problem high on the agenda. Tuna became a national priority, complete with extensive studies from the Institute of Oceanography and the National Fisheries Commission. A drastic solution was decided upon: a rehabilitation of the old tuna monopoly, this time not managed by the dukes of Medina Sidonia, but under state leadership. A mega-merger of all five big *almadrabas* and the associated canning industry into a state consortium to hold sway over tuna fishing with a united front from then on.

Thus, in 1928 the Consorcio Nacional Almadrabero was born, the National Almadraba Consortium, a state monopoly which would take control of all tuna fishing from the *almadrabas* along with the associated Spanish canning industry [150]. The Consortium became one of the greatest food companies in Spain, an industry with thousands of employees and entire towns and villages depending on it economically, a tuna juggernaut of Soviet proportions. Technically the *almadrabas* continued to function much as they had during previous centuries, but now the eight enormous fish-processing factories could produce around 100,000 kg of canned and salted tuna per day. On average there were 600 people working on each *almadraba*. They formed complete factory colonies, with the fishermen, the women in the cannery and their children cared for in their own hospitals, shops, hotels and schools. The consortium succeeded in producing a sizeable quantity of fish internationally. Almost three quarters of European consumption of bluefin tuna was fished from the Spanish south coast.

The Consortium did good business in its initial years. With its schools, hospitals and social facilities the state monopoly brought prosperity to the fishing villages. The Spanish tuna juggernaut also proved capable of winning back some ground on the unfavourable price agreements with the Italian trading houses. It successfully brought an end to the 're-canning' of the Spanish tuna in smaller Italian cans. Exports grew as never before. A new historical record was broken with an export of seven million kilos of tuna in 1929.

Nevertheless, the state monopoly soon proved to be a giant with feet of clay. When the stock market crash of 1929 plunged the world economy into a recession of unprecedented proportions, the brief prosperity Spain had built up over the previous years was soon a thing of the past. One of the first victims was dictator Primo de Rivera himself. He was banished from the country by Alfonso XIII in 1930. The king himself followed soon afterwards. In 1931 the Second Spanish Republic was established, which governed the country until the army rebellion under General Franco in 1936. The Consortium was not spared by the crisis either. Export of bluefin tuna to the Italian market—the only stable market for Spanish bluefin tuna—collapsed in 1933. A more fundamental problem also became visible behind the collapsing international trade: the tuna stocks turned out to have been decimated once again by successful overfishing. On the south coast of Spain the nets began to empty and in Italy, Portugal and Tunisia, too, catches were drastically reduced. The Consortium continued under the rapid succession of Republican governments, but it was on the decline. Eleven years later, when the subsequent Civil War was over, only four *almadrabas* remained active. The legendary *almadraba* of Zahara de los Atunes ceased activities in 1935 due to lack of tuna.

The Consortium, now under Dictator Franco, succeeded in hanging on for a few more decades. In 1972 the tuna monopoly was dissolved for good. In the final years not even one million kilos of fish were caught per season. The new canning industry of the twentieth century had allowed the Spanish *almadrabas* to rise again like a phoenix from the ashes. After the Phoenicians, Carthaginians and Romans, and the dukes' tuna monopoly, it was the cans that saved the fishing industry from oblivion. But despite the growing demand, in the end the classic fishing method ran up against a hard truth: fewer and fewer tuna were caught. The story from the other side of the ocean at this point was completely different: the American tuna fleet, which fished with boats dragging a new sort of net, had succeeded in catching 83,000 tonnes of tuna in 1937, many times more than the Spanish *almadrabas* in their heyday. The classic fishing methods had degenerated into an expensive collective heritage that was unable to stand up to new competition. If no new market was found with higher margins for the bluefin tuna, then the *almadraba* would be condemned to death.

The Tuna Machine

11

> *Or speak to the earth, and it shall teach thee: and the fishes of*
> *the sea shall declare unto thee.*
>
> Job 12:8

Let us pause in our quest for tuna to examine some technicalities, since besides its long history of economic and cultural heritage and its status as an icon of sustainable transition, there is also a biological and technical reason why such fascination surrounds the bluefin tuna. And since we have entered the centuries of new technological inventions, it is a good moment to pay attention to these often undervalued features of tuna. Let us call it the tuna machine. To understand the tuna machine we must first travel to Barcelona. At the end of Las Ramblas, Barcelona's magnificent promenade through the Gothic district, is the city's Maritime Museum. The royal shipyards which house the museum date back to late medieval times and are worth a visit in themselves for anyone interested in maritime history. The hallways, attractively covered with Gothic stone arches supporting the roof structure, reveal an ambition for naval power. Inside we find a replica of the colossal, 60 m long galley El Real ('The Royal'), a faithful replica of the flagship of commander John of Austria, who defeated the Turkish sultan in the Battle of Lepanto (1571), as mentioned in the first chapter. Maybe Miguel de Cervantes had been aboard El Real to receive the letter of recommendation from the hand of the commander himself. Cervantes own marine division probably sailed a similar ship.

It is difficult not to be impressed: propelled forward by 290 oarsmen this ship could accommodate 400 soldiers and mariners. It was built in the sixteenth-century shipyards of Barcelona, but the design is substantially modelled on the *penteconters* with which the Phoenicians travelled their trade routes 2500 years previously and which were later copied by the Romans. Even the ram's head on the prow had survived the millennia. Here we stand eye to eye with a larger and more modern

© Springer Nature Switzerland AG 2019
S. Adolf, *Tuna Wars*, https://doi.org/10.1007/978-3-030-20641-3_11

version of the boat which distributed the salted tuna in amphorae throughout the Mediterranean.

But the feature that brings us here stands in the garden at the entrance of the museum and is much smaller. It is a beautiful submarine, with a pretty polished-hardwood hull, the attractive line of a torpedo and cute little portholes. The spyholes in the bow were added so that someone could give the helmsman instructions for maintaining the correct course. After all, you needed to know where you were going and the periscope had not yet been invented.

Here we stand before a faithful replica of Ictíneo I, the wooden submarine completed in 1859 by Catalan inventor Narcís Monturiol (1819–1885). Ictíneo was the first really built submarine in the world [207]. Monturiol based the name on the Greek word *icthus*, which means 'fish' and *naus*, which means boat: the 'fish boat'. Better still: the tuna boat, as it is highly probable that Monturiol had tuna in mind when he designed his Ictíneo.

Monturiol was an enlightened spirit in nineteenth-century Catalonia. The son of a Catalan bootmaker and a follower of the French communist Étienne Cabet, was known as an idealist. He rose in the Catalan bourgeoisie through curiosity and self-study. The enlightened bourgeois elite were familiar with trading traditions, and like the Basques and Galicians, the Catalan people have a long history as sailors and fishermen. Narcís Monturiol spent the summers in the Catalan coastal town of Cadaqués. In those days Cadaqués still had its own traditional *almadraba* in which bluefin tuna were caught en route towards Marseille. A century later the Catalan artist Salvador Dalí immortalised the local tuna fishermen in his 1967 painting *Tuna Fishing*. Cadaqués' *almadraba*, a spectacle which left a deep impression on Dalí, was discontinued not long after he painted it due to lack of tuna. But when Monturiol spent his warm summers on the beach, the nets were still fully active.

Like Dalí, Monturiol must have been impressed by the powerful giant tuna which was brought ashore. After all, the inventor was intrigued by fishermen and the sea. He had great sympathy for the coral divers who still had a flourishing business in Cadaqués in the nineteenth century. He saw the village divers disappear into the depths to resurface with the valuable red coral. Red gold, as the coral from the Mediterranean was known. The coral divers, who descended under their own breath, led a difficult and dangerous existence. Things would often go wrong during their journey to the seabed and more than once a coral fisher who was unable to reach the surface on time paid for his dive with death by drowning. The coral came at a high price.

That needed to change, Monturiol felt. After all, science offered modern techniques and machines that could help the coral fishers' progress. It was a question of time before humans would be able to travel, work and live under water, he believed. The underwater world was ripe for exploitation. Monturiol thought about it, drew up sketches and discussed his idea with his friends and contacts from the Barcelona business elite. The Ictíneo project, a submarine for coral fishers, was born. Ictíneo got off to a flying start. Monturiol even succeeded in enthusing the government in Madrid, and by issuing shares accumulated enough capital to develop the prototype of his coral fishing dive boat.

The design was a closed cylindrical wooden construction, 17 m in length, three metres wide and three and a half metres tall. A double-walled hull ensured extra strength, so that a depth of 50 m could be attained. That was deep enough to fish for coral, Monturiol had calculated. An ingenious system ensured that you could keep breathing in the submarine. The vessel offered space for 16 people, who had to drive the boat's propeller with their own strength on a kind of bicycle. The Ictíneo was launched in Barcelona's port on 23 September 1859. It was a huge success: before a surprised audience the wooden colossus sank to the port seabed and—at least as amazingly—then returned to the surface with a crew who had survived the submersion alive and well. Never had such a thing been seen before. Fifty test dives followed, reaching depths of up to 30 m.

Propelling the submarine proved to be a problem. Monturiol searched for a solution. The result was Ictíneo II, now with a steam engine. A major drawback: as a result of the engine, the temperature inside the submarine rose to sauna levels. Monturiol went back to the drawing board, but money ran out for further development of Ictíneo II. The company developing the submarine went bankrupt. In 1868 the prototypes were sold off.

Monturiol would never get over this blow. He tried again with a design for a cigarette machine and a new method for preserving meat, but these inventions were also unsuccessful. The inventor became depressed and was soon forgotten. He died deeply disappointed and convinced that his fellow Catalan people had never properly comprehended the Ictíneo's potential.

Greater success was achieved by the French author Jules Verne. Two years after Monturiol was forced to sell his boats, Verne published a new novel, *Twenty Thousand Leagues under the Sea*. The submarine, the Nautilus, described by Verne, exhibits remarkable similarities to Ictíneo. It's likely that the Frenchman was aware of the craft and he may even have been present at the port of Barcelona during one of the test dives of the miraculous fish boat. Verne in fact succeeded in solving the propulsion problem: the Nautilus was able to reach the unprecedented speed of 50 knots. The Frenchman's book was a great success.

Thus Ictíneo became the prototype of the perfect submarine. There had been other forerunners—such as the unlikely underwater rowing boat of the Dutchman Cornelis Drebbel in 1620 and David Bushnell's one-man Turtle in 1776—but Ictíneo was unmistakeably the greatest success. And no wonder: Ictíneo was nothing less than a streamlined wooden bluefin tuna. What better model for a machine than this perfect fish?

The physical side of the bluefin tuna inspires nothing but lyrical admiration. Among admirers of tuna that can take on a euphoric tone. Tuna is the perfect fish. Speak to the *almadraba* fishermen and they never stop talking about it. The same goes for fisheries biologists and tuna researchers, as well as Japanese sushi chefs. Their admiration goes far beyond ordinary culinary, biological or scientific interest. It is a declaration of love for the perfect underwater machine. Tuna dynamics, tuna swimming style and tuna design: bluefin tuna is power plus speed converted into beauty.

For marine biologist Boris Worms unconditional love was kindled when he was invited to a documentary shoot about bluefin tuna on Prince Edward Island off the east coast of Canada. Worms was already famous as the lead author of a notorious article in the American magazine *Science* in 2006 [256] in which, along with an international group of fisheries biologists, he warned economists that fishing for wild species of fish, including tuna, would collapse worldwide by 2050 if no measures were taken to ensure sustainability. Worms went diving with food, looking for a shoal of the wild species that was under threat, coming face to face with giant tuna weighing 1000 lb. 'They were eating almost out of our hands. It was an amazing experience,' he remembers. 'Tuna are simply beyond comparison. The high-speed control. The power, the speed. You're impressed for the same reason people like cheetahs and not sleepy lions. Sharks are like lions: majestically big but slow, lazily floating in the water. Tuna are like torpedoes. Tuna is the cheetah of the ocean,' Worms said when I met him at a conference in Amsterdam. 'The herring floated away a bit and there was the tuna, snapping it up at full speed. They knew exactly what they were doing. Not one fish would touch me. One moment there was a herring between me and a bluefin tuna, which came at it. He had made a turn and obviously couldn't see me as he swam full speed at my face, on a collision course. In a split second I looked him in the eyes; there was this terror in his gaze. Then he made another turn and just swam underneath me without touching me for a moment.'

'Tuna is pure energy,' Spanish marine biologist José Luis Cort testifies. Tuna is a wild, untameable predatory fish that travels enormous distances; it can shoot down into the depths at lightning speed, disappearing into the ocean and suddenly surfacing again in places no one would foresee. 'Only in recent decades have we come to know anything more about the bluefin tuna, but most of it remains unknown. That's because the fish is a difficult creature to study in controlled conditions.' Of course you can catch them and keep them in a tank or cage net, but that's expensive and conditions for keeping fish in captivity are complicated, Cort tells me.

The book *Tuna, Physiology, Ecology and Evolution* is effectively the scientific bible for bluefin tuna, with articles by tuna researchers from Anglo-Saxon countries such as Barbara Block, John Gunn and Carl Safina [43]. It is known as the obligatory reference work for tuna research. The introduction, however, looks more like a poetic declaration of love. 'Watching a tuna swim, you can't help but be impressed by its power, grace, and speed,' write specialists Barbara Block and Donald Stevens, the volume's editors. 'If you've tried to catch a tuna on a fishing line, you have undoubtedly discovered its tremendous strength. Its superbly streamlined shape reflects its ceaseless activity.'

This tuna bible makes one thing clear: everything to do with tuna—from streamlining to the way in which internal tuna mechanics are propelled—is for the optimal use of its energy in the ultimate machine. With its cylindrical body, it can reach maximum speeds beyond those of almost any other fish. But the tuna can also keep up moderate cruising speeds for weeks at a time, powering across distances of thousands of kilometres of ocean. The tuna alternates between a torpedo and an ocean liner.

One of its most fascinating traits is that from a state of repose, the bluefin can burst into top speed in a fraction of a second to shoot after its prey. Scientists measure the tuna's speed in relation to the length of its body, noting top speeds between six and ten times their own length per second. For a tuna of two metres in length that means a speed of more than 70 km per hour. Larger tuna would theoretically even be capable of reaching speeds over 100 km per hour. The bluefin tuna is a swimming Maserati, a nervous racehorse capable of spectacular acceleration.

The secret of the speed is in the special propulsion, with a number of innovative technical intricacies in its design. The first thing that stands out is the extreme efficiency with which the tuna's body is capable of taking up large quantities of oxygen from the water through its system of gills. The fish has this in common with the shark and swordfish. Scientists call it *ram ventilation*, a sort of turbocharger but with gills, requiring a constant stream of water past the dense system. This is also the reason that tuna always swim open-mouthed. It may make the fish look a little clueless, as if they swim around in a state of permanent surprise, but it enables them to absorb more oxygen from the water than most other fish.

In order to take in sufficient oxygen, tuna must remain in constant motion. Stillness and floating in the water is fatal to this fish. The tuna is therefore always swimming around. It even swims in its sleep—sleep being something that scientists assure us it really does, despite the lack of eyelids.

The strong tuna puts up a serious fight when caught in a net, as it immediately starts to suffer from lack of oxygen. And that's not all: when tuna is not in motion, its hydrostatic balance soon becomes out of kilter. The two side fins have a function comparable to the moving flaps on an aircraft wing: they provide upward pressure. When motionless, the tuna therefore not only suffocates, it can also sink to the seabed due to lack of upward pressure [46].

The fascination with bluefin tuna is not new. From ancient times on the bluefin met with warm interest from nature researchers. From the start, its ability to travel enormous distances by sea caught people's attention. In his *History of Animals* Aristotle (384–322 BCE) described how tuna swam into the Mediterranean from the Atlantic via the North African coast in spring, eventually mating in the Black Sea and then returning via Greece, Italy and Spain [9].

In fact not all his observations were equally accurate. Aristotle was convinced that Tuna saw badly through its left eye. He came to this conclusion from the fact that the fish always enters and leaves the Black Sea with the bank to its left. The Greek philosopher's thinking was that in this way its good right eye was always directed at the open water, as a possible source of danger. Clever Perhaps even more spectacular was Aristotle's observation of a gadfly which he believed plagued the tuna under-water. It might put us in mind of one of the 72 parasites which are found on tuna, but no, Aristotle really is talking about an insect. 'The tuna is driven mad with the stings of this fly,' according to the Greek philosopher. That's why a tuna sometimes jumps out of the water. He advises against eating tuna in summer, because at this time the tuna gadfly is active, making the fish less fat from all the jumping out of the water. We now know that the ancient Greeks, like the Japanese much later, loved fatty tuna.

The Roman procurator of Gaul, Spain and Africa, Gaius Plinius Secundus (23–79 AD), better known as Pliny the Elder, devoted a separate entry in his *Naturalis Historia* to tuna [196]. 'An exceptionally large fish. I have seen a tuna that weighs more than 800 lb, with a tail more than a metre wide,' the Roman magistrate noted. This sounds probable enough, but then Pliny's imagination goes to town on the bluefin tuna. He writes that the great fish has a small enemy: the anchovy. 'A very small fish called an anchovy kills the tuna by biting through a particular vein in its throat; the little fish attacks the tuna exceptionally aggressively.' This turns the world upside down, but makes a good story.

Pliny reveals knowledge of some other points of interest. The male tuna has no ventral fin and the only place they spawn is the Black Sea. He also repeats what Aristotle says about the problem of the worse left eye. 'They see better with their right eye, although neither eye sees particularly well.' Pliny also warns that eating tuna always leads to terrible diarrhoea. That sounds reasonably reliable when it comes to fresh tuna: high fat content and lack of refrigeration probably led to considerable spoilage in Roman times.

The Greek geographer Strabo (64–19 BCE) was the undisputed tuna specialist of antiquity. The fish is discussed in detail in his description of southern Spain in his book *Geographica* [229]. Perhaps Strabo—Greek for 'cross-eyed'—felt a certain kinship with tuna due to his own eye problems. We can be certain that Strabo was fond of fish. Although his scientific books in general do not overflow with emotion, 'the cross-eyed one' can hardly control himself when it comes to bluefin tuna, particularly those found in the Atlantic Ocean beyond the Strait of Gibraltar. Enormous shoals of plump, fatty tuna come swimming from the Atlantic into the Mediterranean, the Greek writes. Why are the tuna here so fat? According to Strabo it is because the tuna eat the acorns from the holm oak found all over the coast of Andalusia. After all, enormous quantities of the acorns which this oak produces end up in the sea. They can be found floating in the water as far as Sardinia. The tuna love them. 'These fish are like a kind of marine pig, because they love acorns and grow exceptionally fat on them, even to the extent that more tuna are born when the sea contains a higher quantity of acorns.'

The echoes of Strabo's acorn story can still be heard today. The monk Martín Sarmiento draw on the acorn diet as a possible remedy against the declining tuna population in the eighteenth century, and in the south of Spain tuna is known as *cerdo del mar*, the 'sea pig'. Like the tuna of Phoenician and Roman times, the Iberian ham is prepared by salting. The preparation process of the *mojama* (Arabic *musama*, meaning 'dry'), of the fermented tuna, has many similarities with that of Iberian ham. *Mojama* can also barely be distinguished from a piece of ham. The fact that the grey-brown Iberian pig is raised on a diet of *bellotas*, or holm oak acorns, certainly contributes to the specific flavour of the meat and also ensures that much of the pig's fat content is unsaturated, like the fat in bluefin tuna. A shame that tuna do not eat acorns after all.

An important question which kept on recurring: is tuna really a fish? Long after Aristotle, Strabo and Pliny, the issue remained the subject of debate. Didn't tuna in many ways more resemble a bird? Or a mammal? Was it not necessary to establish a separate classification for tuna?

The question is not as crazy as it seems. Fish are cold-blooded creatures. When the water gets colder, the fish gets colder; when the water gets very cold, the fish gets very cold and often sluggish. But that does not apply to tuna. Tuna are warm-blooded. The exceptionally large tuna heart enables high performance. Scientists refer to this as *high power output*. The heart has a high stroke volume and that means it almost always works at its maximum rate. Tuna also has high blood pressure for a fish. The tuna heart pumps at seven times the power of other teleosts, enabling it to take up much more oxygen in its blood and also to transport that blood rapidly and in large quantities to the muscles [43].

This metabolism keeps the fish permanently in the starting blocks, speeding up smoothly and without losing time when it is necessary to flee from an orca or go after herring. And an even more ingenious feature: tuna can control its body temperature as if with a thermostat.

Tuna Turbo

Then there is the tuna turbo, as you might call the unique energy quality in the form of a heat exchanger grouped around the red muscle fibres. Researchers have discovered a construction in the red muscles of the body which they call the *rete mirabile*, 'wonderful nets', the name conveying their admiration [43]. The *rete mirabile* is a dense network of blood vessels in which the cold blood of the tuna is first warmed by the heat released from the permanent muscular activity of the red muscle mass. It is a biological heat exchanger avant la lettre, comparable with the condensing boilers currently used by every sustainably equipped household. The energy is not lost, but keeps the entire machine nicely up to temperature and ensures efficient performance.

It is suspected that this is the crux of the high performance of the tuna physique. Other fish have to manage without such a turbo system. They see their muscle warmth disappear hopelessly into the ocean via blood circulation in their gills, which are in contact with cold water. The heat exchange mechanism also contains a thermostat to regulate the degree of heating. The tuna thus closely regulates its temperature as needed. Depending on the temperature of the water it finds itself in and the speed at which it is moving, the tuna stokes up the heat a bit or cools things down. Its body temperature is not kept at the same level as it is in mammals, but the effect comes pretty close. Tuna thus has an extremely efficient internal heat exchange mechanism, a kind of biological air-conditioning system which existed long before Mitsubishi entered the market.

At least as fascinating is the special manner of propulsion which makes tuna one of the fastest fish species of the oceans. It is no exaggeration to speak of a 'tuna style' of swimming. It has been discovered that the red muscle mass of the tuna, unlike that of other fish, is mainly grouped around the spinal column. This gives rise to its characteristic rigid swimming movement. The tuna appears barely to move but makes swift progress through the water. The contracting muscle mass causes a fast, backward vibration in the body, eventually concentrating in the swelling of the tail fin, leading to propulsion. This convex tail fin gives extra mass in forward movement. With its rigid swimming style the tuna may be less manoeuvrable, but it can shoot through the water like a torpedo where necessary.

So using bluefin tuna as a submarine prototype is not a bad choice. The fish is a wonder of hydrodynamic design. Its tiny scales, which barely stand out on the metallic skin, make the body extremely smooth. Studies show that with its perfectly streamlined shape and low friction, tuna comes close to being the optimal high-speed fish of these dimensions. The thickest part of the body on the bluefin is a little further back than on other fish. Their convex, rounded appearance might bring to mind associations with the outdated aerodynamics of 1950s American cars, but that appearance is deceptive: it is this bulbous shape which enables the fish to swim more efficiently. Examination of a slice of tuna on the market immediately reveals that the tuna is not round but elliptical. This shape removes the need for the body to twist back and forth too much for propulsion, the narrowing at the tail prevents further energy loss and the streamlined ridges in the tail stop unwanted turbulence from the water streaming past the body. The row of little fins on the back and belly improve balance and reinforce upward pressure. Finally a surprising feat: tuna can adjust its body shape. When speed is required, it switches into high-speed mode: the large dorsal and ventral fins then click inwards into special cavities, while the pectoral fins are pressed flat against the body. The perfect torpedo shape.

Robotuna

The bluefin tuna's superior mechanism still inspired imitation long after Ictíneo. In fact when it comes to biomimicry tuna is at the vanguard of underwater robotics, the development of swimming robots to carry out tasks underwater. The prototype of the robot tuna was launched in 1995 by the Massachusetts Institute of Technology. According to its inventors, Charlie, named after the cartoon character Charlie the Tuna from the famous canned tuna brand StarKist, had to be a submarine-like fish, with a fairly sturdy body. The inventors had reasonable success: Charlie was recognisable as a tuna. The problem remained that the steering electronics were the size of a substantial freezer unit. It was difficult to hide this away in the hull of the mechanical tuna, so it was placed outside the test swimming pool, with cables attached to the fish.

But Charlie was just the beginning of the developmental trajectory. Commissioned by the American navy and Homeland Security, researchers from the Franklin W. Olin College of Engineering and Boston Engineering developed the prototype of an autonomous tuna robot, the Robotuna, that was supposed to be able to swim through the ocean on its own [206, 216]. Although the scientific experiment was the priority, the navy hoped eventually to develop a submarine capable of exploratory dives with new, extremely efficient, energy-saving propulsion. The Robotuna, like its biological model, is propelled by a wave motion transferred to the tail fin via a mechanical backbone. The imitation of the tuna even goes as far as making use of synthetic muscles in a type of plastic that contract under the influence of electric impulses. After the Robotuna came Robotuna 2.0, in which the propulsion electronics were small enough to fit into a coffee cup. That opened up new possibilities. More powerful computers with specific algorithms to simulate the tuna's muscle led to the most recent versions: the GhostSwimmer and the BIOswimmer. These swimmers designed by the Boston Engineering group, came even closer to the tuna original. Mechanical engineers copied the exact dimensions of the fish and with the help of biologists the function of the skeleton and muscle groups was analysed. This had great advantages for the mechanical replica: the motionless hull of the fish was able to fit all the computers. The engineers then developed the software of a 'genetic algorithm' in which every component of the swimming motion was imitated and tuned to

(continued)

the action of the tuna. In this way a robot was created which remained underwater autonomously, was very agile and could reach high speeds. Ideal for tasks such as inspecting drilling platforms and other underwater research, as well as war and security purposes of course. Engineers have long dreamed of entire shoals of Robotuna, together with drones in the air and navy ships enabling an entirely new form of operational marine actions. The American navy is clearly betting on tuna for a new era of sea warfare.

Tuna Mapping

12

What does the bluefin tuna do with its turbo propulsion, collapsible fins and internal air conditioning? It swims over enormous distances, from one continent to another and from the surface of the ocean down to many hundreds of metres below. We know that the tuna already has been swimming into the Mediterranean via Gibraltar since the times of the Neanderthals. But where else it heads for exactly in its life of roughly 30 years is a well-kept secret. What routes does it follow on its long journeys? How long does it stay in certain places? Are there really different groups of bluefin tuna in the Atlantic Ocean? These puzzles still have not been solved.

From the end of the 1960s tuna experts have been at work fitting bluefin tuna with electronic equipment to offer more insight into their migratory pattern and daily activities. Tracking and tagging was first used with albacore tuna in the Pacific Ocean around Hawaii and on Atlantic bluefin tuna off the coast of Nova Scotia. Initially the equipment consisted of small radio transmitters which emitted a signal with a range of a few kilometres. With a boat travelling a couple of days behind the tuna, part of its route could be mapped out, and information collected about water temperature, internal body temperature and swimming speeds.

One of the first tagging results in the Atlantic Ocean already gave a direct indication that bluefin tuna was capable of high performance. A tuna tagged near the Bahamas was found off the coast of Norway within 50 days. The fish had travelled 6500 km at an average speed of 130 km per day, including breaks for eating, diving down deep and so on.

From the 1990s radiographically monitored tuna made way for digitised tuna. A quick operation fitted the bluefin with a small digital computer with sensors in its abdomen, to record a much larger quantity of data over a much longer period of time than was possible with radiographic monitoring. Using this method American tuna researcher Barbara Block succeeded in turning up lots of new information on the living conditions and behaviour of tuna. Temperatures, depths, speeds, light and geographical position, feeding and reproductive habits: everything was recorded by the little computer in its belly. The method is anything but simple and certainly has

© Springer Nature Switzerland AG 2019
S. Adolf, *Tuna Wars*, https://doi.org/10.1007/978-3-030-20641-3_12

drawbacks: batteries stop working, the computer gives up under too much pressure at great depths, the sensors are disrupted by quick changes in temperature. A great deal also depends on the cooperation of fishermen when they find a digitally tracked tuna in their nets, and that is not always forthcoming, Block complains. When fishermen involved in illegal fishing find a little computer they will most likely throw it overboard if they are afraid that their catch will be traced. You never know what might happen to all the data recorded on it and there is a good deal of illegal, unreported and unregulated fishing (IUU) in the tuna sector.

Tuna informatics has meanwhile developed into a science in itself. The latest off-shoot involves pop-up tags, floating mini-computers attached to the tuna's dorsal fin with a little hook. At a pre-programmed point in time the tags are detached from their hosts and float to the surface, where they can be traced by a satellite signal and collected up for further data-analysis. The pop-up tags only work for a short time, but are worth the money because their information can be collected to greater effect.

The series of radiographic and digitised bluefin tuna projects, run by organisations such as the WWF, Monterey Bay Aquarium, the Pacific Community (SPC), the National Oceanic and Atmospheric Administration (NOAA), regional tuna organisations such as ICCAT and WCPFC, and a large number of universities, have turned up a wealth of information in recent decades. We now know that tuna is a creature of certain habits. Every day at sunrise and sunset the fish dives into the depths. This may be a kind of reconnaissance mission to see what there is to eat in the deep water. The timing lends itself to the purpose: in dim light it is easiest to distinguish their prey—herring, mackerel, sardines, anchovies and squid—against their transparent surroundings. A bluefin tuna in search of prey can dive to depths up to a kilometre, which is also useful when it wants to shake off its natural enemies—orcas and mako sharks.

It is believed that deep sea diving is also used by tuna to probe the earth's geomagnetic field to determine its location, or to compare temperatures at different water levels. Sharks and swordfish do the same for these reasons.

The bluefin tuna is a proven cosmopolitan among fish. It is designed to swim in both subtropical oceans and cold seas. During the day it likes to sunbathe for hours at a time in the surface water, turning a silvery belly upwards to warm itself. This may be a way of speeding up digestion. The bluefin can go for weeks without food while in search of feeding grounds, but once it has arrived at a rich fishing ground it can exhibit unprecedented voraciousness and avails itself of as much fish as possible [43].

Thoroughly fed and equipped with a substantial reserve-layer of fat, the Atlantic bluefin returns full of energy to spawn in its native waters in the Mediterranean or the Gulf of Mexico. Its brother in the Pacific commutes between the western breeding grounds (the Sea of Japan, the southern islands of Okinawa and the Taiwan Strait) and eastern feeding zones off Mexico. The distance of thousands of kilometres between its feeding and spawning grounds is travelled as quickly as possible, the fish probably making use of prevailing currents to move more efficiently.

Once back at the spawning grounds, hormones take over control of the bluefin tuna. The shoals seek out warm water where reproduction can take place, and a less hostile environment exists for the tuna larvae. The latter is in fact largely doomed to failure: of the many millions of eggs generated by each mating of each tuna pair, only a few will grow into sexually mature tuna. Annually changing conditions in the Mediterranean, the Atlantic and the Pacific also lead to stronger and weaker cohorts of new tuna. Research shows that there may be a 5–7 year cycle in which the population of bluefin tuna drops significantly and then grows again. The precise reason why one year's cohort is so much better than another remains unclear.

Bluefin Tuna is a transcontinental fish: its migration route runs from the Black Sea via the Mediterranean to the Gulf of Mexico in the west. The Atlantic bluefin tuna can pass five continents on its journey. A strip of sea 10,000 km wide runs diagonally across the Atlantic, from west to east, and considerable quantities of tuna are always caught here (Fig. 12.1). The tuna spends a couple of weeks at the spawning grounds. In the Mediterranean this involves the waters around the Spanish Balearic Islands, the area around Sicily and off the Libyan coast, and, as we have only discovered during the last couple of years, also the area between Turkey, Cyprus and Lebanon. These are the same waters in which the Phoenicians had their home ports before they went after the tuna in the western areas of the Mediterranean Sea.

In the Pacific the distances are no less spectacular. For the first 2 years of its life the bluefin tuna stays around the spawning grounds on the western side, near Japan and South Korea, then crosses the Pacific to the west coast of the US and Mexico to spend the following 2 years there. It then swims back to the spawning grounds to reproduce. The migration subsequently moves back and forth to the southern hemisphere, towards Australia.

The bluefin tuna's reproduction is a puzzle that has received particular interest from tuna researchers. After all, if we knew how the tuna spawned, we would have the key to a more effective artificial reproduction. Attempts to get tuna to spawn by injecting it with hormones have so far been unsuccessful. Since it became possible to fit tuna with electronic devices it has become apparent that they are only stimulated to exhibit breeding behaviours in high water temperatures. Block, for instance,

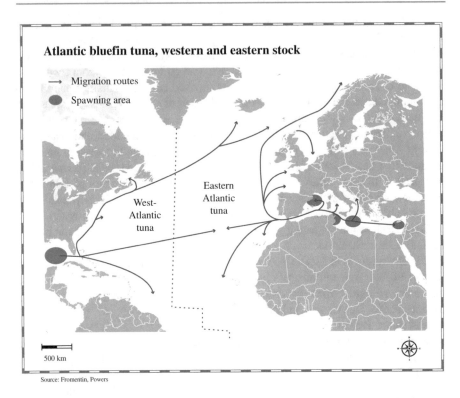

Fig. 12.1 The Atlantic Bluefin Tuna is a born migrant. Travel routes through the Atlantic cover thousands of miles between the continents (Infograph Ramese Reijerman)

discovered that in its spawning grounds tuna constantly dives down into deep water and resurfaces at night, but what exactly they are doing is not clear. Some of the tagged tuna never turned up at any spawning ground, which also fed the idea that perhaps other spawning grounds remain undiscovered along the Gulf Stream in the Atlantic Ocean.

During reproduction the female tuna produce millions of eggs. This quantity seems designed to safeguard against tuna extinction, and is probably behind the enormous recuperative powers of the fish population. If measures are taken to guarantee sufficient sexually mature tuna, a decimated tuna population can come back strong again. Only a tiny fraction of the countless millions of eggs will ever grow into mature tuna, the rest are fodder for those around. Once fertilised, a tuna larva measures two to three millimetres in length, largely consisting of an ugly head with bulging eyes. The tuna larvae, which remain in the same place for the first half year, have a large appetite from the outset and grow surprisingly fast. After 2 months on average the tuna already weighs 170 g, the weight of a large herring; a month later it has grown to 430 g, as much as a large mackerel. On leaving the spawning grounds the tuna already weighs more than a kilo and has reached the size of a small cod [43].

Nevertheless, the vast majority of larvae never reach a gram in weight. It is estimated that there is a one in 40 million chance of a tuna living to sexual maturity (now considered to be 4–5 years old for bluefin tuna on the eastern side of the Atlantic and its brother in the Pacific, while until recently most researchers assumed that stock on the western side of the Atlantic took as long as 8–9 years to become capable of reproduction). As larvae and young tuna they are a much sought-after prey for all other predatory fish in the sea. But once the perfect underwater machine has reached adulthood, only orcas and sharks remain fast enough to be taken seriously as natural enemies. Humans remain the greatest danger of all.

Despite all research the bluefin tuna remains above all an unpredictable mystery. The enormous distances it travels during its life, its fickle behaviour in search of places where food can be found and its relatively long life all make it difficult to chart the places where the fish stays. Digital technology for monitoring tuna on its journeys is expensive. However, research on bluefin tuna is becoming ever more urgent. Both from a biological viewpoint and from an economic perspective, people are in a hurry to learn more about tuna in order to maintain the species. Never before has science had so many clever tools at its disposal. At the beginning of the twenty-first century an army of tuna scientists worldwide have been activated to gather and compare their knowledge. More than ever before it was indispensable as a basis for managing the endangered stocks.

The Mystery of the North Sea Bluefin Tuna 13

People who think they know something about the sea because they've read a book about the biology of fisheries, well they only know what's in that book and nothing about the sea. They know nothing about the sea, man! The fisherman receives the best lessons from the fish.
Retired captain of Barbate's almadraba, twenty-first century

When did we notice that something had gone wrong? With tuna—as so often with other environmental issues—it crept up on us. At first you don't notice at all. Then people start to talk about a problem. Often this is followed by a stream of confusing information, so that it's not always easy to gain a clear picture and it's often the subject of dogged polemics, against a background of significant financial and political interests. Sometimes this is followed by timely intervention, as in the case of the worldwide ban on propellant gases for the ozone layer. But more often it is suddenly too late. As with cod on the Canadian and American east coast, for instance, or coral on the Great Barrier Reef, and perhaps soon krill in Antarctica. Things go wrong underwater because we don't see much of it from above (Fig. 13.1).

When I lived in Madrid I first began to notice that something was really going wrong with tuna. Anyone who liked fish was always well served in Spain. In the early 1990s it was still a feast walking past the stalls on the daily markets and large shops. Under the cast iron arches of Mercat de la Boqueria in Barcelona, for example, or the granite alcoves of the fish market in Santiago de Compostela. Even the displays of the San Miguel market in the historic centre of Madrid emanated the salty aroma of ocean around an endless assortment of fresh fish. There were conger eels metres long, sea bream, horse mackerel and red scorpionfish. At the top of the display were the larger items: large hunks of albacore tuna, red bluefin tuna, and pale swordfish and dogfish. Brown-grey crayfish tried to escape their wooden baskets, beside water tanks containing large live spider crabs and dark

© Springer Nature Switzerland AG 2019
S. Adolf, *Tuna Wars*, https://doi.org/10.1007/978-3-030-20641-3_13

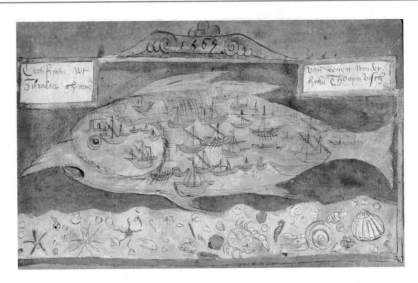

Fig. 13.1 In his 'Fish Book' the sixteenth century Dutchman Adriaen Coenen drew a bluefin tuna that was had been found on the shores of Ceuta. Mysteriously enough somebody had engraved a scene on the tuna skin, of one of the large sea battles that were fought around the Street of Gibraltar (Visboek Adriaen Coenen by courtesy of the Dutch Royal Library)

blue lobsters. There were fresh anchovies to pickle or salt, fresh cod and stalls with stockfish. There were baskets of soft-shell clams, mussels, oysters, cockles, winkles, scallops and the curious barnacles, prehistoric little claws perilously plucked from the rocks in northern Spain. A child's head would easily have fitted into the enormous, gaping mouths of the monkfish.

Less than two decades on, little remained of the colourful still life of wild fish on the Spanish markets. The rich collection was slowly supplanted by a limited quantity of farmed fish. Sea bream and salmon dominated the shelves. The better quality salted anchovies—once a cheap staple—were selling for the price of caviar. Red mullet, John Dory and wreckfish had disappeared. The monkfish were now so small that they could not even hold an apple in their mouths. Bluefin tuna, too, largely disappeared from the beds of freshly crushed ice.

Even to those who had never seen the statistics it was clear that something was slowly but surely changing on the wild fish market, indicating a change in fishing and in the oceans.

Bluefin tuna disappeared from a number of places in the world, initially without much fuss or commotion. To say the least, that is remarkable for such a big fish. It had somehow happened under my nose without my noticing. I discovered, more or less by accident, that in the early 1980s there were still bluefin tuna in the North Sea. I could hardly believe it: bluefin tuna off the coast where I had grown up? Had I unknowingly swum year after year in the same grey sea as the giant tuna? It was almost unbelievable, yet true. But now they had disappeared. A tuna population suddenly disappearing without notice: how did that happen? And could it be reversed?

Dutch herring fisher Floor Kuijt was a sturdy blond figure with a blushing face who appeared considerably younger than his pensionable age. When I spoke with him in 2008 in the Dutch fishing village of Katwijk, he still remembered the bluefin tuna well. We drank coffee at his house behind the beach boulevard. Kuijt told me about his life as a fisherman. Like many youths from his village, at the age of 14 he left for the North Sea to fish for herring. His first memories, however, go further back, to 1952, when as a child of five he was allowed to accompany his father on the boat.

The fishing municipality of Katwijk breathes a different atmosphere from that of the Spanish tuna fishing village of Barbate or the Japanese tuna town of Misaki. This is mainly because Katwijk does not have its own port. Katwijk doesn't smell of fish. The fishing boats are moored in distant harbours of IJmuiden or Scheveningen. In the heart of the little village, not far from Kuijt's house, just behind the dike, there stands a church that acts as a fortress against the threatening sea: Andreaskerk, named after Saint Andrew, patron saint of fishermen. The tight-knit Protestant fishing community could be found here every Sunday, where faith without fanfare prevailed. No Shinto shrine with a bearded god of fish, and certainly no Catholic image of Our Blessed Virgin the Virgen del Carmen, that papal patron saint of fishermen who is carried to the sea in Barbate. The Reformed village of Katwijk has to make do with the fish of the Old Testament, with stories from the time when the Phoenicians were on the point of setting sail for the western Mediterranean. It was an exacting god who ran the show there: the sea was cruel, man an insignificant plaything for the waves, and they paid dearly for the fish. We read in the Old Testament, Exodus 7:21: 'And the fish that was in the river died; and the river stank, and the Egyptians could not drink of the water of the river; and there was blood throughout all the land of Egypt.' Or take Ezekiel 29:4: 'But I will put hooks in thy jaws, and I will cause the fish of thy rivers to stick unto thy scales, and I will bring thee up out of the midst of thy rivers, and all the fish of thy rivers shall stick unto thy scales.' Or Nehemiah 13:16: 'There dwelt men of Tyre also therein, which brought fish, and all manner of ware, and sold on the Sabbath unto the children of Judah, and in Jerusalem.' Thus Tyre, along with Sidon the most important Phoenician trading city, entered Katwijk's Sunday sermon via the first direct translation of the Bible into Dutch, the Protestant *Nederlandse Statenbijbel*. And with them came tuna, the fish that had been so important in the Mediterranean from biblical times on.

Kuijt could still remember it vividly. He saw the bluefin tuna on his first trip with his father. One morning he found a pair of enormous fish on deck. 'They were massive things,' Kuijt remembered. Not as big as the common mink whale nicknamed Kees, sometimes sighted in the North Sea. That might have been ten metres long. But to a 5-year-old boy the bluefin tuna was impressively large too. 'Perhaps two or three metres long,' Kuijt estimates. They were caught at night by the fishermen between hauling in the nets. Just for fun, to pass the time. 'First they threw some herring or mackerel overboard to lure the fish in. The tuna would then rush at it,' Kuijt remembered of the hunt. Then a hook was thrown overboard, generally no more than a bent and sharpened stove poker, attached by a short line to two sturdy buoys. The tuna bit at the hook, then shot off, pulling the two buoys in a spectacular sprint over the waves.

The fishers didn't reel in the line, as it would have snapped like dry straw under the force of the fleeing giant tuna. They would wait until the tuna was exhausted and allowed itself to be gradually pulled in.

Once on deck the fish was slaughtered for eating. If there were Norwegians or Danes nearby, the tuna was saved for exchange purposes, as the Scandinavians could trade it in their home ports. The herring fishers of Katwijk could not do anything else with it: the fatty tuna spoiled quickly and once ashore in Holland no one was interested in the fish.

Until the 1970s bluefin was a regular guest in the North Sea, where the herring was among its favourite prey, especially the fatty herring which swims into the North Sea ready to spawn. Fishermen also consider this herring a desirable catch. No wonder that fishermen and tuna met in the herring breeding grounds [35].

When Floor Kuijt decided to go to sea himself, he soon found a career as a helmsman and captain. By then the days of fishing for tuna with an old stove poker had long past. Tuna had suddenly disappeared from the North Sea. In retrospect it was a warning signal. At precisely the same point the herring population in the North Sea collapsed. For the first time it was necessary to call a halt to herring fishing. This caused a shock in a country that may not have been full of big fish eaters but where raw herring was seen as a local delicacy and a source of national pride. Until the end of the nineteenth century the fish was a major cheap staple in smoked form, which had the advantage of a shelf life lasting several months. In the twentieth century the *maatjesharing*—soused herring caught from May onwards, with a minimum fat content of 16%, fileted, gutted, brined and matured—came to be seen not only as food but also as part of the national identity. The Dutch pick up the raw fish by the tail, run it through chopped onions, tilt their heads back and lower it into their wide open jaws. Before the Japanese became known for their sashimi, the Dutch would boast that they were unique in eating raw fish. It is a source of distinction and national pride (Fig. 13.2).

Tuna has been swimming around northern Europe since the Stone Age. Bluefin vertebrae of the period 7000–3900 BCE have been excavated around Copenhagen. The Danes found shoals of bluefin around their coasts. A tuna was caught off the coast of Holland during the sixteenth century. Norwegian naturalists describe how their compatriots in the eighteenth century could catch 15–20 tuna per season with a harpoon.

The Norwegians in fact started hunting tuna on a large scale in the 1920s—not to eat but out of irritation [161]. Bluefin tuna destroyed their nets in order to stuff themselves with the herring and mackerel trapped inside. The tuna was known as an awkward pest. Fishermen attacked them with hooks and harpoon cannons, but the tuna were quick and always came as an entire shoal to rob them of their fish. Eventually the solution was found in fishing away all the tuna in one go with purse seines, which were drawn in around the entire shoal. Only in this way could the tuna be effectively kept at bay. Savvy Italian businesspeople, who had heard of the tuna plague in the north, set up canneries in Norway and Denmark. This suddenly

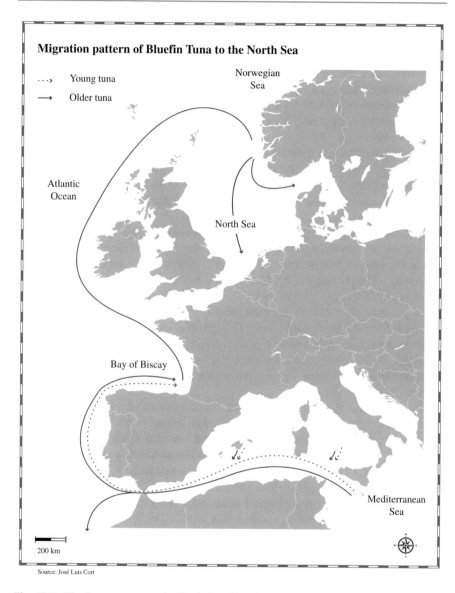

Fig. 13.2 Bluefin tuna routes to the North Sea. Old migration route that disappeared but is being recovered by new generations tuna (Infograph Ramses Reijerman, based on Cort and Nottestad)

gave the North Sea fishermen a market for the tuna which had previously been more or less worthless.

Only after World War II did tuna fishing really get going, thanks to new technologies. The new hydraulic winch made setting out and particularly hauling in the purse nets a good deal easier. In a couple of years a Norwegian fleet of 470 small purse-seine ships sprung up. The warm summer of 1952 was a record season. Large shoals of tuna swam right along the coast and even into the fjords. The calm sea and beautiful weather provided optimal conditions for tuna fishing. Eighteen thousand tonnes of bluefin were caught, the vast majority by the Norwegian fleet. By today's standards that was an enormous catch for one country, more than the entire Mediterranean annual catch quota established 50 years later for all bluefin tuna. Nowadays this would supply a substantial proportion of annual consumption on the Japanese sushi market. Over the 1950s the catches fluctuated around 10,000 tonnes.

Bluefin seemed to do it extremely well in the Northern waters. There are many indications that the bluefin tuna even used the British Channel for its migration route. The hungry giant was a kind of plague for the local fishermen. In the bay of Trebeurden, an Asterix-like coastal area at the northside of the French Brittany region, the local fleet of sardine fishers were complaining in the summer of 1946, the first summer after World War II, that an unknown fish was stealing away large amounts of sardines caught in their nets. A doctor in one of the villages nearby, a certain Dr. Miroux, got interested in the case. He fabricated his own catch device combining a Mauser gun and a harpoon and went to sea with the sardine fishermen. To everybody's surprise Miroux shot a 175 kg, 2.5 m bluefin tuna. Being an engaged sports fisherman himself, Miroux went looking for fellow sportsmen experienced in buefin tua fisheries. Next season, in 1947, a group of invited sports fishermen of the US came over, fully equipped with gears and outboard engines, to help their French colleagues to hunt the bluefin. Between 1947–1953 bluefin was caught each year by sport fishermen, some years with a poor catch, sometimes with excellent catches [101].

After 1963 the catches of bluefin in the North Sea suddenly dropped. The fishermen found more and more larger adult fish in their nets at the expense of the younger year classes. Younger tuna disappeared from the North Sea. According to French marine biologist and tuna specialist Dr. Jean-Marc Fromentin, the tuna's migration routes changed and from then on the tuna remained on the western side of Ireland and Great Britain rather than venturing into the North Sea [107, 108].

Why did the bluefin tuna disappear so suddenly from the North Sea? Was it the intensive fishing which suddenly soared after World War II? Did something happen to the water temperature? Had the North Sea lost its attraction for the tuna once and for all, now that the herring population had been overfished and decimated?

Fromentin studied centuries of tuna statistics and data from the library of the duchess of Medina Sidonia. From the catches in different places he deduced that after reproducing near the Balearic Islands and Sicily the bluefin tuna swam back to the Atlantic in search of food. He believes that the North Sea, the Norwegian Sea and surrounding areas formed an important section of the tuna's migration route until the 1960s. Since it is estimated that three quarters of the menu of the North Sea tuna consisted of herring, it seems obvious to assume that the reduction in herring stocks in the 1980s was the main reason why the tuna stayed away. But that is not quite right: the North Sea herring that was worst hit was mainly the group that spawned in spring, when the tuna itself was supposed to be busy reproducing in the Mediterranean, a trip of several thousands of kilometres down south.

Perhaps water temperature also played a role. The 1960s saw a substantial drop in temperature. Although tuna can survive in fairly cold water, Fromentin suspects that the cooling made the North Sea less attractive.

But a truly conclusive explanation for the disappearance of the North Sea tuna still eludes us. It is most likely that a combination of factors caused the bluefin tuna in the North Sea to disappear: mass fishing by the Norwegians, reduced food supply, few new sexually mature fellow tuna and water which grew increasingly cold.

No one in Western Europe at the time was losing sleep over the sudden disappearance of bluefin tuna: it happened practically unnoticed. The threat to the marine environment might have been gaining interest in public opinion, but it was mainly other animals which attracted attention, such as seals. Brigitte Bardot devoted herself to the battle against clubbing seal pups to death. Dolphins also enjoyed warm attention, thanks to the popular television series *Flipper*. Whales, too, did well. Marine mammals with their cuddliness factor won hands down against chilly fish. Even if the general public had been aware of the threatened existence of bluefin tuna in the North Sea, the fish still probably would not have stood up to this competition.

Cousteau from Scheveningen

Tuna might not have been particularly well known as a consumer fish in the north, but it had already been recognised as a remarkable fish swimming in northern waters as far back as the seventeenth century. This was largely thanks to Dutchman Adriaen Coenen, who put tuna in the spotlight. It would not be too much of a stretch to call this fisherman and beachcomber from Scheveningen, a fishing village near The Hague, an early forerunner of Jacques Cousteau. In 1577, at the advanced age of 63, Coenen embarked on his life's work: the *Visboek* (Fish Book, Fig. 13.1) [155]. In this weighty 800-page tome Coenen documented the wondrous fish and sea creatures he had found in his nets and on the beach over his lifetime. The book was richly illustrated with countless drawings. Coenen's masterpiece, later supplemented with two separate books on whales, brought him considerable status, not only among scientists of the day, but also with the general public. Like Cousteau, Coenen understood that there was money in revealing the underwater world. His book could be viewed, along with a collection of dried fish, for a modest fee at annual fairs in Holland.

On 18 July 1554 Coenen discovered a tuna for the first time at a fish market in The Hague. No one seemed to want to buy the giant tuna, so the market trader had the creature prepared at an inn, De Kroon, where Coenen succeeded in obtaining a piece. The tuna made such an impression on Coenen that he depicted it in various places in his *Visboek*. Later he was required to explain this special fish to a high-placed administrator in The Hague who had also bought a tuna.

(continued)

Clearly there was something magical and mysterious about the fish. In his book Coenen mentions a painted bluefin tuna found in 1565 on the distant beaches of Ceuta. Six witnesses who found the fish were so surprised by images on its skin that they decided to have an official record of the incident drawn up by a notary with an office in Gibraltar.

'Fifteen or sixteen days ago a tuna was found on the coast off Ceuta, on which a large number of ships had been drawn with many sloops, masts, oars, oarsmen, artillery, and an armed galiot preparing to attack one another. It was represented in a very natural and realistic style, as if the picture had been miraculously drawn in the skin and flesh of this tuna, the like of which we had never seen before. Truthfully recorded in Gibraltar, 13 May 1565, with my signature as witness below, Johan Frutuoso.'

The last tuna was fished from the North Sea in 1985, following several years in which the species had not been sighted there at all. During the same period, something similar occurred on the other side of the world, off the coast of Brazil. To their surprise Japanese longline ships had discovered large shoals of bluefin tuna, which they immediately decided to capture. A small fleet of longliners hauled in 5000–12,000 tonnes of tuna per year. In precisely 10 years, between 1960 and 1970, all bluefin tuna were fished from the Brazilian waters. As far as is known the tuna never returned here. Again, there was little public commotion, except perhaps among the Japanese fishers and a handful of tuna experts.

To many experts, the disappearance of the bluefin tuna from the North Sea is typical of how little we really know of this migratory fish and the thousands of kilometres it travels along routes no one truly knows. The solution to these kinds of mysteries has gained in urgency. Proper management of the tuna stocks will only ever be possible if more is known about the composition of the populations and how and why the groups of tuna move around.

The migration to the North Sea was part of the migration of groups of tuna from the Mediterranean to the Atlantic. In 1929 a tuna was caught in the Mediterranean with fish hooks in its mouth that were used by fishers in the northern Atlantic Ocean. It was the first evidence of the prevailing theory that the tuna in the Mediterranean was the same as that which passed through the Strait of Gibraltar each spring from the Atlantic Ocean. In her research with radiographically tagged tuna in the 1990s American tuna researcher Barbara Block argued that two tuna populations swam in the Atlantic Ocean: one that bred in the Gulf of Mexico and another that spawned in the Mediterranean. According to one of the theories the two groups meet in the fish-rich regions of the Atlantic. But the pilot tagging studies remain restricted to relatively small quantities of fish. The sudden disappearance of tuna from certain areas seemed largely to point to the tuna population and their migratory routes being more complicated, with subgroups that have their own migratory routes and behaviours.

The North Sea is an important link in the quest for tuna. Those who had the key to the disappearance of the bluefin tuna from the North Sea could perhaps contribute to a policy that would make the species sustainable. If the French tuna researcher Jean-Marc Fromentin was right and the tuna in the North Sea had followed a route along the Spanish coasts, along the Bay of Biscay 1000 km south of the North Sea. This sea lies in the triangle formed between the northern coast of Spain and Brittany in France and is famous for its storms, but also for its rich fishing waters. There we find the *bonito del norte* (*Thunnus alalunga*), as the Spaniards call albacore or white tuna, and the Atlantic bonito (*Sarda sarda*), a mackerel-like cousin of tuna. Both albacore and bonito are canned, but can also be found fresh on the Spanish market.

For the Basques, traditionally a fishing people on both sides of the border between Spain and France, the sea has always been a rich fishing ground for tuna, including bluefin. Only in the 1940s did problems arise. Albert Elissalt, a ship owner and manufacturer of canned tuna in the French-Basque port town of St-Jean-de-Luz, like many of his French and Spanish colleagues after World War II, had earned a great deal of money catching young bluefin tuna. Demand from the canning industry continued to grow, but the Basque fishers could only provide a limited supply to the factories [244]. There was a technical problem. For centuries the Basques had been using sailing boats and even rowing boats to fish for tuna. They used a form of trolling: pulling lines and hooks camouflaged with horsehair, straw and feathers. A little tin blinker above the hook was supposed to catch the fish's attention. It didn't take much to get the hungry tuna to bite.

In the twentieth century the Basques had exchanged their rowing and sailing boats for small cutters—initially running on steam, later diesel—but the traditional hooks with horsehair remained the preferred fishing method.

Elissalt realised that this method yielded too little. Far more tuna was caught on the west coast of the United States, he had read. He decided to go in person and see how those Americans did it there. In August 1949 he travelled to San Pedro, Los Angeles' large fishing port, which had grown to be the most important landing site for the American canning industry. There he found a large pole and line fishing fleet which fished with rods and live bait. The Americans were said to have derived their traditional technique from the Japanese and fishermen in Hawaii. But the Frenchman also saw that the use of purse seines was on the rise. This was the same technique which the Norwegians had so successfully applied in their battle against the tuna plague, but which evidently had not been previously noticed by French and Spanish colleagues. This was all the more surprising given that the technique was an application of the traditional circular net, previously used by the Basques themselves.

Catch volumes in the US were spectacular, '. . .more than 100 tonnes of tuna in a single haul', Elissalt noted enthusiastically in letters home. They even used light aircraft and helicopters to trace the shoals of fish. That greatly impressed him. If these fishing techniques were introduced in Europe, success was guaranteed, the canning factory owner believed.

Right after his return, Elissalt began to apply some of the lessons from the New World to tuna fishing in the Bay of Biscay. First, pole and line fishing was enthusiastically embraced by the French and Spanish fishers. They lured the tuna towards the boat with live bait and then reeled it in en masse, just as their American colleagues did. The results of the new technique turned out so much better that within 2 years the entire fleet had switched to pole and line fishing.

From 1949 the new fishing method was fully operational. Never before had so many young bluefin tuna been reeled in from the water. According to counts undertaken by Spanish tuna scientist José Luis Cort, between 1949 and 1960, this effectively led to a veritable massacre in the Bay of Biscay and off the coasts of Portugal and Morocco. The haul largely consisted of bluefin tuna traditionally found in the Bay of Biscay: small immature tuna, 1–4 years old, weighing under 35 kg. He calculates the number of juvenile bluefin tuna fished from the water in that period at 6.5 million. The Bay of Biscay accounted for the largest share. It was mainly these not yet sexually mature bluefin tuna which were caught here.

Elissalt the ship owner proved to be a visionary in several respects. He introduced the successful pole and line technique. Later, he was proven right in his prediction that the purse-seine nets and light aircraft that he had seen in the US were the future. A couple of decades later these fishing techniques were introduced on a large scale in the Mediterranean to catch the bluefin tuna right at their breeding ground.

Is the solution to the mystery of the disappearance of the North Sea tuna finally here, on the green coasts of the Spanish and French Basque Country? José Luis Cort believes so. The former director of the Oceanographic Institute of Santander on the Atlantic north coast knows the Bay of Biscay well. He is also a specialist in juvenile bluefin tuna. After years of studying the migratory movements of the tuna, he developed a theory that young larvae from the spawning grounds around Sicily and the Balaeric Islands reach sexual maturity in the Bay of Biscay. Once mature, in summer this group migrated to the North Sea in search of food. Since the juvenile tuna were fished from the Bay of Biscay in extremely large numbers between 1949 and 1960, in subsequent years there was an increasing shortfall in the maturing year classes capable of reproduction. It would be several years before the effects worked their way through the whole population. Fewer and fewer tuna from this group grew large enough to complete the journey to the north and return to spawn in the Mediterranean. In 1963 the decline was so significant that the supply of North Sea tuna began to collapse [63, 64].

Floor Kuijt was on top of the situation—quite literally, since, like the bluefin tuna, he was after North Sea herring. When I spoke to him in 2008 he had just stopped working as a captain in the herring fleet. Although it would hardly have been surprising to see him get up mid-conversation, walk out the door, cross the square and boulevard to the beach, to quickly pull in a net of herring with his strong hands. Kuijt was a fisherman born and bred, his family is as much a fixture of Katwijk as the dunes, the beach and the Reformed Church. His surname means 'spawn' in Dutch. For centuries, generation after generation, his family had been fishing.

But during his long fishing career he saw the definitive dawn of a new era. Stocks were under pressure. For the first time in history, new technology and increased scale created the fleet capacity to completely eliminate entire stocks, of herring as much as tuna.

Kuijt started out on a lugger, fishing for herring off the English coast and Sandettie bank. Luggers were engaged in passive fishing techniques: the little boats dropped their nets three metres into the water like curtains, forming a loose, floating version of the *almadrabas*. At night, when the herring had swum into the nets, the catch was pulled in. Using this technique the herring caught in the North Sea were more or less stable at around 650,000 tonnes per year until the start of the 1960s. Then the fishing techniques changed. The ships actively went after the herring with trawls and drag nets. The capacity of the boats grew quickly and the equipment used became increasingly efficient. Research showed that the herring population was dropping rapidly. Fishermen from the Netherlands, Denmark, Norway and England charted in increasing detail the regions of the North Sea where the fish came to breed. In 1965 there was a record catch of 1 million tonnes of herring. Eighty per cent of that was young herring, leading to a rapid drop in herring reaching sexual maturity. Right after that record year came the great blow. The population of North Sea herring collapsed. In 1975 the catches had dropped to 300,000 tonnes. According to herring experts the fish were on the verge of extinction. Circumstances led to a decision to take a unique step: an absolute ban on fishing for herring. The moratorium lasted from 1977 to the end of 1980.

Kuijt remembered the ban well. The herring fishers were furious. There was particular dissatisfaction about unfair treatment. 'They should have closed the entire North Sea, but fishing was permitted in the Skagerrak,' he told me. The Danes simply continued fishing for herring between Jutland and Norway. That caused bad blood. Kuijt, too, initially decided to disregard the ban. 'I was young,' was his excuse for illegally breaking the moratorium. The second time he set out, it was definitively over. The navy was stationed outside port. The moratorium was now effectively enforced.

In retrospect, the ban on herring fishing signalled a turning point, Kuijt believes. When the fishermen resumed North Sea fishing a few years later, some recovery was evident. 'But not to the extent we had expected,' Kuijt says. The moratorium had saved the herring from its doom, but it turned out to be more complicated than had initially been assumed to maintain the population.

'Of course it helps to close everything down, but the stocks of fish only recover properly when you get a strong year class,' Kuijt explains. Why does herring succeed in reproducing better 1 year than another? Kuijt puts it down to a range of conditions: the number of fish larvae in the breeding year, feeding conditions in subsequent years, and overfishing of the still young herring before it can reproduce, the latter being a key factor.

In Kuijt's view, the overfishing, juveniles included, was largely due to rapidly changing technology. The luggers previously used to catch the herring had been replaced with large trawlers, with more powerful engines. These in turn were replaced by even larger freezer trawlers with even greater horsepower. Fishing became an accelerating race of ever increasing scale in the form of greater freezer capacity in larger ships with more powerful engines and ever better navigation and fishing equipment. Things went the same way as with the tuna fleet. Larger scale meant reduced costs per fish caught, but also higher expenditure on fuel and rising investments in ships. In order to cover those costs it was necessary to fish more, leading to a vicious cycle of increasing scale, efficiency, capacity and catch. In the years when Floor Kuijt began as a fisher, a *Kantje*—90 to 100 kg of herring—per fisher per day was sufficient to earn a decent living. When he retired more than 40 years later, the crew of 15 had to catch daily around 2800 boxes of herring, 25 kg each, to make a normal wage. That's more than 45 times as much per fisherman.

Of course, the Netherlands is rich; the fishermen became much more demanding of their income. 'In the past you prayed for food, now for a Mercedes,' Kuijt observed. But it was mainly increased need for fuel and depreciation of the boats that raised the costs. Larger boats, increased energy usage, everything contributed to making the catch ever bigger. It was no different at other fisheries. And certainly when it came to tuna, with its ever growing purse seiners and longliners.

For Floor Kuijt the conclusion was clear. New technology was not necessarily a good thing. It depended on how it was applied and regulated. The reasoning error which people in fishing make time and time again in his view: when the catch threatens to reach its economic ceiling, new methods and increases in scale are developed to cut costs and catch exponentially more. The fish initially adapts to a new assault on the population, but in the end the yield drops. 'The effectiveness you introduce in the end works against you,' Kuijt explained. The greater the capacity, the greater the pressure to fully exploit it. 'That means: the greater capacity of your ship in the end is more of a sign of how badly you're doing.'

The result was a desert in the North Sea. 'You chase the fish away or make a graveyard of the place,' Kuijt summarised the tendency in fishing. 'We can't find a balance between prosperity and greed.' Another approach was needed. Some kind of management was required, he thinks, for instance by allocating fishing regions to be managed by the fishers themselves, in combination with a catch quota. 'Give the fishermen their own stretch of sea, like a farmer ploughing his own land.' We needed to think more creatively, Kuijt believed. By developing meshes which further limit bycatch, or turning our attention to the ships' energy consumption. By developing new methods whereby drag nets leave the seabed in the breeding grounds undisturbed. By fishing selectively so that the sexually mature fish have a chance to reproduce.

Kuijt didn't know that the latter was seen by tuna scientists as the reason why bluefin tuna had disappeared from the North Sea. But at sea he had seen with his own eyes how the herring had declined and drawn his conclusions, which were not so different. It was necessary to develop more selective, sustainable fishing. And the fishers had to take their share of the responsibility, Kuijt felt. 'We have to own up. We caused the overfishing. Now we have to help solve the problem.'

The good news from the conversation with herring fisher Kuijt was that this experienced fisherman realised that something had to happen to ensure sustainable fishing, and that this was at least partly the responsibility of the fishing industry itself. It was to be hoped that the new generations of fishermen would take these lessons on board. Then herring as well as bluefin tuna might have a chance in the North Sea, we said, half in jest, as we parted company.

At that moment I didn't really believe in the possibility of a comeback for North Sea tuna. Who did? Years passed with effectively no return of the tuna to the Northern waters.

But then, as suddenly as they had disappeared, bluefin re-emerged in the North Sea.

I found a small one on a fish market in Zeeland, the southern coastal province of the Netherlands, in 2013. That must have been a coincidence, a lone individual that had taken a wrong turn for one reason or another and instead of ending up in the warmer part of the Atlantic found itself in the North Sea. But 3 years later another one washed up on the beach and a year later a large male, a metre and a half in length.

He had already been considerably chewed, but there was no doubt about it: bluefin tuna on a Dutch beach.

The comeback began to look more robust. There were more and more reports of a bycatch of tuna captured with the herring in northern waters [68, 97]. Even further north, off the coasts of Iceland and Norway, large schools of bluefin tuna appeared. It began to resemble times gone by [233]. There was another reason to be happy: the management of overfished stocks in the northern waters finally started to pay off in a remarkable way. The improved herring stock, was one of the great successes of the European Common Fisheries Policy. And that did result not only in improved yields for herring fishers, but also in the return of their old fishing competitor bluefin tuna. In 2017 the Dutch division of the WWF joined forces with the Technical University of Denmark and the Swedish University of Agricultural Sciences to work on a study of bluefin tuna in the North Sea [262]. A tagging project was set up. With the help of anglers, bluefin tuna was caught off the Swedish and Danish coasts. After carefully fitting them with a tag—tuna is a nervous fish and cannot sit still for long—the fish were set free in the hope of gathering further information on their migratory behaviour in the North Sea. It was the first time in the history of Scandinavia that they had carried out research into bluefin tuna. The case reached the television news. So tuna was back again. This time perhaps to stay. Something was really improving for Atlantic bluefin tuna.

The *Almadraba* Revisited

14

I had personally experienced the previous downfall of bluefin tuna at close quarters. It was a cool spring morning in the mid 1990s. A nautical mile and a half away was the deserted Zahara de los Atunes, where it all started, this affair with bluefin tuna. The first warm sunbeams had cast a clear morning light on the tuna palace of the dukes of Medina Sidonia onshore. The dilapidated walls stood out sharply against the new hotels and apartments which had mushroomed in recent years on the edge of the long beach. On the corner, invisible behind the dune with its green pine trees lay the excavation of Baelo Claudia with its ancient salting pits. The captain's whistle sounded high and shrill above the calm morning sea. The *almadraba* could begin.

The fishers on board felt a close affinity with the bluefin tuna or *atún rojo*, 'red tuna', as it is called. They had grown up with stories of Phoenicians, Carthage, the Romans and the dukes of the fish, and talked about them not like history but as close friends they just had met the previous day. They knew that from the next bay, where the ruins of the Roman town now lay, tuna was once exported to all corners of the Mediterranean Sea. Behind us, now hidden by the thick morning mist in the strait, lay Tangier. At the breakwater of Barbate's harbour we had passed the Reina Cristina, a Japanese reefer. Soon the haul from our ships would disappear into the hull of this boat. Beheaded, gutted, quartered, the fish would be transformed into rock-hard fillets at temperatures below −60 °C, to be offered on the Tsukiji fish market in Tokyo in a couple of months' time.

* * *

The *almadraba*: some effort was required if you wanted to partake in the oldest surviving fishing tradition of our civilisation. People gathered punctually at six in the morning on the quay in Barbate. Coffee was served from a shipping container housing a café on the quay. The mood was pessimistic. It wasn't looking good for fishing, the fishermen agreed. A lukewarm offshore north wind had stirred up the sandy bed and clouded the water, a phenomenon familiar to everyone engaged in *almadraba* fishing.

© Springer Nature Switzerland AG 2019
S. Adolf, *Tuna Wars*, https://doi.org/10.1007/978-3-030-20641-3_14

'The tuna likes to see where it's swimming. It's not a fish for cloudy water,' Rafael Marquéz Guzmán explained to me over coffee. Today he had swapped shifts at his garage in Barbate to help with the catch. Guzmán (as far as we know unrelated to the famous Guzmán mentioned in Part I), like most people here, was not a fulltime fisherman anymore. But his father, grandfather and great grandfather had served on the *almadraba* all their lives. Like most of the approximately 100 men distributed over our little fleet of motorboats, flat-bottomed barges and dinghies ready to depart, their knowledge and experience of tuna stretched back generations. The *almadraba* runs in your blood, said Guzmán. Someone with the surname of a tuna empire was immediately credible.

The mood among the crew was tense, as it is before a football match with their favourite club from Cádiz. 'The fishermen are paid a percentage of the catch,' boatswain Carlos de la Cruz Día explained the air of mild excitement that prevailed. At 50 he was the eldest in the fleet, a small, agile man under a baseball cap and a yellow oilskin jacket. He once worked as a fisherman off the coasts from Cádiz to Larache, in search of anchovies and sardines. This was before Morocco temporarily ended its fishing agreement with Spain. He now lived in Isla Cristina on the border with Portugal, where he spent the rest of the year fishing for Venus clams and razor clams. Clams are definitely different, the boatswain explained. They are dull, in a way. With clams you go fishing. With tuna it's hunting. Every year he has come to Barbate specially, a journey of three and a half hours, to join the *almadraba*. He showed me the scars on his forearm where he had been struck by a tuna head. Lucky it wasn't the tail. A tuna's muscle mass can break bones like matchsticks. When he started out, more than 30 years ago now, the *almadraba* was rather harder work. There were fewer machines on board, no diesel engines to winch in the nets. The whole thing was pulled up by hand (Fig. 14.1).

Once on the water, while heavy pulleys, buoys and cables were dragged back and forth around us, there was sufficient time for stories. About the addiction of the small army of fishermen to the battle with the biggest fish in the ocean. About the force of a blow from a tuna's tail, that great, robust fin on a thick, bulbous muscle mass. About the aggressive swordfish hidden among the tuna, which shoot out in a rage to fight for their lives. And of course the orcas, who at this time of year lie in wait at the entrance of the strait. They return to their usual spot year after year. Some orcas were easily recognisable by their fins. The fishermen would name them.

Orcas knew how to swim into the nets of the *almadraba*. Once inside with the imprisoned shoal of tuna, they would have a feast. That was considerably less strenuous than chasing tuna at top speed. After eating they would neatly swim back out. It used to be war between fishermen and orcas, but now there is more respect. Sometimes they would leap from the water around the boats with their enormous black and white bodies. It could be a game, or it might be intimidation, you never knew with orcas.

The bluefin tuna was suddenly of interest again. Spanish television crews came down in the fishing season to report on the *almadraba*, newspapers wrote extensive features on the subject. Now that the bluefin tuna was threatened, the realisation began to dawn that the days of the ancient fishing tradition of the *almadraba* were

Fig. 14.1 The Almadraba. Bluefin tuna appeared on the surface. The next moment it started to rain fish (S. Adolf)

also numbered. Every cloud has its silver lining: the *almadraba* was rediscovered as a piece of cultural heritage. They laughed about it on the boat. There were several trades which claimed to be the oldest profession in the world, without immediately being elevated to cultural heritage. But there was also a conscious pride behind the jokes. The fishermen knew that fishing had finally gained the status it deserved, and that inevitably reflected on them too. Civilisation had always gone after the tuna, and now that large-scale fishing and the fish's survival was at stake, it was this artisanal fishing technique that could guarantee a new sustainability, without the large bycatch characteristic of industrial fishing, and without catching the undersized fish which have yet to reproduce.

The *almadraba* was a living myth, bloody, earthy and impressive. It belonged to a culture which inhabited the coastal villages. The immense size of the nets alone appealed to the imagination. The figures spoke for themselves. Hundreds of heavy anchors on Barbate's quay in winter, reddish-brown with rust and covered in barnacles, bear silent witnesses. It took a month to set up the network of hundreds of metres of vertical nets. Four hundred anchors, each half a tonne in weight, hold the wall of nets in place, including a labyrinth of chambers from which the tuna cannot escape. Thousands of buoys keep the structure upright. Twelve hundred rolls of net cable, and 150 tonnes of lead-weighted cable for the seabed are required to keep the thing in place. Setting out and fixing, the nets used, the knowledge of currents and tides: centuries and generations of experience lie in the water here.

It was addictive. The garage brought in the money, but Guzmán happily continued with the *almadraba* on the side, at least as long as there was tuna. He had a 10-year-old son, but would he ever come fishing on the *almadraba*? He doubted it. 'I don't think our family will produce a fifth generation of fishermen.' The bluefin tuna has been overfished. The entire fishing industry was in crisis. Anyone wanting work in the fisheries here on the coast of Cádiz could really forget it. Under dictator Franco, when the state tuna canning industry El Consorcio held sway over the *almadrabas*, there was still a proper industry which kept half the village in work. Bluefin tuna was a cheap canned product. No part of the fish was thrown away. Even the big heart found its way in small pieces into a tapa, accompanied by the dry Manzanilla sherry which is popular here. In the aftermath of the dictatorship the catches from the *almadrabas* suddenly dropped substantially. The falling yields made the expensive *almadrabas* less and less profitable over time and were a nail in the coffin for the canning industry. A few small factories for tuna were active on a modest scale in the village. In the second half of the twentieth century other types of tuna had displaced bluefin in the cans. Most of the catch now disappeared into the freezer ships headed for Japan, which was a good thing, because that was where the big money came from. Sashimi and sushi, the Japanese cuisine that had conquered the world, drove up tuna prices, particularly those of bluefin tuna. Tuna is to sashimi as tomato is to pizza, an indispensable ingredient. Hundreds of millions of euros go into it, and the *almadraba* took its cut. The fishermen were well aware that without the Japanese their tradition of millennia would long have disappeared.

The question was how long it could continue. According to marine biologists the Atlantic bluefin stocks that migrated this way had been under great pressure from overfishing for years.

If expectations on departure were rather gloomy, the mood notably improved on the way to the nets at the sandbank off the beach of Zahara de los Atunes. The wind had changed to a southerly direction. On arrival at the lightship, the seawater turned out to be nice and clear. The tuna had a good view. There was nothing to stand in the way of a decent catch.

Three men had kept watch by the nets overnight. Some 20–25 m above the seabed we bobbed around the rectangular chamber where the tuna was steered inside along a wall of long vertical nets. The fishing fleet was now complete: the scout ship, two large motor boats, three flat-bottomed barges with masts for winching the nets on board, four smaller boats, three dinghies and the main ship, with its storage space filled with ice. The boats were anchored in a rectangle around the chambers of the *almadraba*. The wind had turned to a more favourable direction. We waited for the tuna to come in.

The captain, a small, greying blond man with sharp features, whistled to signal the attack. The men in the boats peered into the water in concentration. The divers who had sunk down into the labyrinth of nets surfaced, hauling a dead bluefin tuna with them. The creature, a couple of metres in length, around 200 kg, was winched up onto the boat by the tail. Was it fatally wounded by an orca or swordfish? Had it got itself trapped in the nets? It was unclear. But there was tuna and that gave the fishermen hope. Tuna never travels alone. The divers descended again. The men remained watching, tense, on the edge of the boats, the air filled with the sweet aroma of smoke from the heavy Ducados cigarettes.

The divers resurfaced, this time with thumbs up. There were tuna in the chambers of the *almadraba*. There was money in the water. The tension was discharged in raw battle cries from the fishermen, shouted out over the water. The fleet prepared for battle. The captain was now giving out rapid commands with his whistle, directing the actions required, like a conductor before an orchestra, signalling with this hands and arms. The nets were attached with cables to the winches, engines started up, chains rattled. '*Adelante*, onwards,' came the rasping cries. On the main ship a protective tarpaulin had already been laid out so that the fish could soon be winched up onto deck without being damaged. The boatswain took up position at the prow to collect up the smaller fish with a net.

Everything then became a confusion of nets drawn up, shouting and swearing; a pandemonium in which only the trained eye could discern any system. By striking the water with oars the tuna was driven from the side chambers of the *almadraba* net into the main chamber or *copo*. The chamber of death. The captain was now continuously whistling out commands. As if to add power to the operation, large explosions sounded from military training grounds on the coast. There was still nothing to be seen in the water, at least to the novice, but the trained fishermen's eyes could see from the crests of the waves that some large mass was rising. Slowly, as if the sea were coming to the boil, the water began to stir between the boats. A dinghy with ten men left the main boat and paddled to the middle of the rectangle formed between the ships.

The bluefin appeared on the surface. At first as a giant shadow shooting back and forth under water, in a panicked search for an escape route. The shadows took shape: gleaming, silver-grey torpedoes, metres in length, large eyes flashing through the water and staring at the approaching surface. It is probably an anthropomorphic perception, but the large eyes of the tuna seemed to reflect panic and terror. The first silver-white bellies turned upwards in their attempts to remain in motion as the nets closed in. The sea had changed into a white, foaming fountain, shooting metres into the air. Dorsal fins appeared on the surface and dived down again.

Suddenly there was a crashing sound of tuna tails beating on the water. The next moment it began to rain fish. Literally. If you think 'raining fish' is just a kind of biblical metaphor, you've never witnessed an *almadraba*. Fatty *caballas*, Spanish mackerel, and *melva*, another relative from the mackerel family, were swept out of the nets and metres into the air by the powerful tails of the bluefin tuna in their death throes. They clattered down onto the decks of the surrounding boats to be cleared up in scoop nets and plastic bags and stored away. It was not often possible to get a meal of fish out of the sea this easily.

The enormous, glistening, torpedo-shaped bodies of the tuna were now clearly visible among the white foam thrown up into the air by the beating fishtails. Nineteen tuna, cried the captain. Two of the largest must have weighed more than 300 kg. Yelling into his mobile phone, the captain reported the catch to the port office in Barbate. The people at the cold storage could start preparations for cleaning and freezing the tuna.

The men stepped from the dinghy into the middle of the boiling mass. Dangerous work, as the tuna was not yet dead. It remained a case of watching for the feared blow from the tail. The tuna in the water were pulled to the main boat with pike polls, bound and winched onto deck. There they quickly disappeared into the hull of icy water. The white fountain of foam between the boats gradually grew red with tuna blood. The beating and splashing of tails quietened, the foam fountains slowly died down. Once out of the water, the tuna did not last long. Between the boats the water was now deep red.

If you stand eye to eye with a classical Greek or Roman statue you can be overcome with the sudden realisation that you are bridging a gap of thousands of years. In an instant, we witness something which was seen in precisely the same way thousands of years ago. The *almadraba* is a time machine of this kind. I stood watching alongside the Roman Pliny (23–79 CE), natural scientist and writer of the reference work *Naturalis Historia*: 'When the tuna dives into the Spanish *almadraba*, the nets are also full of mackerel' [196]. That might sound trivial at first, but really, here at sea, just off the beaches of Zahara de los Atunes we could see the exact same thing with our own eyes. The cheering of the men bringing in the great nets, the tuna in a fountain of foam and blood fighting for its life, the view of the shoreline: history loses its abstraction and turns into a palpable and perceptible experience.

'Abundant and wondrous is the haul of the fishers when the army of tuna sets off in spring,' writes the Greek poet Oppian, favourite of the Roman emperor Marcus Aurelius, in the second century CE. He describes tuna fishing in remarkable detail in the *Halieutíka*, a treatise on fishing dedicated to the emperor and his son, Commodus [182].

'The masses of tuna enter our sea from the wide Ocean in spring, driven by restlessness to breed. First it is the Iberians who throw themselves upon the shoals and attempt to catch the tuna; then at the river Rhône the Celts and inhabitants of Marseille chase after them; and third come the inhabitants of Sicily and the surroundings of the Tyrrhenian Sea. From the immeasurable depths there, they turn back and swim out through the entire ocean.'

'Here exceptional quantities of tuna are found, unimaginably large and fatty. The fish is salted, stored in amphorae and transported back to Carthage. This is the only tuna which the Carthaginians do not export; they eat it themselves for its excellent flavour,' a text attributed to Aristotle tells us.

Still further back in time there are the myths. The Greeks believed that darkness, chaos and destruction reigned in the impenetrable waters beyond the Pillars of Hercules. There was only space here for gods and heroes. Here lay Tartessos, the land where Zeus battled with the giants. This was where Heracles, later known by the Romans as Hercules, carried out his mythical labours and fought the sea monsters. According to Plato, Atlantis was devoured by the waves here. The Jewish god Yahweh killed the sea monster and the many-headed serpent Leviathan. The Old Testament states, 'Thou didst divide the sea by thy strength: thou brakest the heads of the dragons in the waters. Thou brakest the heads of leviathan in pieces, and gavest him to be meat to the people inhabiting the wilderness' (Psalms 74:13–14) [52].

The Hebrew word *thanin* is translated here as 'dragon'. The tuna monk Martín Sarmiento wrote that the *thanin* can be a dragon or a sea monster, but was more likely a collective term for large fish. It is a small step from Hebrew *thanin* to Greek *thynnos*, and from there to Latin *thunnus*. *Thunnus thynnus*. The sea dragon struck dead by Hercules and Yahweh was none other than the giant tuna [52]. The god fish, food for mythical heroes. They killed the tuna by 'breaking' its head with a sturdy club, providing the coastal inhabitants with food. But it was also a feared sea monster, a demon of power and magnitude from the deep, with the head of a dragon and a tail which could easily break your bones. Tuna was a fish monster that was closely connected with the fate of human mankind.

The blow was dealt, our Leviathan had been reeled in and lay in the icy water of the hull on the main ship cooling quickly. The fleet formation around the nets was broken up. We set course westwards, back to Barbate's port. At the cold store on the quay the teams stood ready to hoist the fish up. A set of digital scales displayed their weight. The largest fish went up first. Two tuna of 425 kg each, almost three metres in length. These big guys were followed by a tuna of 380 kg and another of 378 kg. The harbour master noted the catch, while the fish disappeared inside and was lifted up on the hydraulic platform.

Up on the platform Takeshi Noguchi and his assistant stood ready and waiting to sample the fish. They looked a little out of place in these surroundings: two Japanese men in their boots, oilskins and a protective hats amidst the Andalusians with their heavy Cádiz accent. Around their necks hung mobile phones with plastic tuna charms attached. Noguchi moved among the fish with an ease indicative of routine. Here was a man who knew what he was doing. In his hand lay a long, hollow steel prong which looked like a piece of kitchen equipment akin to an apple corer, but rather longer. Noguchi pricked the tuna's belly, pulled it carefully back out of the flesh and inspected the colour to assess fat content. The Japanese like fatty tuna for their sushi and sashimi. The higher the fat content, the higher the price.

Noguchi said something to his assistant and indicated with a nod that the tuna had been approved. Soon it would go to the reefer, the transport boat anchored a little further out, to take the tuna to Tokyo. Then the harbour boys of Barbate started on the heavy work. They were dressed in oilskins and hairnets, hygiene requirements from clients in Japan. A single hair on the tuna can cause scandal and loss of value in Tokyo. A power saw started up and in flowing strokes the heads and fins were separated from the bodies. Their bellies were opened with large, sharp butchers' knives and the entrails, a couple of arms full, were cut from the trunk. The large gill system was removed. Within a couple of minutes the impressive silver Leviathan was transformed into an anonymous, torpedo-shaped hunk of flesh, ready for freezing. Two tuna from the catch were destined for the Spanish market. The rest were bought by Tokyo Seafoods, the company for which Noguchi worked. The tuna would be frozen in the boat's hull. At the start of June the freezer ship would sail for Tokyo. There the tuna would be displayed on the market with a colourful label: 'Wild tuna from the *almadraba* of southern Spain'. Connoisseurs would be lining up to buy it.

Afterwards we drank coffee in the bar on the way out of Barbate's port. Outside, the harsh Levante wind had now risen and sand was blowing over the boulevard. Tuna master Takeshi Noguchi sat at the table, legs wide apart, with his assistant Takeshi Homma, and their youngest attendants Tatsumi Yoshzya and Makoto Soma. Noguchi exuded natural authority like a dyed-in-the-wool samurai from a Kurosawa film. For 26 years he had been purchasing tuna from all over the world to take back to Japan.

He knew the Spanish coast like the back of his hand. They were good people, those Andalusians, you could have a good laugh with them. But when it came to work they could learn a thing or two from the Japanese, Noguchi felt. 'We like to get things done quickly and efficiently. But that's difficult here,' he said with a thin smile. Arriving on time was not to be taken for granted in Andalusia. And then there were all those holidays: you were barely done with one village fair when the next procession with a virgin came around.

In any case, you had to come here to Spain for the best tuna. Beautiful red flesh, plenty of fat. That quality of bluefin could not be found in many places around the world. Sadly things were also declining here, Noguchi thought. The tuna flesh was often brown and that meant loss of flavour and a lower quality of sashimi. What was the cause? Noguchi shrugged. The tuna wasn't always handled right and wasn't cooled fast enough. Some scientists talked about rising water temperatures. The fat levels were also falling, and fat content was highly prized by his clients. That was probably to do with reduced food at sea, Noguchi suspected. The bluefin tuna is a voracious predator and the supply of food simply followed certain cycles at sea. Fewer herring and anchovies meant less fatty tuna.

Another factor important for quality—Noguchi could not explain it to Barbate's fishermen often enough—was that the tuna must be killed promptly. If the fish tries to escape, its body temperature rises. Tuna is by nature a somewhat nervous fish that quickly becomes distressed, and fear and stress hormones are not good for the quality of the flesh. That is the big advantage of the tuna caught and fattened on tuna farms, found all over the Mediterranean Sea. There they shoot a bullet through the creature's head and you're done. It's a good deal quicker than dragging them up and fishing them out of the *almadraba* nets. Nevertheless Noguchi remained unconvinced about fattened farm tuna. After it is thawed, the meat rapidly loses its colour, the flavour is different. 'Tuna is a predatory fish and needs to swim free in the ocean,' Noguchi felt. 'Swimming around in circles in a space of 50 m^2 isn't good for them.' You could immediately tell that the *almadraba* tuna tasted considerably better. 'Hai!' came the affirmative from the assistant and two attendants over their milky coffees.

Noguchi bought the tuna from the *almadrabas* in Zahara de los Atunes and Conil de la Frontera. His business had a freezer ship in Conil's port, the Reina Cristina. The fish would then be distributed throughout Asia, some with Tokyo's Tsukiji market as its destination. But his company delivered most bluefin tuna directly to Japanese sushi restaurants. 'It's worth the effort of coming down here to the south of Spain. You can strike some good deals,' Noguchi said. There were few places left in the world supplying wild tuna in such large quantities as here in the south of Spain. High quality and at low prices too.

In Japan tuna was expensive. The most expensive cut is the belly, the *toro*. That year 7000–8000 yen was paid per kilo of bluefin tuna *toro*. At the exchange rate at the time that was 45–50 euros per kilo for the better pieces at least. The tail was the cheapest cut, at 2 euros per kilo.

Of course the Japanese tuna master was also worried about the future. For Noguchi there was no doubt: the tuna catch had reached its ceiling. And how else it might develop still remained to be seen. The bluefin tuna as a species is endangered when the fish is no longer able to reproduce. You didn't need to be a biologist to understand that. In Noguchi's view the core of the problem lay in industrial bluefin tuna fishing with purse seine nets, large bag-shaped nets which could fish out an entire shoal in one go, without allowing them to spawn. The fish was then fattened at large tuna farms. There were no proper checks or regulations for the tuna farms at sea. The ICCAT, the International Commission for the Conservation of Atlantic Tuna, had limited itself to rules in the form of catch quotas for bluefin tuna, but the quotas are sloppily observed. Some tuna traders have taken no notice of the rules or the official fishing certificates. There was barely any obstacle in the way of tuna pirates.

The tuna master glanced at his watch. Back to work: consultation with company management on the home front. Japan wanted to know what bluefin tuna would be arriving in their ports.

The Tuna Monger

The Japanese adventure in Spain sprang from simple beginnings, with a street fishmonger specialising in tuna. Anyone who came across him in the 1960s with his wooden fish cart, in the villages and towns on the coast of Murcia where he sold his sardines and salted tuna, would hardly have guessed that he was a multimillionaire in the making. Ricardo Fuentes had a profound influence on the history of the bluefin tuna. Spain had emerged from the Civil War poor and backward, under Franco's brutal dictatorship. Only in the 1960s did the country gradually break out of its isolation. Tourism began to pick up, blonde girls in bikinis from Sweden, Britain and the Netherlands appeared on the beaches. Spain grew more liberated and prosperous. Ricardo Fuentes started up his own shop stall selling salted fish on the daily market of his hometown of Cartagena, the town once used by Hannibal as a bridgehead for Carthage in the war against Rome. There he sold the specialities for which his region had been famous since the Phoenicians: bluefin tuna roe and *mojama*, the salted tuna with the Arabic name which looks and tastes like raw cured ham. Fuentes' business did well. He opened a small factory with a warehouse on an industrial estate outside the town. Along with his sons in 1984 he founded the company Grupo Ricardo Fuentes e Hijos, which became the basis of an international tuna empire of 40 operating companies and tuna farms spread over the entire Mediterranean and a turnover which in the first decade of the twenty-first century grew to 250,000 euros. Forty years after he had started out with his fish cart, this tuna trader conquered Tsukiji market. Many of the traders could hardly pronounce his name, but they tasted the bluefin tuna from Ricardo Fuentes of Cartagena.

As early as the 1980s Fuentes was one of the first in Spain to realise that there was a possible market for tuna in Japan. He made contact with Japanese people in the Mediterranean in search of bluefin tuna for their sashimi and sushi. In the eyes of the Spaniards, who had lived in isolation under Franco for many years, the Japanese appeared to come from another planet. They ate raw tuna, and they paid prices which the Spanish fishers could only dream of. The fish farmer's sharp nose detected an opportunity. Japan had money, there was a large market for tuna and their traditional supply of bluefin in the northern Pacific was running low. The Japanese traders were searching high and low for alternative supply channels, and they were prepared to invest substantially.

The early years were not easy. The Japanese may have paid high prices for tuna, but they made quality demands which the Spanish initially did not understand. The tuna was destined for raw consumption and had to be of the highest quality, requiring very precise handling. Every blemish could influence flavour and therefore price. If the flesh was just a shade darker, because the fish had not been cooled immediately after being caught, the tuna was classed as *yake*: unfit for raw consumption. The Spanish had never been

(continued)

accustomed to handling bluefin tuna with care. The *almadraba* was a blood-bath. The fish's brains were beaten in with sturdy clubs and long hooks were attached to the tuna at random in order to hoist the creatures up. There was no question of cooling in transit. Earlier, when everything was salted or cooked and canned, it didn't much matter anyway how the fish looked. The first tuna from Spain to reach the central market in Tokyo barely received a second glance from the Japanese. The Spanish bluefin tuna promptly disappeared into category C, the lowest on the quality scale, just high enough not to sell for cat food.

Ricardo Fuentes soon began to experiment with new techniques to mini-mise loss of quality when transporting the fish to his new Japanese clients. The Japanese helped by bringing their advanced freezing techniques to Cartagena and a new freezer tunnel was installed on the Fuentes industrial estate. At first this was just for freezing the *ventresca*, the fatty tuna belly. That was the part which went off the fastest. *Toro* as the Japanese call this cut of tuna belly, was the most in demand in Tokyo. When the technology improved, entire tuna were frozen. This gradually raised the quality of the Spanish tuna arriving in Tokyo. Export to Japan, still only ten per cent of Fuentes' turnover in the early 1980s, began to grow rapidly. 'The great advantage in the early years,' says David Martínez, director of the Fuentes group, 'was that Ricardo Fuentes also still made *mojama*.' Whatever was rejected by the picky Japanese buyers could easily find its way in salted form onto the Spanish market. The best tuna therefore disappeared to Japan and the Spanish market was more than satisfied with the fermented leftovers. In this way Fuentes gained an advantage over competition from abroad which had to work without such stable sales. This head start allowed Fuentes to operate for a long time without too much competition. His six tuna vessels soon proved insufficient to satisfy demand. The Fuentes group began to buy up tuna from French, Italian and Tunisian fishers.

Tokyo's Pantry

15

Bathed in the dazzling glareOf the lights, the tunas seemAs if they're floating.

Kozaburo Omura

Anyone wanting to see the end of our tuna story on the southern coast of Spain must head for Tokyo. It was a little after four in the morning, making the entrance to Tsukiji market even darker, with its narrow alleys of shops for dried fish and seaweed and restaurants selling omelettes and noodle soup. Tokyo was still asleep, but the working day here was already in full swing. The first delivery vans were racing by. Small, manoeuvrable forklift trucks were starting up their fuming diesel engines. Traders parked their cars and shot past the burning lanterns into the Namiyoke Inari Shrine. The wooden building with its arbour breathed an oasis of peace amid the surrounding industrious activity. A large lion's head guarded the entrance against evil spirits. Restaurateurs could offer their prayers here to a stone egg the height of a man, tuna traders had to make do with a sturdy slab of granite set up for the protective gods of sushi and sashimi. After washing their hands the visitors stomped into the shrine in their boots and pulled on thick cords to sound the bell and awaken the Shinto gods. Prices were good today and the turnover was worth the effort. Inside the labyrinthine market, Katsuji Suzuki made his way, clad in boots, taking long, confident strides to the narrow, slippery alleys between the fish stalls. He was loudly greeted from the shops, stalls, filleting tables and freezers. They consider Suzuki a good surname here on the market: it means seabass in Japanese. This 60-something-year-old, as tall as a tree, with wavy blue-grey hair, is an old celebrity on the market, so our walk was repeatedly interrupted. At one stall sesame biscuits were served, at another lengthy jokes were exchanged over the quality of fish he had recently bought. The personnel stopped filleting the tuna for a moment and smiled, long, flat knives in their hands.

© Springer Nature Switzerland AG 2019
S. Adolf, *Tuna Wars*, https://doi.org/10.1007/978-3-030-20641-3_15

Walking over the smooth cobbles we passed endless rows of stalls where the brokers offered the merchandise they had bought at auction that morning. Tsukiji resembled an enormous nature cabinet, like the ones you might associate with eighteenth-century natural scientists. An endless collection of all conceivable fish and shellfish in the world. Everything that swims and is edible can be found here. It smelt of salt water and fish. The display cabinets were full of fish which just a few days before had been swimming around in the Atlantic, the Mediterranean, the Indian Ocean or the Pacific. Eels and conger eels tried in vain to wriggle out of their polystyrene boxes. Lonely tiger prawns lay neatly displayed in wood shavings for inspection. Octopuses floated elegantly in glass tanks of seawater, yellow trumpetfish lay in ranks on their bed of crushed ice. There were enormous spider crabs, small varieties of mackerel and piles of elegant boxes in which sets of sea urchins were carefully and securely packed. Giant black mussels, cockles and clams lay on display next to red seabass, brown spotted rays and orange squid. A little further along, swordfish was expertly sawed into slices. Next to that came a display cabinet with large red steaks. You could still come to Tsukiji for whale.

But that's not what Katsuji Suzuki had come for. After a couple of hours' sleep, after midnight when the last client had left his little restaurant Ginza Maguroya (the Tuna Shop), the sushi master drove to Tsukiji with a clear aim: getting there early. At half past four it's still quiet in the megapolis. Once its 35 million inhabitants had woken up, the same journey could take him an hour. Having parked his car on the premises he walked straight to the two large auction halls on the outer edge of this largest fish auction in the world. We were now at the epicentre of the global high-grade tuna trade. And Suzuki was delving for bargains. Hundreds, sometimes thousands of tuna—frozen and fresh—are laid out daily for inspection on the boards in the two auction halls. For those with a sharp eye, there might be something special: a top quality fish missed by the buyers that could be nabbed at a good price. That was Suzuki's speciality (Fig. 15.1).

Anyone who has anything to do with tuna ended up at Tsukiji sooner or later. The 'mother of all fish markets' was situated where the Sumida River flows out into Tokyo Bay. You'll rarely saw active fishing ships along the quay at Tsukiji. Pollution put an end to fishing in the bay decades ago. The Japanese fleet of large industrial freezer ships has bases in countless ports in the smaller towns along the coast. The fresh fish come from here or from Narita airport, transported by truck to the market. The Japanese are the biggest consumers of tuna in the world, and Tsukiji reflects that. The location in the heart of Tokyo alone, a vast area belonging to the most expensive land in the world. A couple of metro stops away is the Imperial Palace and the gleaming business district where the Japanese multinationals have their headquarters. The emperor, the economy and fish: the pillars of Japanese society are brought together here in an area of a few square kilometres.

From the seventeenth century the fish market of Edo, as Tokyo was called until 1868, was located elsewhere. But after the severe earthquake of 1923, which destroyed the majority of the city, the authorities decided to move the market. The large grey market halls with their semi-circular shape were designed to be a magnificent specimen of functional architecture in the Bauhaus tradition. Now

Fig. 15.1 An early morning coffee with Mr. Suzuki and a bluefin tuna at the Tsukiji market (S. Adolf)

they present a somewhat worn and listless facade. For decades, apparently interminably, there has been talk of moving the market to a more accessible location outside the centre. The ground on which Tsukiji stands is polluted and would be an interesting prospect for more profitable real estate projects. But the move is cursed. The operation has been continually postponed. First there was opposition from the traders and their clients. Then there was something wrong with the new building. Next, pollution was discovered in the new location at such high levels that it exceeded the municipal safety standard.

But finally the market relocated two kilometer eastwards to the Toyosu island area. The 'New Tsukiji' opened the 11th of October 2018 in a huge, state of the art, climate controlled interior market building of 40 ha, almost doubling the space of what was already the biggest fish wholesale market in the world. Four floors full of all the fish in the world you can eat, including a restaurant area and a roof garden. Sanitized and brand-new, it looked like a 'cutting-edge' market space, ready for the twenty-first century. Visitors were not allowed to cross the slippery market floor, with its files of frozen tuna and trading boots where you smelled the salty scents of the world oceans. They could watch it from a gallery on the upper floor, behind well secured windows. It looked more professional perhaps, but not all the Tsukiji-veterans seemed to be happy to leave behind the rat-infested 83-year old market space. The New Tsukiji lost its soul, they complained. The shrine of the guardian spirit of Tsukiji, the well-respected Sui Jinja, was tucked away in a lost corner of the new building. The stubborn Shinto god from the granite monolith at the old market's entrance finally lost the battle against the reallocation.

Japan cherishes Tsukiji. *Tokio no daidokoro*, 'Tokyo's pantry', the market is called. The brokers offer their wares on hundreds of stalls lit by the warm glow of old-fashioned lamps. Even the cold neon light on the auction floor contributes to the authentic atmosphere. The bell for the bidding rounds, the secret language of gestures used by the bidders and the dancing auctioneers have not yet fallen prey to the sterile screen technologies of the digital era. The brokers who participate in the bids through auction houses set the dynamic of this market. Their relatively small scale means that they attract a close-knit public, with purchasers ordering remotely or coming to take a look for themselves.

The Tokyo fish market is the largest transshipment location for fish in the world. A daily average of 50,000 traders, restaurateurs, fish retailers, tourists and market staff come together in the 230 ha area of halls and side buildings. Around 450 different species of fish and shellfish are traded here. In one morning the market serves an average of 14,000 clients. Restaurant owners, fish retailers and sushi masters visit the often narrow stalls to study the range that the seas and oceans have to offer, and tuna is the fish to which the market owes its fame. A daily turnover of 300 tonnes of tuna was no exception. Although Japan has hundreds of fish markets, Tsukiji, which processes around a sixth of the total wholesale turnover, is by far the largest. The record turnover was 890,000 tonnes of fish in 1987. Trade weights subsequently decreased to around 650,000 tonnes in 2016, which translated to a daily trade value of 2 billion yen or 16 million euros in 2016. In total around 60,000 people make a living from activities directly relating to the market.

As island inhabitants the Japanese never live far from the sea. The islands are spread over a distance of 3000 km. The coasts are enormous. The surrounding seas are deep and subject to cold and warm currents, making them ideal for the plankton development fundamental to rich marine life. No wonder that the sea has always played a crucial role in Japanese cuisine. The surrounding seas hold 2000 species of fish, hundreds of different shellfish and a whole array of edible seaweeds. At 27 kg of fish per capita the Japanese were overtaken by the Chinese (38 kg) at the start of the twenty-first century but remain among the world's biggest consumers of fish. Portugal and Spain top the European fish-eating league (not counting Iceland) with 60 kg and 40 kg per person respectively. In the United States people only eat 20 kg of fish per person. In Germany and the Netherlands the figure is as low as 15 kg.

Everything edible from the sea can be found in the Japanese diet. From seaweed to urchins, from crabs to sea cucumbers and from whale to tuna. Fish in Japan is far more than just food: it is a birth right and cultural heritage, part of the collective identity. When you get your hands on that, you impinge on Japan, as revealed by the country's firm resistance to bans on the consumption of whale meat, or the tenacity with which they refused to engage with worldwide criticism of traditional fishing and dolphin slaughter, publicised in a 2009 documentary about the Japanese fishing village of Taiji [200].

Nevertheless Japan has not evaded the trend started in the economically most developed countries during the second half of the twentieth century: fish consumption has dropped in popularity and consumers are not willing to pay as much for it. This contrasts with the countries with middle classes, particularly China, where fish consumption has grown spectacularly. Tsukiji's turnover weight initially dropped after 1987. Harvard food anthropologist Ted Bestor describes it clearly in his reference work on Tsukiji [40]. In 10 years the turnover dropped by 20%. That was undoubtedly related to the cracks in the property bubble and the economic crisis Japan found itself in. At the same time it also seems characteristic of a broader global trend in developed countries. Fish is complicated food: it contains bones and tends not to be a quick snack. Until 2005, when China became the largest importer of fish, Japan was the market leader. In 2005 Japan still imported 11% of world fish production. Tsukiji is a barometer of the Japanese economy—but at the same time it is closely bound up with Japanese international interests in areas such as ecological sustainability of fishing and the disputes arising from that regarding international waters and fishing methods.

Among the broad range of fish on offer, tuna is central. According to the Japanese Fisheries Agency at the beginning of this century it was estimated that 80% of bluefin tuna caught globally ended up on a plate in Japan. And when it comes to buying the giant bluefin tuna or *kuromaguro*, Tsukiji is traditionally the number one place in Japan. In the first decade of the twenty-first century eighty-five per cent of Atlantic bluefin tuna officially recorded as caught in the Mediterranean therefore ended up at Tsukiji or in the cold stores of the big Japanese multinationals. In 2006 Japan officially imported 23,000 tonnes of fresh and frozen bluefin tuna for a total value of 350 million dollars. The *kuromaguro* in particular has almost sacred significance in Japan. The fish is associated with wealth and good fortune. In the periphery of Tsukiji visitors can choose from countless restaurants offering sushi and sashimi fresh from the market. The Japanese are certainly keen: there are often queues at the doors of the most popular spots, such as the famous Edogin sushi bar.

Next-door to the north of the market an entire neighbourhood is set up with all kinds of fish paraphernalia, sets of sushi knives, frighteningly realistic plastic model sushi for restaurant owners, fish tools, seaweed, dried and fermented fish and little plastic tuna charms to hang from your phone. This is also the location of the restaurant Sushizanmai, open 24 h a day and part of a chain run by one of the Tsukiji dealers. In the adjacent neighbourhood, less visible to the general public, are the exclusive restaurants where the financial and political elite like to pick up their sushi. The powers that be feel at home in this atmosphere of tradition, the high-quality image of the market and the idea of fresh *kuromaguro* within reach.

In this almost magical atmosphere of money and good fortune a separate place is reserved for the annual auction of the new season's first bluefin tuna, which takes place in January. It is a ritual which always reaches the international press due to the record prices. A photo of the smiling buyer, generally a restaurant or sushi chain owner wielding a large butcher's knife over the open belly of a fatty bluefin tuna, now belongs, along with the concert of the Vienna Philharmonic Orchestra and ski jumping at Garmisch-Partenkirchen, to the list of iconic New Year traditions to receive worldwide interest. The mega-bucks paid are precisely recorded and constitute a source of pride. In January 2009 a fresh bluefin tuna of 128 kg from Japanese waters took the honour. Two rival sushi bar owners came to an agreement and paid almost 10 million yen (80,000 euros at the time) for the fish. That's 625 euros per kilo. The year 2001 saw a record price of 20 million paid by a trader from Hong Kong. At the time the fish was worth more than its weight in gold. For many Japanese people it was a definitive signal that China had awoken as a major economic power. From 2012 Kiyoshi Kimura of Sushizanmai was the lucky winner, beating his competitors with the highest price six times in a row. Mr. Kimura established a record price of 155.4 million yen (1.8 million dollars) in 2013 for a bluefin tuna weighing 220 kg, giving an average price of almost 8200 dollars per kilo. In subsequent years Kimura landed his prizes considerably more cheaply: in 2017 the price had dropped by more than half to 3000 dollars per kilo.

You might reasonably wonder whether there was a connection between the price of the first bluefin tuna and the state of affairs in the Japanese economy. Was the bluefin perhaps a sort of economic gauge indicating how things were going? Mr. Kimura himself liked to emphasise that he was supporting Japan with the high prices: things were not going badly for the nation. But after detailed research *The Economist*—which as the organiser of conferences on sustainability in the oceans was aware of the issue—came to the rather disappointing conclusion that this sadly was not the case [40]. A correlation between the price of the first tuna and growth in gross domestic product (GDP) turned out to be largely absent. It smelt a little fishy: an expensive marketing event providing the highest bidder with a load of publicity. The fact that the price subsequently underwent a downward trend was perhaps more to do with something completely different: the negative publicity around overfishing by Japan, which had brought the stocks of bluefin tuna in the Pacific to the edge of the abyss.

We had not come for the first bluefin tuna but for an average trading day at Tsukiji. Preparations began around midnight. The small forklift trucks brought the

Fig. 15.2 The deep frozen torpedos of tuna, and an unfrozen part of the tail for the traders to taste and judge (S. Adolf)

frozen tuna in and out of the cold stores. Tuna carcasses, frosty white with condensation, were unloaded from lorries and delivery vans and thrown onto the floor from the cargo boxes. Old lorry tyres broke the fall of the cylinders of flesh, often weighing hundreds of kilos, which were then pulled over the floor of the large hall with pike polls (Fig. 15.2). There the fish lay ready for inspection and auctioning. Endless rows of uniform white, torpedo-shaped fish carcasses in an area equivalent to half a football pitch. An army of tuna, fins and tails removed, a gaping crescent-shaped hole at gill level. Between the boards on the floor hung a white mist of freezing dew. Fresh bigeye tuna was collected in the hall next-door. The metallic fish were a different shape and size, multi-coloured sticky labels providing information on origin and weight. A long cut had laid the belly cavity open to remove the innards, but the fish were otherwise complete and easily recognisable. Sometimes there was still a nylon line hanging out of their wide open mouths, retained from the longline ships that had caught them a couple of days before.

Katsuji Suzuki inspected the range on offer on the floor with the eye of a connoisseur. He shone a torch into the open abdomen. Using a sharp fishhook with a wooden handle he carefully pricked at the tail, which had been sawn open, and prised little pieces of flesh from the frozen mass to taste. On the wooden display tables lay thawed sample slices of the frozen tuna which had regained their original red colour. From time to time Suzuki noted down on the scrap of paper on his clipboard which tuna interested him. The tuna was inspected for flavour, colour and shape. Blemishes, scars and wounds could all reduce the quality of the flesh. The fat content was crucial. Suzuki brushed his fingertips along the open tail and rubbed them together to judge the fat content. The shape of the tuna was carefully inspected. The ideal of beauty here was like that of a Rubens painting: curvaceous and plump. The skin had to be flawless. Too many sores on the surface potentially indicated a long death struggle, with all the loss of quality that entailed. At Tsukiji an army of men, along with the odd woman, did this work day in day out. They all looked more or less identical: boots, small fish hook in hand, torch stuck into their belt, clipboard at the ready.

From half past five onwards there were continuous auctions at the two tuna halls. Restaurateurs and fish retailers had given Tsukiji's dealers their orders. The auctioneers rang their bells and started on a ritual which for outsiders is largely incomprehensible. Even most Japanese people cannot follow the jargon in which bidding proceeds. Every auctioneer had his own style: sometimes the bids were beaten out in a monotone staccato, sometimes it sounded like a fast tuna rap with high shrieks, with the auctioneer dancing back and forth before his audience of dealers to the rhythm of the bids. The bystanders looked on in concentration when a fish interested them, otherwise gazing vacantly. The bidding was completed with quick hand gestures. They got through the fish on offer at breakneck speed.

In a corner of the market hall for frozen tuna lay the bluefin from the *almadrabas* of Zahara de los Atunes, Conil and Barbate. Unlike the complete frozen tuna torpedoes, here I found chunks of tuna cut lengthways along the backbone into four large concentric pieces. Broker Kitano Kubota was pleased with his pieces of 'wild tuna'. He liked to buy in Spain. You knew you were getting quality then, said Kubota, as he sawed the frozen tuna into manageable pieces and placed them in his freezer. He already had a buyer for this fish. It was destined for a wedding dinner here in Tokyo. A special tuna for a special occasion.

Sushi master Suzuki wasn't as lucky that day. His trading competitors had paid close attention and had not missed any bargains. After the auction Suzuki shrugged and disappeared from the floor, on his way to the brokers' market where he buys the rest of his sushi fish, among the hundreds of stalls at the heart of the market halls. We walked past the long wooden benches where the fresh tuna lay staring blankly as it waited to be carefully filleted. Here it was a mixture of slaughterhouse and market. The tuna had their heads removed, a delicacy consisting of fine pieces of flesh from the top of the head and under the eyes, which some restaurants have made their speciality. In the corners of the stall the tuna heads lay piled high. The tuna was carefully cut from the spine with long, flat knives and divided into manageable chunks of different qualities. From a little further on came the screeching of the electric band saws with which the frozen tuna was cut into slices. The filleted fish was carefully arranged on display, disappeared into the freezer or was packed in ice and polystyrene for transport. The air between the stalls was heavy with the metallic scent of fresh tuna blood.

The Tuna Shop

Katsuji Suzuki knew tuna better than he knew his best friends. For years he had travelled the world seas as a skipper on a Japanese longliner, in search of bluefin, yellowfin and big-eye tuna. But in the end his heart really lay in the cuisine. Now he ruled as sushi master behind the bar of his restaurant on the edge of the central business and entertainment districts of Akasaka and Minato-ku. If you don't know his restaurant it's not that easy to find, hidden away in the basement of an anonymous office building. Ginza Maguroya was just big enough for three tables. When they were full, there was always the sushi bar, with its traditional glass display cases where the raw ingredients of Suzuki's sushi art lay on show. There was red octopus, herring roe and fresh sea urchin. Tuna roe was displayed beside various prawns and lobsters, mackerel, scallops and squid. There was also fresh tuna meat, in various sizes and species, from all the world's oceans. Everything was neatly recorded in characters on the little wooden boards behind the sushi master's back.

For fish lovers, Tokyo is a feast. In the Michelin Guide, it has long been the top city for starred restaurants. Two thirds of the renowned chefs specialise in Japanese cuisine, with sushi restaurants taking up an important place. In 2017, 92-year-old sushi master Jiro Ono was the oldest three-star chef in the world, with his chic restaurant Sukiyabashi Jiro, near the gardens of the Imperial Palace. Spain's top chef Ferran Adrìa, a member of the culinary premier league himself, praised the elderly master as the producer of the best sushi in the world. The most spectacular sushi and sashimi is undoubtedly to be found in the restaurant of the Tokyo Ritz-Carlton on the 45th floor of the Tokyo Midtown Tower, with a panoramic view over the city and far beyond. The décor of Katsuji Suzuki's Ginza Maguroya had considerably fewer pretentions. In the windowless basement room of the Sankaido building, photos on the wall show the longliner which Suzuki had once captained, loaded with frozen tuna at sea. But sushi lovers felt liberally compensated for the humble surroundings by Suzuki's culinary expertise. With his large hands he kneaded the rice and adorned it with expertly cut fish. He was assisted by two sushi masters, each with two aides who knew the tricks of the trade. From the kitchen you could hear the distant sounds of opera music.

Along with the Japanese veteran of the world league of tuna scientists, marine biologist Peter Makoto Miyake (1933), we inspected the best tuna the restaurant had to offer. A bluefin tuna from the cold waters of the southern Pacific, red big-eye tuna, bright red yellowfin, albacore with its pale flesh. All 'wild', not fattened on farms. They were served up raw and skilfully cut into sashimi on little glazed ceramic plates. A little bowl in which the soy sauce could be mixed with a pinch of the bright green wasabi paste of Japanese horseradish provided the right dipping sauce. The pièce de résistance today was an authentic bluefin tuna from the northern waters of Japan. It was the same fish that had been worth its weight in gold at Tsukiji at the last new-year auction. Katsuji Suzuki pointed with restrained pride to the slices of sashimi. This tuna had swum around freely in a cold sea, Professor Miyake explained. That's why the flesh was marbled with fine fatty tissue, just as you find in the better-quality Iberian ham. The fat content is essential, Miyake told me. It made the bite of the tuna subtly crisp, the flesh melted on the tongue and left behind a flavour between fish and butter. From behind the display case Master Suzuki received the compliments with a smile. It had been among the most expensive tuna available on the market, he told us proudly, but by striking at the right moment he had been able to procure the showpiece at a relatively cheap price. He keeps his most beautiful tuna frozen to serve little by little to the greatest connoisseurs among his clients.

The small restaurant is fairly busy this evening. The tables are soon all taken. The incoming guests are boisterously welcomed by the entire staff, a fixed Japanese ritual. 'Still, the good times are a thing of the past,' says Katsuji Suzuki from behind his display cases. The collapse of the housing market in the 1990s put an end to a long period of growth, with clients who could afford top quality tuna. A long, hard economic recession followed. Political and business elites went under and Japan had immense difficulty wrestling itself free of a crisis lasting decades. Fishermen went on strike because they were at the end of their tether. Political leaders came and went. In 2011 the tsunami and the Fukushima nuclear disaster dealt the country deep wounds, not least due to the possible nuclear pollution of local fishing grounds. Prime Minister Shinzo Abe, elected for a second term at the end of 2014, introduced a special stimulus package in an attempt to breathe new life into the economy. Certainly, the Japanese were still rich, but faith in future economic growth just would not recover. That was reflected in attendance numbers at sushi restaurants in general and tuna consumption in particular. More expensive sushi restaurants like Ginza Maguroya struggled with a clear drop in custom.

This all worked to the advantage of the cheap sushi restaurants, such as the *kaiten-zushi*, where little dishes of sushi and sashimi pass by on a conveyer belt. Here sushi was transformed into a form of sophisticated fast food. Together with the prefabricated supermarket sashimi, bluefin tuna from Mediterranean fattening farms served as a cheap ingredient here. Since these farms had been introduced to the Mediterranean in the 1990s the import of bluefin tuna had rocketed: from 3000 tonnes in 1995 to around 35,000 tonnes from 2005. The high volume and lower quality makes fattened bluefin tuna relatively cheap, with prices half to a sixth of those for a 'wild' bluefin tuna. The fattened fish facilitated a market for the general public, maintaining its exclusive image of luxury fish, but remaining affordable for the mass consumer. The popularisation of sushi and sashimi also seems to have caught on in restaurants in big cities in the rest of Asia. Aside from the 127 million Japanese consumers there is an army of potential sushi eaters in the starting blocks in Shanghai, Guangdong, Beijing, Dalian, Hong Kong, Taipei, Singapore, Kuala Lumpur, Bangkok and Ho Chi Minh City. It is only a question of time before the little plates with affordable raw tuna, served on rice with seaweed, appear en masse on conveyer belts in these locations too.

Tuna professor Peter Makoto Miyake grimaced at the mention of bluefin tuna from fattening farms. He wouldn't touch the stuff. 'You can easily pick out fattened tuna. The fat tastes strange and bears no resemblance to wild tuna like that of the *almadrabas* in southern Spain,' said Miyake. In his view it is comparable to the difference between wild salmon and fatty farmed salmon. Miyake refused to eat farmed salmon. You wouldn't find him in those cheap sushi chains either.

Popularisation, sharp price competition and a Pacific Ocean population on the point of collapse according to scientists: how could a future be guaranteed for the bluefin tuna? Behind his bar, sushi master Suzuki patiently turned out little balls of tuna. He didn't have to think long about it. 'The bluefin tuna will never disappear,' he said resolutely. It sounded like an irrefutable command. 'The general public are now eating farmed tuna. That's where the bulk comes and the quality is always rising,' Suzuki hoped. Japanese scientists were perfecting techniques for growing tuna, he was certain, and the future lay in artificially bred, farmed tuna. The first fully farm-reared tuna has meanwhile been on the market for years. The big Japanese multinationals will never allow this trade to disappear, and scientists are getting better and better at genetic reproduction methods. 'In 20 years,' Suzuki predicts, 'we'll be farming freshwater tuna.'

The Japanese Goldfish

Sushi has not been around as long as people tend to think. Japan's most successful contribution to global cuisine was invented in its current form in the nineteenth century. Food anthropologist attribute the invention to sushi chef Hanaya Yohei (1799–1858) [40]. In his shop in the city then known as Edo, Yohei developed the so-called *nigiri-zushi*, or 'hand-kneaded sushi', a snack consisting of a thin sliver of fish on a small piece of rice, pressed into shape and soaked in rice wine vinegar. Yohei had been inspired by the much older, original sushi. According to food historians this could be traced back to sometime in the seventh century. The original sushi, strictly speaking, was a classical method of preserving fish: just as the Phoenicians and Romans used salt for the fermentation process, the Japanese used rice. The rice was literally packaging and was thrown away on consumption. Sushi chef Yohei turned this concept on its head: the fish was not even given the chance to ferment, the vinegar kept the food fresh and in so far as there was still concern over incipient spoiling, that was easily overcome by the pinch of horseradish paste buried under the fish.

Yohei's sushi was a great success.

Impatient in line
clients press hands together
Yohei kneads sushi.

(continued)

So reads a popular haiku. Yohei's shop was in business until the 1930s. From here sushi went on to conquer the world, at first as a popular street food in Tokyo, prepared in itinerant carts. This, unfortunately, has nothing to do with tuna. In fact tuna was not the first fish to be used for this [62]. On the contrary, all tuna species, in particular bluefin, were highly unpopular and despised by the sushi-masters as an inferior kind fish basically unfit for human consumption, according to author Trevor Corson. Originally it was even considered *neko-matagi* even not worth to feed your cat with, because of the strong taste, the bloody meat and the quick decay.

The rise of modern refrigeration techniques in the 1970s and the new air traffic from the US East-coast to Tokyo put a new slant on the range of sushi available. Suddenly it became possible to introduce this fattier fish into the dish that had previously spoiled too fast. This coincided with a changing consumer taste to softer, smooth meaty dishes. Tuna soon became the star of the show, and of all the species of tuna the Japanese particularly wanted bluefin in their sushi [146].

Walking into the little museum in the gardens of the Imperial Palace in the centre of Tokyo, you will find among the works of art on show from the Imperial collection a Japanese print drawing with a selection of fish from the Sea of Japan. Emperor Hirohito (1901–1989) is considered by many to be a war criminal for his involvement in World War II, but throughout his life he was also an enthusiastic marine biologist. One drawing shows a beautiful, enormous octopus, a garfish, a longnose stingray, a cofferfish and a mackerel, and, directly below the trumpetfish, the unmistakeable image of the bluefin tuna. The Imperial prints make it clear that bluefin tuna was a longstanding acquaintance of the Japanese. It traditionally belonged to the fish roaming the Sea of Japan and the western Pacific. However, it is unclear how far back in history Japanese tuna fishing goes. It appears that it has only been fished for a few centuries. Japanese tuna historian Fumihito Muto found documents belonging to the ship owners of the Edo period (1603–1867) and government registers from the Meiji Restoration (1868–1912) which show that over the centuries there was large-scale tuna fishing on the east coast of Japan. The Japanese used circular nets, a kind of predecessor to the purse nets, but made with straw.

Large-scale tuna fishing really got going in Japan, as in Europe and the US, at the start of the twentieth century. The introduction of motorised boats enabled the Japanese tuna fleet to travel ever further from their coastal waters in search of tuna. Catches rose from 1000–5000 tonnes annually in the early twentieth century to almost 20,000 tonnes in 1930. At the end of the 1930s record catches of 47,000 tonnes were even recorded. World War II, followed by the occupation by the American troops, more or less brought tuna fishing to a standstill. But from 1952, after General Douglas MacArthur, commander-in-chief of the Allied Forces in Japan, lifted the ban on fishing in territorial waters, Japan soon developed into a world-class fishing nation. The Japanese longliners ventured ever further into the Pacific. In 1954 one of the Japanese fishing boats was located at the Bikini Atoll when the Americans were detonating their experimental hydrogen bombs. The 23 people on board the boat were brought to Japan seriously ill from the radiation. The fears for radioactive tuna caused a crisis in the market which lasted for months. Concern caused by a radioactive accident with tuna fishers also led, quite unexpectedly, to the birth of Godzilla, an ancient dinosaur like creature. According to the story it was woken from hibernation by the harsh explosions of the same American nuclear test bombs that made the tuna fishers sick and created this atomic mutant. Again tuna, the source of the biblical Leviathan, was a source for a mythical deep-sea monster that spread fear. Godzilla swam to the coast and headed into the cities to terrorise the people.

But the real Godzilla never left the ocean. It was the monster of the ever expanding Japanese fishing fleet, crisscrossing the world to satisfy its insatiable hunger for tuna [31, 127]. The Americans in a sense, woke up a new era fisheries at the Asian side of the Pacific. The ancient fishing knowledge, coupled with great economic ambitions and technological ingenuity, led to unparalleled industrialisation of fishing at this side of the world. The Japanese longliner ships were capable of towing fishing lines 30 miles long, with thousands of fish hooks attached. The large ships increasingly functioned as floating factories, where the fish caught were immediately cleaned and frozen. In the latter case a leap in quality irreversibly changed the nature of fishing. Salting had enabled the Phoenicians and later the Romans to transport tuna from the coasts of the Mediterranean to the corners of the known world at the time. The invention of the tin can opened the way to a large industry of canned tuna.

And now the third revolution which prepared the way for tuna to conquer the world came in the 1970s from the East with the new Japanese conveyor-based freezer system which froze the tuna to a temperature of $-70\ °C$. This almost completely halted the process of decomposition of the delicate tuna flesh, safeguarding quality to the utmost and changing fishing boats into travelling fish factories and storage with a virtually unlimited operational radius. The technology did not remain limited to Japan but was rapidly exported: fishing fleets from countries such as Spain, Russia, South Korea and Taiwan were now able to catch unlimited fish all over the world and go in search of fishing grounds and new species of tuna, regardless of where their home ports and sales market were. Tuna processing, trade and consumption created a global value chain that was unprecedented in its extent and complexity. For the perishable bluefin tuna this change was particularly advantageous. The use-by date, previously an essential restriction, became less of a priority. As a torpedo-shaped lump or frozen in loins, it was barely recognisable as fish anymore. It turned into a commodity that was traded in bulk quantities for the world market, just like cocoa, grain crops, coffee, oil and metals.

While Ricardo Fuentes was walking the streets of Cartagena in Spain with his salted fish, the Japanese Godzilla was searching the world seas for more tuna. The fleet of freezer ships had to attain sufficient volume. Bluefin tuna become increasingly rare in Japanese waters, the Pacific did not produce enough fish to supply the growing market demand for raw sushi and sashimi. In the 1960s and 1970s, when the Japanese fleet gradually spread over the world seas, the longliners also turned up in Port Lincoln, a flourishing tuna port on Boston Bay in the south of Australia. There they fished for southern bluefin, the Australian relative of the bluefin. Port Lincoln enjoyed a reputation as the tuna port of Australia. A large proportion of the 14,000 inhabitants found work in a flourishing canning industry. From the beginning of the 1970s Toyo Reizo, the tuna-trading subsidiary of the Mitsubishi Corporation, began to buy the fish off local fishermen. The tuna was shipped on ice to Tokyo. It was a good outcome for Port Lincoln's fishermen: tuna was becoming scarcer and fishing had become too expensive to provide raw materials for the low-priced canning industry [31, 127].

The Japanese longliners had meanwhile left the Pacific in their search for new tuna stocks. They discovered some tuna in the Atlantic off the coast of Brazil. It emerged that tuna also swam further north, off the east coast of the United States and Canada. As elsewhere in the northern Atlantic, the big fish did not have a great reputation here. The fishermen saw bluefin more as a worthless pest that broke their nets and ate their fish. Only anglers on the American east coast saw the bluefin tuna as a desirable prey. They liked to be photographed next to the captured giant. But in the eyes of commercial fishing that was at most a childish pastime.

The Japanese found an ideal base for their search for tuna in the American fishing location of Gloucester, Massachusetts [81]. In size and industriousness Gloucester had much in common with tuna ports elsewhere in the world, such as Barbate and Misaki. The proximity of the rich fishing grounds of Georges Bank and the coastal regions of Nova Scotia and Newfoundland meant there was a lively local fishing industry. In the twentieth century Gloucester had grown into an important port for catching cod. It was also one of the ports where purse-seine fishing had been developed in the nineteenth century, a method which later enjoyed such success in Norway and the Mediterranean, as well as the rest of the world via Australia and Japan. Gloucester, however, was also one of the main players in one of the greatest ecological fish disasters in history: the implosion of the cod population in the waters off the American and Canadian west coast. The species became so overfished that the massive stocks collapsed dramatically at the end of the 1980s. For a long time it was even thought that cod had been more or less eradicated for good on these fishing grounds. The implosion of the cod population developed into the ultimate nightmare of both fishers and sustainable fishing activists globally: it proved possible to overfish a species so extensively that the population, once it slumped below a critical biological threshold, was definitively eradicated.

For the fishermen of Gloucester the cod implosion generally meant unemployment. The Japanese interest in tuna unexpectedly opened up new opportunities. Adventurers suddenly discovered that serious money could be earned from tuna fishing. A supply of bluefin tuna was set in motion towards Japan. The Japanese airline JAL soon started up regular flights transporting tuna packed in ice as fast as possible to the Tsukiji market. In Gloucester the tuna had suddenly changed from an unpopular net destroyer into a 'golden fish'. In fact, it was this Japanese interest in the local bluefin tuna that created a firm basis for the popular reality television series *Wicked Tuna* about the commercial tuna fishermen from Gloucester that started much later, in 2012, on the National Geographic Channel [176]. The Japanese economy was growing spectacularly, with astronomical tuna prices. In 1991, however, that came to an abrupt end, after Japan's speculative property bubble burst, plunging the country into recession. Prices for bluefin tuna came under pressure. At the same time the rapid invasion of Japan had caused the sea off the American east coast to be fished clean. There was no question of a complete implosion of the tuna population as there had been with cod, but tuna was becoming scarce and expensive. If Japan wanted to guarantee the supply of affordable tuna to still the insatiable hunger of the sushi market, then it would have to find other fishing grounds where tuna could be obtained more cheaply.

So Godzilla turned his head from the American Eastcoast even more eastward, to the other side of the Atlantic, where people had been fishing bluefin tuna since ancient times.

Gurume Bumu

16

While Japan searched the world seas in order to safeguard the supply of tuna, on the demand side there was also a development with far-reaching global consequences. In short, everyone wants to eat well. In the growth economy that characterised the final decades of the twentieth century in the economically developed world, food culture gained in status. Eating and cooking became a global hype. Not only did food cultures fuse, but traditional divisions also became blurred: on the one hand fast food was winning culinary ground, while on the other the higher eating culture was becoming democratised. Like *haute couture*, the walls around *haute cuisine* were being broken down. An extensive middle class turned out to want to obtain products which had previously been labelled luxury goods for the 'happy few'. It was often more a question of status than of taste, but that didn't matter: everyone now had the right to show their status, in the kitchen and on the table too. Exclusive products such as caviar, foie gras and truffles were no longer reserved for a privileged elite.

The Japanese call this culinary democratisation *gurume bumu*, a Japanese corruption of the phrase 'gourmet boom'. Japanese restaurants, previously an exceptional and elite phenomenon, cropped up all over the world. In parallel with this development, Japan in turn began to adjust its eating patterns to those of the rest of the world. Fast food advanced, as anyone walking through Tokyo would soon see: McDonald's and Kentucky Fried Chicken won their place in the big cities, as did pizza chains. The middle class preferred to eat out quickly and cheaply in restaurants to eating at home around the kitchen table. Those who did eat at home were increasingly served ready meals, often with meat instead of fish. The Japanese are still among the greatest fish eaters in the world, but around the turn of the century only 40% of the animal protein consumed came from fish. According to figures from the Japanese Ministry of Health, Labour and Welfare, between 1995 and 2004 there was an unprecedented drop in fish consumption among all age groups. Until 1976 prices for fresh fish were 40–60% lower than those for fresh meat, but by the early 1990s that difference had practically disappeared. Once its inedible head and spine had been removed, the price of fish per gram worked out considerably more

© Springer Nature Switzerland AG 2019

S. Adolf, *Tuna Wars*, https://doi.org/10.1007/978-3-030-20641-3_16

Fig. 16.1 Gurume Bumu. Tuna sushi and sashimi became industrialized, fatter, faster, cheaper (By courtesy of Jonathan Forage/Unsplash)

expensive. Around the turn of the century, two thirds of Japanese children preferred meat to fish, and many Japanese parents found it easier to prepare [40, 146].

To the extent that fish was still eaten, it changed in character. Rising prosperity in post-war Japan brought obesity. The fatty tuna belly or *toro*, once a waste product used as cat food, became the ultimate sushi fish. Fatter, faster, more convenient. 'Japanese cuisine became industrialised,' food anthropologist Ted Bestor concludes. With new forms of mass distribution and large-scale manufacturing technologies the food supply became standardised. What had previously been bought as raw fish from the fish monger around the corner was now bought pre-cut at the supermarket (Fig. 16.1). Ready-prepared sushi on a plastic serving dish, supplied and produced by large companies, often the same ones in charge of ships, fish farms, storage and distribution, enabling them to control the entire value chain.

The crisis which struck Japan at the start of the 1990s had repercussions for culinary culture. There was less money, but the Japanese did not want to give up the status of sushi. The solution was found in *kaiten-zushi*, cheap restaurants where sushi went by on a conveyor belt. The dishes were presented on plates of different colours, each indicating a specific price. At the end of the meal the bill could be simply calculated according to the colours of the empty plates. These were restaurants for the general public, aimed at supplying affordable sushi of reasonable and reliable quality.

The new, large-scale markets also meant the need for a new, controllable supply of affordable tuna. In the Australian Port Lincoln [146], Japan found the commercial answer to this new challenge of the industrialised sushi market: the tuna farm. After freezer technology and the *gurume bumu* this was the third change to make its mark on bluefin tuna fishing. The idea emerged after the rise in demand from Japan rapidly shrank bluefin tuna stocks in the Pacific Ocean. In 1984 the Australian government was even forced to implement a strict limit on the catch to save the species from extinction. From then on only 14,500 tonnes of tuna could be caught per year. The situation turned out to be serious: in the following years there was so little fish in the sea that the quotas were not even reached. The Japanese fishing cooperatives came up with an idea for compensating for the declining catches. Was it possible to keep tuna, like salmon and trout, in captivity?

The plan met with a great deal of scepticism, but research began nevertheless. Port Lincoln, still the number one tuna port, was chosen as the place to set up the first tuna farm, with Japanese assistance. The largest tuna fishing company, in the hands of an Italian immigrant, decided to collaborate on the project. Technically this was not real rearing of fish from eggs, but involved fattening existing tuna. Wild-caught, juvenile, small tuna were kept alive and captive for months in cage nets, where they were fed on fish. This appeared to work: you put in tuna weighing around 20 kg and after a few months of feeding the fish that emerged had doubled in size and weight. Admittedly the quality was not that of wild-caught Atlantic bluefin tuna, but it was good enough to supply the growing demand for tuna for the sushi and sashimi market. The farmed tuna was thus middle-of-the-road: not the best quality, but reliable and affordable. It was just the tuna for the bulk market of sashimi and sushi that the Japanese consumer needed at the time.

The tuna farm was a successful innovation which rapidly took on in Port Lincoln [81]. The tuna farms were largely in the hands of speculators among the local immigrant population, unattached youths who earned millions through export to the Japanese market thanks to the new fattening method, and became the city's nouveau riche. People talked of a tuna 'gold-rush'. One of the 'tuna cowboys' who threw himself into tuna fattening was Dinko Lukin, an immigrant from Croatia. Lukin came up with something no one had previously thought of. Instead of fishing the tuna from the water with lines and hooks, it was much better to catch them with the purse seine or purse (bag-shaped) net. It was a technique that had been used for other types of tuna. The advantages of such a purse net were legion. With the use of a light aircraft it was possible to detect the large shoals. Then you could capture an entire shoal in one go. Catching the tuna with purse seiners meant they stayed intact. There would be no nasty wounds from the hooks, and the fish could stay in the water to be transferred in one go into a transport cage and transported to the farm. That saved a great deal of stress in this naturally nervous fish. Purse seine nets were not new. The ancient Egyptians mentioned them, as did the Greeks. The *almadraba de tiro* was in fact an early form of purse seine. The Basque people had used circular nets. In the years following World War II in Norway these nets where used against tuna. They were reintroduced for tuna fishing in the Bay of Biscay. The use of purse nets was greatly boosted by use of motorised power blocks which used hydraulic technology to set out and close heavy nets. Now the combination of powerful purse seine boats with tuna farming in Port Lincoln proved extremely efficient. Never before in history had bluefin tuna been fished out of the sea so quickly. The Japanese sushi market could now be much better supplied than previously, with bluefin tuna captured in water cages, fattened in controlled conditions, and fished out—or rather, harvested—systematically. Within a couple of months Port Lincoln had adopted the new fishing innovations from the Croatian immigrant.

The success of the tuna farms in Port Lincoln got the Japanese tuna traders thinking. Why shouldn't this efficient approach work for Atlantic bluefin fishing? Already in the 1980s some tuna ranching activities were started in the Strait of Gibraltar next to Ceuta, right at the opposite site of the rock of Gibraltar [174]. There were other locations along the Mediterranean coasts which would be excellent for setting up the fixed, circular net cages for tuna farms. They could set up these new production techniques for bluefin tuna in joint ventures with their business partners in the Mediterranean. It might require some initial investment but in the end it was a good deal more efficient than the *almadraba*, with all its complicated, labour-intensive fishing techniques [226]. Ricardo Fuentes was one of the first, together with his Japanese customers, to venture into tuna farming. From 1996 Fuentes started up three joint tuna farm ventures [204]. Here forces joined in different alliances. Tuna Graso became a large fishing and processing company—now also a farm owner—on the Mediterranean market. It was born from a joint venture between the Fuentes group and the Japanese trading house Mitsui. With the Maruha Corporation Fuentes founded the company Viveratún; and a few years later Atunes de Levante was founded with the Mitsubishi Corporation. 'Typical Fuentes', say the tuna fishers and traders in Spain. He was a relatively small player—fisherman and farm owner—compared with the powerful Japanese trading houses, but by avoiding supplying one buyer exclusively, he retained the possibility of playing the Japanese off against one another. That was particularly clever because the traders in the tuna value chain had taken up powerful oligarchic positions on a global scale. By keeping 51% of the joint ventures in his own hands Fuentes, however, succeeded in retaining control for himself. This was the new, borderless reality of the world economy: a Spanish fish trader started up tuna farms with Japanese partners following a concept developed by a Croatian in Australia, all for the sake of exploiting bluefin tuna, a fish swimming in the various world seas, for the Japanese market. Everywhere, from the Mediterranean to Australia, bluefin tuna was now tracked down with aircraft and fished out in entire shoals in purse-seine nets. The tuna then remained in the farm nets for up to 6 months, fed on herring, sardines and squid. When the fish had achieved sufficient fat content, it was slaughtered, frozen and transported to Japan.

Ricardo Fuentes' initiative soon gained a following in the Mediterranean. It was in full swing when I first went along with the *almadraba* fishermen in southern Spain in the 1990s. The circular nets of the tuna farms cropped up everywhere, first in Spain and Croatia, then Italy, later eastwards and off the coast of the North African countries as well as around Malta. The Fuentes group took the lead and spread all over the Mediterranean with daughter companies in Croatia, Sicily, Tunisia, Cyprus and Malta. In under 10 years the fish monger from Cartagena had established the greatest empire ever to have existed in the Mediterranean in the new tuna era. Besides tuna farms, Fuentes also owned purse-seine boats, aircraft for tracking down the tuna, tugboats for transport, freezer installations, cold storage warehouses and trucks. The annual production rose to many thousands of tonnes of bluefin tuna. Eighty per cent of that was destined for the Japanese market. It was as if there had never been a crisis in bluefin tuna. Japanese longliners and the new large-scale purse-seiner-ranching activities grew at an explosive scale in the Mediterranean. And with not many management tools or compliance of the few rules, a new phenomenon took off in an alarming pace: massive illegal, unreported and unregulated (IUU) catches to satisfy the fast growing appetite of the Japanese Godzilla [226]. If you didn't know better you would think that bluefin could be caught in large quantities and fattened on the Mediterranean tuna farms for eternity.

Part III

The Final Tuna Trap

I'll make him an offer he can't refuse.—*Michael Corleone* in The Godfather

The Fall

17

'Disappointing this year,' said Diego Crespo in 2008 in response to an inquiry as to how his *almadraba* was doing. We stood on the quay of the port of Barbate and watched the boats which had just returned from Zahara de los Atunes. Those asking the fishermen for figures kept on receiving gloomy answers. Around 1200 tonnes were caught in the four *almadrabas* on the south coast in the 2008 season. In 2007 that was still 1350 tonnes. Diego Crespo is the director of the *almadraba* of Zahara de los Atunes and also president of the association controlling the last working *almadrabas* on the Spanish south coast. 'Of course,' he says, 'you never know with tuna. It swims closer to the coast 1 year than another. There are stronger and weaker year classes. The catch can depend on the clarity of the water, the temperature and the strength of the currents.' But altogether the trend in catch figures was a source of worry a decade into the twenty-first century. Fewer and fewer bluefin tuna were swimming into the nets.

Crespo kept his office where the river Barbate runs into the sea, not far from the spot where in Roman times the port of Baessipo loaded up the tuna for shipping. A gate formed the entrance to a spacious patio surrounded by largely empty, dilapidated halls and buildings. From the 1930s until the early 1970s this was the main building of the massive tuna processing of El Consorcio, with its own hospital, communal areas and a canteen. In a place that had once blazed with industriousness, now the paint was peeling and the windows were broken. From the outset the *almadraba* had been a battle for the Crespos. Almost literally. In the 1970s Barbate had been split into two camps: the Crespos on one side and the Ramírez's clan on the other. Everyone in Barbate, and later even as far as Japan, was aware of the discord between the families. In the San Paulino church opposite the town hall the front rows of pews were reserved for members of the two clans, who avoided one another's gaze during mass. At the Virgen del Carmen festival, when the image of the patron of fishermen was carried from the church for her annual boat procession through the port, the families gave one another as wide a berth as possible. The Crespo and Ramírez families were irreconcilable, like the Montagues and Capulets of Romeo and Juliet. But unlike Romeo and Juliet, of course, everything hinged on tuna.

© Springer Nature Switzerland AG 2019
S. Adolf, *Tuna Wars*, https://doi.org/10.1007/978-3-030-20641-3_17

The Crespos had set off for Morocco in 1943 to set up the *almadraba* near Larache, the former Phoenician tuna town of Lixus. At home in Barbate that was impossible due to El Consorcio's state monopoly. When the Spanish protectorate in northern Morocco ended in 1956, the family returned. At the start of the 1970s a new opportunity arose when the state monopoly was lifted and the Crespo and Ramírez families both entered the race for *almadraba* concessions for Barbate. Along with his four brothers, Diego Crespo—father of the current Diego—succeeded in trumping the Ramírez family with a better offer. Two years later the Ramírez family struck back. Through good relations within the Ministry of Agriculture, Food and Environment, they laid claim to the Barbate concession. Harsh words were spoken. The feud deepened. In the end reason won out and a compromise was agreed. The Crespos left Barbate's *almadraba* concession to the Ramírez family and had to be satisfied with the *almadraba* off the beach of Zahara de los Atunes, next to the ancient palace of the dukes of Medina Sidonia, which had fallen out of use in 1936, but now served as a good alternative.

The Crespo family, children of the founders, still manage Zahara's *almadraba*, but were also shareholders in the *almadrabas* of Tarifa and Conil. Barbate's *almadraba* was sold by the Ramírez family after the turn of the century. The new owners included the family of Ricardo Fuentes from Cartagena. After the success of the tuna farms, his son Paco and his brothers had decided to buy up a number of other *almadrabas*. On the opposite shore, in Morocco, the Fuentes Group ran four *almadrabas*, including that of Larache, where Crespo also once held sway. So fishing for bluefin tuna remained in the hands of a limited number of families who knew one another very well and were constantly running into one another.

Diego Crespo remembered how difficult the 1970s had been for his family. The organisation of the *almadraba*, compared with the period during which it was under state monopoly, was slimmed down and more efficient, but on the other hand the market for canned fish was not exactly flourishing. Bluefin tuna was yielding low prices and the company ran at a loss or at best broke even during the first few years. For the Crespos, who had invested all their savings in the *almadraba*, there was no way back. Then, out of the blue, the first Japanese longliners boats appeared on the coast, in search of bluefin tuna in the Mediterranean. In 1978 a small quantity of bluefin tuna was sold to the Japanese for the first time. During the 1980s exports grew further. The Japanese *reefers* or freezer transport ships became a familiar sight in Barbate's port. Japanese fishers went along on the *almadraba* ships to teach the fishers how to handle their tuna. The brutal slaughter became a thing of the past, as did the relaxed cigarette after drawing in the catch. 'For the Japanese, it was mainly about speed,' Crespo reminisced about the culture change. Any dawdling meant loss of quality.

The Japanese trade breathed new life into the fishing industry. The old *almadraba* boats were replaced with newer models and equipped with tanks where the tuna was immersed in ice water. Its size—250 kg on average, but sometimes as much as 400 kg—as well as the fact that tuna often came in by the dozen at a time, meant that the freezer capacity of the cold stores in Barbate were insufficient for deepfreezing the fish quickly to the core. That changed when the Japanese brought their superior freezing technology to Barbate. Japanese interest represented salvation for the ailing tuna industry. Because the *almadraba* is expensive. It can quickly involve a hundred fishermen. A quarter to a third of the material for nets and cables is replaced annually. It is only thanks to the high price paid in Japan for the exclusive wild tuna that this artisanal fishing method could be kept running.

Thus the oldest industrial fishing technique of the West was saved by the rising demand of a culinary revolution in the East. 'Without Japan,' Diego Crespo said, 'the *almadraba* would have disappeared.' As was already clear, however, there was in fact a whole other side to the Japanese interest. Besides the *almadraba* with its high-quality tuna, the Japanese were even more the engine behind the introduction of purse-seine fishing elsewhere in the Mediterranean, fishing the shoals of tuna en masse and systematically so that they disappeared into the tuna farms to supply the Japanese mass market for sushi. Even more worrying was the explosive growth of a black market, where massive amount of IUU bluefin tuna found its way to Tokyo. That began to have repercussions for the catches. For the first time since the tuna monk Martín Sarmiento had studied sustainable fishing, overfishing of bluefin tuna began to be a serious problem. For Crespo there was no doubt: the introduction of purse-seine fishing threatened to decimate Mediterranean tuna stocks. The introduction of the purse nets increased catch capacities exponentially. In the past a fishing boat would have left its port, tracked down the tuna, caught it and returned to the port to unload. That took days. With the farms it was much simpler: you caught the entire shoal, transferred them to a transportable net and took them to the farm. The boat was ready to continue fishing within half an hour, meaning ten, 20 or 50 times the capacity.

As chair of the association of *almadrabas*, Crespo argued for a return to small-scale fishing to save the tuna, thus also limiting the tuna farms, where the Fuentes family played such a prominent role. Crespo regularly met Paco Fuentes in Barbate over those years. The issue of the Japanese intruders was carefully avoided.

Since the 1980s fishing in southern Spain had been battling with a crisis that mirrored the worldwide problem of overfishing. Barbate, a port town with a population of 23,000 in the province of Cádiz, had once been the base for the Atlantic fishing fleet of Andalusia. Fish was the cork that floated the local economy. But in the 1960s and 1970s, after their own coastal waters had first been fished clean, the fleet had to search at greater and greater distances. They began with the Canary Islands, then moved on to the Moroccan coast in search of squid, sardines and anchovies. Morocco, dissatisfied with the fees it was receiving for its licences, cancelled the fishing agreement with the European Union in 1999. The fleet in Barbate largely came to a standstill. After difficult negotiations a new agreement was reached, but the fishing industry was never what it had once been. Unlike the long-distance fleet of large purse seiners and longliners built by their colleagues in Galicia and the Basque Country, the Andalusians had far fewer modernised industrial vessels. Many fishermen in the south gave up. Their ships were scrapped under European funding, and they retired early. The speculative construction industry which gripped Spain in the 1990s offered better employment opportunities for the Barbate population. When the housing bubble burst in 2008, it turned out to have been fool's gold. Barbate was struck again. Forty per cent of the population was unemployed, with the rate even higher among young people. Those who were able, sought their fortunes elsewhere. The city had fewer and fewer inhabitants.

A decade into the twenty-first century the Andalusian region officially had only 10,000 fishermen left, half of whom earned a living on an artisanal scale. Spaniards remained big fish eaters, but the continued scrapping of the fishing fleet meant they were no longer able to supply their own consumption. The fish was obtained from ever further away. The festival of the Virgen del Carmen, patron of fishermen, turned into a tourist attraction. Many young people in Barbate looed to smuggling for salvation, a long tradition here on the edge of Europe. *Chocolate*, as they call the hash, is transported in fast little rubber boats from Morocco and dumped on the beaches. The 'easy money', *dinero facil*, is an enticing idea: what their fathers had to work a month to earn, the young people earn in an evening by helping smuggle hash. In the port, beside the fish market and ice factory, outside the fishing season hundreds of rusty anchors from the *almadraba* nets bear silent witness to what remains of artisanal tuna fishing. It increasingly gained the status of an open-air museum. For years there was literally a tuna museum opposite the port, where a local canning factory owner had collected some photos and models of tuna fishing in its heyday. You could buy an expensive piece of *mojama* or can of bluefin tuna in olive oil. In a small building beside the marina the regional government set up a modest exhibition, which closed after a couple of years due to lack of resources. Any tuna still caught with the artisanal *almadrabas* disappeared into the freezer ships headed for Japan. The main question was how long it could continue. 'The scientists have sounded the alarm,' said Diego Crespo. 'If we go on like this, our fishing industry will disappear due to overfishing. So we have to find another way.'

They had to find another way. At the start of the twenty-first century bluefin tuna was confronted with an existential crisis. The warnings of Martín Sarmiento dating back 250 years had once again become relevant. Could bluefin tuna be saved from extinction?

Heart for Tuna

Barbate on the Costa de la Luz in the province of Cádiz is not a tourist attraction. Surrounded by pristine nature reserves and military exercise zones, the town has never had the option of building hotels and golf courses as an alternative to the ailing fishing sector. But the surprisingly flashy restaurant El Campero, a mere 100 m from the port, is a concept that draws tourists from far away. They come to find out more about the culinary secrets of bluefin tuna. No part of the fish is thrown away here. Since Roman times it has been known that pretty much all of the tuna can be eaten. There is *ventresca*, the fatty piece of belly meat, or there are *morillo*, *mormo* and *contramormo*, parts of the head that melt on the tongue. There is tuna roe, tail and loin. The tapas menu includes smoked and stewed tuna, tuna sausage, tuna tempura and tuna carpaccio. Sushi and sashimi appear too: El Campero was one of the first restaurants in the entire region to immerse itself seriously in Japanese cuisine. Even tuna heart, decoratively served in small black pieces on a white plate with a little oil and sweet red pepper, is served as a snack here.

'We exclusively use bluefin tuna from the *almadraba*,' the proud chef and owner José Melero Sánchez, Pepe to friends and acquaintances, tells me. In the kitchen he is just working on a new experiment: gazpacho with tuna. 'Our bluefin tuna is the best in the world, the Jewel of the Strait of Gibraltar. This is a magical, exceptional fish,' says Pepe Melero. The story behind El Campero says a great deal about the evolution of bluefin tuna on the Spanish coast in recent decades. Melero learnt the trade in his father's little café in Barbate in the 1970s. Those were different times: they sold the villagers cups of coffee and the traditional, cheap tuna *tapas*: salted tuna, tuna roe, grilled offal and stewed tuna with onions. The *morillo*, now one of the most expensive cuts of fish, was then given away free due to its fatty quality, for clients to feed their cats with. But Pepe's tuna tapas were a big success in Barbate. In 1994 he took a gamble and started up a restaurant with a large kitchen and an army of cooks. The menu was mainly based on the classically caught bluefin tuna, firmly rooted in the local cuisine, but with new, innovative dishes with clear Japanese influences. In El Campero tuna gained culinary momentum. The restaurant grew to be a reference point for the bluefin from the *almadraba*. Pepe Melero remembers how the Japanese cooks from the reefers in Barbate's port used to ask if they could use his kitchen to prepare their tuna. For both parties it took some getting used to initially. The Japanese watched in disgust as the tuna was cooked through in pieces; the Andalusians in turn saw the Japanese eating raw tuna. 'We were shocked,' Melero remembers. But mutual curiosity soon won out. The Japanese taught him their special cutting techniques to achieve tuna at

(continued)

raw sashimi quality. In turn Melero cooked traditional Spanish recipes for them and taught them how to use salted tuna. Fusion dishes arose from this cross-fertilisation: for instance dipping the sashimi in a sauce of sherry, salt and olive oil, instead of soy sauce and wasabi.

From a fish which generally disappeared into cans to an export product for the Japanese market, in recent years re-evaluated as a gem of local cuisine. El Campero developed bluefin tuna into a high-quality product for which customers were willing to pay the price. You won't find fattened, farmed tuna here, on principle but also because of the flavour. 'The colour, the taste, the fat: there is a gulf of difference in quality between *almadraba* tuna and farmed tuna,' says Pepe Melero. It is this quality, to a large extent determined by the fat content and structure of the various cuts of tuna, which determines its application in various dishes. 'For tuna tartare we use the tail, for the *tataki* the white part of the belly, *solomillo* and *lomo negro* are best for carpaccio. His favourite part of the tuna is the little *contramormo*, a cut of a couple of hundred grams from the head of the tuna. In all his new culinary experiments he is keen to emphasise that his restaurant will never betray the local cuisine. '*Atún encebollado*, grilled tuna with onions, is a dish I learnt from my mother. That will never disappear from our menu,' says the chef.

Like the fishers from his village, at the end of the first decade of this century Pepe Melero was worried about the future of the bluefin tuna. The fish was becoming less and less available, with the meagre catch disappearing to Japan and prices continually rising. For its kitchen El Campero consciously opted for bluefin tuna from the *almadraba*, a fishing method which Melero believes can be termed as sustainable because of its small scale and lack of bycatch. But it became increasingly clear that the tuna would not survive this way. Policy was needed to guarantee sustainable fishing, and that policy needed to be as international as the tuna itself.

Misaki, a little port town 50 km south of Tokyo, might be Japan's equivalent of Barbate. After all, when you say Misaki in Japan, you're talking *maguro*, or tuna. The port at the entrance to the bay has traditionally been the base where the Japanese fleet lands to unload. Misaki's fish market is famous among the residents of Tokyo as a tourist daytrip, an outing to see the ospreys in the harbour, to enjoy the view of Mount Fuji, to eat in the local fish restaurants and make a boat trip to the island out in the bay across from the little town. Visitors could head for the market hall on the port to buy tuna.

Down in the austere hall of Misaki's fish market, people lugged frozen tuna back and forth. Here we found bigeye tuna from the coasts of India and China, as well as albacore and yellowfin. But there was also bluefin: one from the south Pacific and a couple of frozen slices from the *almadrabas* of Zahara and Conil. The fish was driven in on a forklift truck and dragged on hooks over the wet polished concrete floor to the pallets to be laid out in long rows for inspection. The wheeling and dealing of Misaki's tuna auction could begin. How long could this continue, auction director Yamamoto Takaichi wondered. As in Barbate in south Spain, the spectre of a future as an open-air tuna museum threatened Misaki too at the start of the twenty-first century. 'If you ask me, I can't be sure if our market will still be going in a month's time,' the auction director said gloomily. In his office above the big auction hall he told me about the Japanese fishing fleet's difficult summer. The ships had held a symbolic 1-day strike to call attention to the problems. Two hundred thousand vessels, from the smallest fishing boats to the big freezer ships, declined to set out their nets. Four hundred thousand employees in the fishing industry went on strike, warning of mass bankruptcy due to high running costs. The government was asked for help in the form of tax measures. The prices which the Japanese were prepared to pay for their fish were too low to cover the costs; the fishers had reached the bounds of their capabilities.

Yamamoto Takaichi could only confirm the situation. The Japanese consumer was no longer willing to pay a price that would keep the market profitable. Their tastes were also changing. 'The younger generation has grown up with hamburgers and meat,' Takaichi sighed. 'They want fatty fish like salmon or fattened farmed bluefin tuna.' Alternative markets had to be tapped to make up for the falling fishing activities in Misaki. Tourism scored high as an opportunity with potential. Special marketing campaigns had brought interest to Misaki as the best place to eat and buy tuna. A dedicated tuna festival would draw visitors in the summer. There were school trips to the auction. In the gallery overlooking the trading floor a class watched, quiet as mice, while their teachers pointed to a wooden model of a tuna sliced open to reveal the tastiest cuts of *maguro*. Visitors could head for the large market hall beside the port for frozen tuna. There are also little plastic tuna charms on sale to hang from your mobile phone, alongside Hello Kitty bibs, allegedly the Japanese copy of Miffy, playing with a big tuna head. *Maguro no kabuto-yak*, grilled bluefin tuna head, is the local speciality.

Early that morning fish dealer Nobu had attended the auction along with his son, in search of tuna. With his fish hook and an experienced hand he picked out a piece of tuna flesh from a sawed off tail. The fish monger briefly rubbed the piece in his hands to bring it up to the right temperature, bit off one end and threw the rest down onto the concrete floor in disdain. This tuna was too lean for his customers. That tuna from Spain, that was another story, Nobu acknowledged. It was the best on the market that morning. A shame it was so expensive. Too expensive for him.

On the way to Nobu's shop we passed the empty high street with its restaurants and their special tuna menus. On the street in front of his fish shop, filleted fish hung

Fig. 17.1 The gods watch over the future of tuna in Misaki (S. Adolf)

in racks to dry in the air. Nobu-san was famous: writers and intellectuals came from Tokyo to the two restaurants he started up near his fish shop. There we ate a sashimi dish of two kinds of *toro*, the fatty fish from the tuna belly, and *akami*, a leaner, redder cut of tuna from the flanks.

Misaki is the place to go for expertly prepared tuna. The question was: how long would that continue? In the temple in honour of Fujiwara no Sukemitse, a hero from the little town who kept the pirates at bay, the traders have hung a large portrait of *Iwaka mutsukari no mikoto*, the god who watches over food. At his feet lie large turnips, melons and other produce from the land. But almost half of the picture is taken up by an enormous bluefin tuna lying at the divine feet. This god watches over Misaki and over tuna. He will have to save them from ruin (Fig. 17.1).

Meanwhile on the other side of the globe the realisation began to dawn that some management measures at both sides of the Atlantic became urgent if any bluefin tuna were to remain [105, 106, 108, 172–174]. The International Commission for the Conservation of Atlantic Tunas (ICCAT), the international regional fisheries management organisation, had monitored Atlantic tuna stocks since 1969. This organisation's scientific committee, a collection of the best tuna scientists from the affiliated countries, had been warning of the worrying state of bluefin tuna from as early as 1981. Initially this mainly related to the western Atlantic tuna population along the American coasts, which came under pressure from Japanese demand. In the Mediterranean things were going in the same direction, now that it had caught the eye of the Japanese market. At the start of 1994 ICCAT agreed to ban longline fishing in the Mediterranean in June and July, the tuna breeding season, but because the affiliated countries barely monitored this and ICCAT did not have its own coast guard, there were few resources to check and enforce the regulations with clear sanctions. So ignoring the ban, longliners fished the Mediterranean Sea for years, sailing under a flag of convenience or without any flag at all.

All the policies to ensure sustainable tuna fishing seemed to become quickly mired in reluctance and fraud. IUU-catch flourished. It emerged that until the end of the 1990s Spain was still bringing a great deal of tuna ashore which failed to fulfil the agreed minimum size. Year in, year out, ICCAT's tuna experts recommended limiting the total catch, to allow the species to recover somewhat. Year in, year out they observed precisely the opposite: more bluefin tuna were caught rather than fewer. The effect was noticeable. From 1.7 million in 1969 the estimated population fell to 900,000 in the early 1980s. After a brief recovery the graph of the tuna population fell sharply again in the 1990s.

In 2008, 100 tuna scientists from ICCAT countries met at the Gran Hotel Velázquez in Madrid. In this hotel with its somewhat threadbare chic décor, with wood panelling and marble pillars in the lobby, they discussed the state of affairs in a meeting room. Among the scientists were veterans such as José Luis Cort from Spain, Peter Miyake from Japan and Alain Fonteneau from France. The report that these elite scientists published read like a requiem for bluefin tuna. Between the lines of the dry scientific prose you could read desperation regarding the largely failing catch quotas that had been introduced since 1998. All those years a substantial proportion of the catch had gone unreported, the scientists concluded. How substantial, no one knew, because the quality of official statistics on the tuna catch was 'very poor'.

At that point 80–85% of the bluefin tuna in the Mediterranean and eastern Atlantic had been caught with purse-seine nets. The fishermen on these boats did not cooperate properly with data collection, so the statistics were extremely unreliable. Even more difficult was the situation with the tuna farms, where there were no checks but the prevailing impression was that substantially more tuna was produced than the official data would suggest. Everyone knew that this was no miraculous biblical multiplication of fish but corrupt tampering on a large scale to launder illegally caught 'black' tuna through the farming administration. Other protective measures, such as closing the fishing grounds and banning the use of aircraft tracking, were also undermined by corruption and evasion or simply not pursued. Checks 'should receive more attention', the scientists complained. It was a diplomatic way of saying that nothing had come of the measures.

The catches in the Mediterranean and eastern Atlantic in 2007 were estimated by ICCAT scientists at 61,000 tonnes. That was more than triple the level they had recommended as guaranteed sustainable for future fishing.

By 2010 the future of bluefin tuna looked gloomy. In a matter of decades the tuna fisheries, trade and industry had transformed into a global chain of unprecedented magnitude and complexity. At the same time an urgent need was growing to come up with some kind management policy to prevent a collapse of the stocks. New forms of governance and partnerships were needed, but this urge seemed not to keep pace with the increasing pressure of a growing demand for tuna and the big economic interests behind it. In the view of American marine ecologist Carl Safina at the end of the decade, bluefin tuna fishing exemplified the tyranny of greed and its devastating consequences for nature, politics and science [212]. ICCAT moved from one impotent resolution to the next, like a ponderous, toothless giant, overtaken right and left by the powerful tuna economy of quick money. If the steadily declining line of population statistics continued, bluefin tuna would definitively disappear from the Mediterranean by 2012. The fish that for millennia had played an important role in successive civilisations, amidst trade, wars and politics, seemed seriously endangered. In the next chapter we will dive in to what seemed to become the end game for bluefin tuna. Could the fish be managed and saved from the tyranny of greed? If so, there was hope for tuna, and maybe for the other wild fish in our oceans. If not mankind would lose an impressive fish and its oldest heritage of industrial fisheries.

Tuna Requiem

18

First the good news. It was the end of July 2013 and orcas were swimming in the Strait of Gibraltar again. And if there were orcas then they were looking for their favourite food: bluefin tuna. That was what made it so special, as it was precisely a year after the predicted Doomsday for the Atlantic bluefin tuna according to statistics drawn up by the ICCAT scientists. Now, in the middle of the summer, tuna was still in evidence. Not just a lost solitary individual desperately trying to find the rest of his shoal. The sea was full of tuna. Even tuna that, having mated, was on its way back from the Mediterranean to the Atlantic. For decades, returning tuna had only rarely been registered during the summer months. Bluefin tuna stocks had been so paltry that hardly any tuna swam out to the Atlantic.

Three years before, by 2010, it had been red alert for bluefin tuna. The ICCAT fishing quotas had been substantially reduced. The NGOs dealing with bluefin tuna did not think that was enough: they believed a total ban on fishing would be best. In 2010, a failed attempt was made to get the Atlantic bluefin tuna onto the United Nations Convention on International Trade in Endangered Species (CITES) list, in order to impose an international trade ban.

But now, suddenly, a different story was going around the whitewashed streets of Zahara de los Atunes, Barbate, Conil de la Frontera and Tarifa, the same story that was circulating on the opposite shore in Morocco, in the fishing villages along the beaches from Tangiers to Ceuta. They were back again: sizeable bluefin tuna measuring up to 2 m in length or sometimes even more. In the ports on both sides of the strait, fishermen jumped into their small dinghies—often with no more than an outboard engine—to meet each other in the middle of the water with one common goal. They threw out their sturdy nylon lines with hooks to catch a few bluefin tuna. The fish were hoisted over the side of their cramped dinghies and sailed home. It did not look very legal. It had a whiff of official-quota-skirting. But checking up on the fishing boats was virtually impossible. A substantial bluefin tuna could bring in many hundreds of euros. So it was worth a try.

The fishermen's biggest problem wasn't the harbour authorities checking their catches, they had told me. They were much more troubled by the orcas which, just

© Springer Nature Switzerland AG 2019
S. Adolf, *Tuna Wars*, https://doi.org/10.1007/978-3-030-20641-3_18

like them, were on the look-out for bluefin tuna. They were lying in wait in large family groups in and amongst the small boats and patiently biding their time until a handline fisherman hooked one. The orcas knew that, once hooked, the tuna would not be able to swim away. Now it was merely a question of gnawing the tuna off the hook, rather like a piece of kebab off its skewer. To the fury of the fishermen. When the whales swiped the tuna right from underneath their noses, it sometimes looked as if they were laughing in their face, with those smirking toothy jaws.

We decided to have a look ourselves and embarked at the port of Tarifa where the Swiss organisation FIRMM arranges special whale watching tours in the Strait of Gibraltar. We set course for the middle of the Strait, where the Spanish-Moroccan fleet of small handline fishing dinghies was drifting around. The orcas' dorsal fins could be seen from afar. Some excitement grabbed hold of the boat filled with day trippers. This was, in a way, like they were in the middle of an episode of *Wicked Tuna* on the National Geographic channel, but live. There were six orcas led by a colossal male measuring around 8 m. It was an impressive sight. I had not expected orcas to be that big. The family plunged towards us, and streaked under the bow, with their clearly recognisable black and white markings. The orcas were on the hunt. Or rather, they were involved in the preparation of a well-planned banquet, with tuna as the main course. Sashimi à la carte. Every now and then a male orca lifted his head above the water. A small Spanish boat from La Linea, the port across the bay from Gibraltar, received its full attention. The line fisherman, a solid man in his forties in blooming health, was so focused on releasing the nylon line through his fat gloves that he was not aware that his trousers kept slipping down, revealing his buttocks in all their glory for all, including the orca, to see.

And then the man hooked something. Together with a colleague he struggled to get the line in. It turned out to be a tough fight: a tuna weighing more than 200 kg did not give in that easily.

This was the sign the orcas had been waiting for. The colossal male launched the attack. The fast approaching dorsal fin rising metres above the water caused panic on the fishing boat. Cursing fishermen picked up speed, reeling in the line pulled taught by the struggling tuna. The orca dived down and struck. A few seconds later the tension in the line released and the fishermen tumbled backwards. On our boat, a FIRMM staff member announced through the intercom in Spanish with a thick German accent that the orca had plucked the tuna from the hook. Polite applause amongst the tourists on board ensued. This had not escaped the attention of the fisherman from La Linea. You could see him ponder for a moment, once back on his feet in his dinghy. Was he an object of amusement not only for the orca, but also for a tour boat full of bloody tourists? Then he turned around, yanked his trousers down as he bent over and smacked his large, pale buttocks. Take that, filthy orca and bunch of ecotourists with your whale watching tour. A brief silence descended on the boat. This is not something you got to see in *Wicked Tuna*.

That evening, the news reported how Spanish customs had stopped a fishing boat from La Linea with a large haul of smuggled cigarettes which was being shipped from Gibraltar to the Spanish beaches. The small boat closely resembled our tuna fisherman's.

Bluefin tuna was doing better. But for the fishermen besieged by orcas they continued their struggle to survive.

During the first decade of the twenty-first century a dark feeling of doom began to grow inside me. In the preceding years, I had developed a passion for tuna. And now I became truly worried. I was not the only one. Many well-informed fish and sea enthusiasts were troubled. Alarming reports were on the increase that the state of fish, coral reefs and global warming were all deteriorating. For me all this doom and gloom seemed to converge in bluefin tuna, the fish I had been looking for years. I knew enough by now to realise that this was not the first time people had formed a sombre outlook about the fate of that extraordinary fish, which had been closely connected to our history for millennia. There was always something up with bluefin tuna. War was never far from the fishing grounds. Power and money were closely bound up in its fate. But from these conflicts of interest, the problem was now shifting unmistakably to the survival of the bluefin tuna itself. No question: the tuna dukes had been worried about the decline in catches as far back as the eighteenth century. And hadn't the bluefin's sudden disappearance from the North Sea and Scandinavian coasts occurred without imminent disaster? But something of a completely different order seemed to be the case now. Tuna had taken over the world and the world now threatened to vanquish tuna.

During the second half of the twentieth century, global consumption of tuna had rocketed. In 2008, ICCAT tuna scientists, as well as NGOs such as the WWF, had warned that there was a serious chance that the Atlantic bluefin tuna would disappear. Trends were set in motion, inducing the sickening feeling that the population was nearing a point of no return, which once passed would inevitably lead to a biological implosion of the species. The estimated biomass of sexually mature fish in the western Atlantic off the American shores was as little as ten per cent of that at the start of the 1970s. The estimated eastern Atlantic bluefin tuna stock, including those fish in the Mediterranean, was ten times as big, but shrank by almost two thirds during the same period. How on earth had it got to this dire point? [141, 142, 166, 173, 212]

We need to begin by looking at the broader picture of fishing during the turbulent first decade of the twenty-first century. This cannot be seen in isolation from economic and social developments during this period. The economic recession post-2007/2008 had struck globally and affected Europe above all. The southern member states of the EU saw their economy contract under drastic austerity policies. Many banks collapsed or had to be bailed out with public finds. Unemployment threatened to take on a structural nature; an entire generation of young people was no longer getting any work. There was political and social unrest, and instability on both sides of the Mediterranean with protest movements and uprisings against the regimes in Libya, Tunisia and Egypt. In that climate, a broader realisation emerged in public opinion around the growing need for greater sustainability of the fish stocks, through better management. Bluefin tuna had played a key image role in this. NGOs such as the Pew Charitable Trusts, WWF, Greenpeace and Oceana saw in the tuna an iconic fish in which many worldwide sustainability problems converged. The British documentary *The End of the Line*, based on the eponymous book by journalist Charles Clover [59], had alerted a broad public to the urgency of overfishing, with the bluefin tuna again taking one of the leading parts. Sea Shepherd, the radical marine direct action group and self-proclaimed fisheries police, released from their nets a small fortune of freshly-caught bluefin tuna into the Mediterranean. There were television series, films and several books about tuna. An international collective of investigative journalists published articles to reveal fraudulent practices around the catch. Just as it was almost too late, tuna had suddenly become cool.

The problem of overfishing had already begun to receive wider public attention back in 1992. During that year, something remarkable happened. Unexpectedly, a huge stock of cod off the North Atlantic coast of Newfoundland had disappeared. The total implosion came as a shock. Since the discovery of the new continent itself, the abundance of cod in these waters ranked as proverbial. The first witnesses described how, at that time, basketfuls of cod could simply be scooped out of the waters. The introduction of new, bigger trawlers at the beginning of the 1950s ramped up cod fishing, after centuries of more or less stable fishing in these rich grounds. Within 15 years it had become possible to catch what in the past had taken over a century of fishing. After some years of record catches, the stocks now suddenly collapsed. By necessity a moratorium had to be enforced after sexually mature cod had more or less completely disappeared for the most important stocks. This brought an end to what had been a rich fishing tradition for 500 years [157]. Other fishing stocks had collapsed suddenly, like herring in the North Sea, anchovies in Peru and later salmon off the California and Oregon coasts. But in the case of cod it attracted more global attention. Mark Kurlanski's book Cod described how there seemed to be a point of no return: 10 years after the introduction of a ban on cod fishing, cod still had not returned to the fishing grounds of Newfoundland. The 'northern cod' thus seemed to follow the ultimate ecological doom scenario. It appeared an entire species could be virtually wiped out through overfishing.

More bad tidings were announced. In a 2006 article in *Science*, a team of 14 ecologists and economists, led by marine ecologist Boris Worms from Dalhousie University in Canada, reported that the Newfoundland cod had not been an isolated phenomenon [256]. Worms had published an article 3 years previously in which he had concluded that the populations of large predatory fish such as sharks and tuna had declined by 90% since the 1950s. Now researchers claimed the oceans were on the cusp of a global collapse of the most important captured wild fish stocks. A collapse is defined as a slump in the original, unfished population by more than 90% within three generations. If there was no change in policy, or rather in the absence of any policy, the most important wild fish species would be below the bottom line halfway through the twenty-first century, according to Worms' calculations. Large predatory fish such as shark, salmon, cod and tuna would be the first to go. Only a new approach of sustainable fish stocks management as integrated ecological systems could prevent a disaster, the scientists believed. 'Unless we fundamentally change the way we manage all the ocean species together as working ecosystems, then this century is the last century of wild seafood,' said study co-author Stephen Palumbi.

While all attention was focused on the disappearance of wild fish, the main conclusion of the article was actually a different one: that the loss in biodiversity threatened the resilience and the capability of the oceans to produce food. That was a bit more subtle conclusion. In a way it recalled the voice of the great German explorer and naturalist Alexander von Humboldt (1796–1859). Von Humboldt—in popularity comparable with the nineteenth century version of Sir David Attenborough—achieved worldwide fame during his life as the man who altered our view of nature for good. During his countless travels and investigations of the continents he saw that nature comprised a complex, connected web of cause and effect. Animals and plants should not be studied as individual species, but as part of a close-knit ecological system. If one tiny thread in the global web unravelled, there would be implications for the entire intricate web. But that was not Von Humboldt's only contribution to thinking about nature and ecology. As Andrea Wulf describes in her biography of Von Humboldt [259] *The invention of nature* the German scientist identified another important link: one between power, politics and economy, and ecological issues. It was to become a recurring theme in Von Humboldt's publications, after his first big journey through America. There he had seen how Spanish colonists had caused disastrous soil erosion because of deforestation and the construction of a river dam in Venezuela. In his essays, he substantiated time and again that economic exploitation and ecological systems were linked in terms of policy and that humanity has the power to rule over his environment with at times catastrophic consequences.

So this notion had been conceived two centuries before, but things seemed more than ever to be coming to a head, and on a global scale. Knowledge of the ecosystem was obviously a primary requirement, but what really mattered in sustainable management and control was the game of power and policy that was supposed to lay the political and economic conditions for sustainability. The *Science* article by the 14 fisheries researchers led to great controversy in the world of fisheries science, in retrospect perhaps even greater than the authors themselves had wanted. In particular the message that wild fish would disappear around 2050 rumbled on for a long time afterwards. The fact that this was an extrapolation based on a number of negative assumptions caught the least attention amongst readers. They were primarily focused on the most alarming message: no more wild fish. 'That was how the press spun it,' said Boris Worms when I called him to account about it years later. In 2009, he published a new article to which no less than 21 scientists had contributed, in which the alarm message was refined [257]. Two thirds of the overfished global population would not be able to recover fully under the current policy, according to the authors, but a combination of measures such as the introduction of quotas, closed fishing grounds and modifications in fishing methods could prevent complete extinction. This would, however, require a global approach in the often economically disadvantaged regions of the world, home to a large share of industrial fishing grounds.

At face value, that seemed a great deal more nuanced and offered some hope: the future of fish had, apart from biological measures, become dependent on economic power games and the way the governance of policy could be shaped. But was this conclusion all that reassuring? Sustainable fishing policies, especially on a global scale, did not have a particularly good track record.

One thing was certain: the issue of overfishing and the future of the oceans could no longer be seen in isolation from economic and social developments. These had accelerated rapidly during the first decade of the twenty-first century because of the two crises that defined the public debate around internationalisation: the financial and economic crisis and the climate debate. In his 2006 film documentary *An Inconvenient Truth*, [122] former US vice president Al Gore had put the topic of global warming on a worldwide map. The apocalyptic message was focused on global warming because of CO_2 emissions threatening to become exponential. If nothing was done, this would have catastrophic and irreversible consequences: melting icecaps, rising sea levels, the shifting of warm and cold gulf streams, an increase in extreme weather conditions, drought and flooding. Coral reefs would die en masse; fish stocks would be hit; flooding, drops in temperature and desertification would set in motion mass migration. It led to a new impulse by the global green movement that has lasted to this day and seems to be gaining in strength rather than diminishing.

The climate discussion also touched upon a number of dramatic consequences for the marine environment, fish and fisheries. But, more than anything else, the discussion drew attention to a deeper problem: that economic power and political organisation can develop a necessary common policy for a borderless, global approach in which common and individual interests are not at loggerheads. In fisheries science, this has been known for years as the 'tragedy of the commons', after the 1968 article by ecologist Garrett Hardin [130]. The term is a metaphor based on a situation in which farmers let their cows graze on a common piece of land. This is not a tragedy in itself, but it can turn into one if the farmers do not stick to the rules and let their cows graze the common land without restriction. While it might be rational self-interest for every individual farmer to put out to field more than the permitted number of cows, the group of farmers, to which he himself belongs as well, will suffer. The common land will be grazed bare. The result: no more grass for any of the cows, the end of the common meadow. And the end of the assumed rationale of economic conduct, which takes its starting point from personal gain, but eventually ends up in a loss for everyone. In this, the tragedy of the commons is closely related to the economic and game theoretic 'prisoner's dilemma'. In decision making, the 'rational' pursuit of self-interest between competing decision makers tends to dominate over cooperation, even if this leads to a less attractive collective result for all the individual decision makers concerned together. Everyone who has shared a kitchen with a communal fridge in a student housing recognises the problem. Unrestricted access to a common resource, such as fish in the sea, or beer in a communal fridge, leads to depletion which ultimately will affect everyone negatively: in the end there is no point in using the fridge if your beer disappears from it. The ultimate consequence: plundering, exhaustion and no use of the communal fridge. The tragedy of the commons is at the basis of the sustainability principle: how can we organise the use of common resources in such a way that we are not all staring at an empty fridge in future? Or at an empty sea? That requires a management policy, governance to guarantee that resources are supplied in a sustainable way. This can be done by introducing right of ownership in common things. Here, this would mean a lock on the beer fridge and you would not get a key until you had paid. Or you could let a central kitchen commission or another form of cooperation between all the stake holders (occupants, visitors) manage the fridge with beer quota to prevent pillaging. Depending on the circumstances, one of these options would have to be selected.

So, it had been thought about, but effectively implementing sustainable fisheries policy was a different matter. Especially when it concerned global interests, with fish moving back and forth across the oceans and production chains stretching across the entire globe. Just as in the climate debate, the issue of economic organisation and political policy to counter overfishing came at a particularly bad time: the world was grappling with the consequences of sweeping economic deregulation that had taken place during the preceding decades. Successful advancement of neoliberalism by the British prime minister Margaret Thatcher and the American president Ronald Reagan, followed by the collapse of the Berlin Wall and communism, had given short shrift to the classic ideas about a government overseeing the management of shared resources. This had resulted in far-reaching liberalisation and deregulation, as in the financial sector, for example. This deregulation and lack of control over financial risks in banks and the financial world in general had laid the foundation for the financial crisis of 2007–2008. In the Great Recession that followed, fortunes evaporated, billons in public money vanished in propping up banks and caused a dramatic downturn in the global economy, with major unemployment, rising poverty and an increase in income inequality. Although it was clear that the root of this crisis lay in a lack of supervision of the common resources (money) and a retreating government, finding new forms of governance in combination with market thinking turned out to be far from easy. During the crisis, the big economic and financial power groups were too busy consolidating their positions. This was not the time to kick up a fuss with discussions of principle about innovative ways to tackle the economy of the future and its 'tragedies of the commons'. It was clear that several global problems were heading straight for us, but as to the solutions, these were much less obvious. Insofar as discussions were held, they had a distinctly ideological and emotional character. Climate sceptics, who, ideologically, were often in the same boat as free market fundamentalists, waged a dogged war, supported by a section of industry that rejected any form of government control and did not want to have anything to do with the common international accord aimed at pushing back warming as a result of CO_2 emissions. From the green side, climate change whistle-blowers kept on at them, warning that conversion to a sustainable, energy-neutral economy was not happening remotely fast enough.

The Battle of Science

The discussions about overfishing did not escape this atmosphere of heated and polarised debate, either. In 10 years the world of fisheries scientists, previously a rather impassive nerdy maths club of biologists, ecologists and economists, had changed into a swirling shark pond in which only the strongest were able to survive. There was a true war between movements, which was being fought tooth and nail. Take the iconic battle between Daniel Pauly and Ray Hilborn.

Marine ecologist Daniel Pauly, professor at the prestigious Fisheries Centre of the University of British Columbia, qualifies as the undisputed standard-bearer of the fight against overfishing in the oceans. In his 2009 essay 'Aquacalypse Now' [188] he passed a devastating judgment on the powerful fisheries lobby which, right across the globe, using billions in subsidies, had succeeded in creating a colossal over-capacity in fishing boats that plundered the oceans. No less than a biological Armageddon was happening here, driven by an economic doomsday machine as efficient as it was ruthless: 'the fishing-industrial complex'. Analogous to the 'military-industrial complex' coined by US president and general Dwight Eisenhower [79], this complex comprised an alliance of distant water fishing fleets, influential lobbyists, associated scientists and representatives of the people. Behind the romantic image of the fishermen, this complex had managed to extract dispro-portionately large amounts of public money which were used to maintain an enormous fishing fleet over-capacity in big industrialised fishing nations. 'In Japan, for example, huge, vertically integrated conglomerates, such as Taiyo or the better-known Mitsubishi, lobby their friends in the Japanese Fisheries Agency and the Ministry of Foreign Affairs to help them gain access to the few remaining plentiful stocks of tuna, like those in the waters surrounding South Pacific countries,' Pauly wrote. In America, with its market in which three big brands in the canning industry call the shots (Chicken of the Sea, Bumble Bee and StarKist) the situation was very similar. At that point the EU, the US, Japan, China and others together landed around 30 billion dollars annually in subsidies in the fisheries sector, approx-imately one third of the global value of what was being fished out of the sea. This maintained a fishing industry which, according to Pauly, functioned like a kind of

© Springer Nature Switzerland AG 2019
S. Adolf, *Tuna Wars*, https://doi.org/10.1007/978-3-030-20641-3_19

Ponzi scheme, comparable to the methods used by the fraudulent stockbroker Bernie Madoff, who had been responsible for a worldwide scandal that covered an estimated $13 billion and resulted in a 150 years prison sentence in 2009. A Ponzi scheme works as follows: you make people believe they have made a brilliant investment by distributing fake profits which you finance with money from new investors. By paying out some of the capital income as dividends, you are concealing the fact that income is not growing sustainably at all, but is entirely absent. As long as everyone believes in it and the money keeps flooding in, all's well.

In the case of fisheries, public subsidies were involved. That maintained an over-capacity in fishing fleets two to four times as big as was advisable for sustainable fishing, Pauly estimated. He also wiped the floor with fisheries biologists, who he believed were being used to gloss over the interests of these industrial fisheries complex whilst failing in their actual duty: researching and controlling fisheries. Their investigations often resulted in calculations regarding fish stocks that were far too positive. It was an absolute fact that fish populations globally were plummeting, according to Pauly. And ultimately this threatened not only fish stocks and fisheries, but the marine ecosystem at large. Meanwhile, the world continued to eat fish as if nothing was happening. 'But eating a tuna roll at a sushi restaurant should be considered no more environmentally benign than driving a Hummer or harpooning a manatee. In the past 50 years, we have reduced the populations of large commercial fish, such as bluefin tuna, cod, and other favourites, by a staggering 90%,' according to Pauly. Only clear policy could avert this plundering. By introducing stricter quotas for threatened fish populations, but also by publicly auctioning individual fishing rights to fishermen, for instance. But ultimately, firm, clear state intervention was the most powerful management instrument for putting an end to overfishing. To begin with, by halting subsidies, but also by allowing countries to rule over their seas and to set up a policy system within the 200-mile zone off their coasts, the Exclusive Economic Zones (EEZs). Introducing large protected marine areas where fishing was banned would enable the stocks to recover.

Pauly's publications and recommendations led to fierce polemic and strife within the world of fisheries scientists. Ray Hilborn was regarded as the most important scientific opponent. This prominent fisheries scientist from the University of Washington warned that the impact of fishing was grossly exaggerated by the NGOs who were making efforts to preserve the marine environment. He felt Pauly had clearly positioned himself in that camp and that his opponent lacked an objective view. Feeling guilty about eating fish was quite unnecessary, Hilborn wrote in an opinion piece in the *New York Times* [134]. He too accepted that particular species such as the Atlantic bluefin tuna were threatened, but there was nothing wrong with the stocks of many other tuna species, he thought. Moreover, not all fisheries are the same: capacity reduction and other measures in mixed fisheries in the American waters had led to particular species being substantially under-fished. Fishing had become a complicated business and it was difficult to combine with maintaining all maritime ecosystems as we used to know them, Hilborn felt. You simply could not have everything. 'We are caught between the desire for oceans as pristine ecosystems and the desire for sustainable seafood. Are we willing to accept some depleted species to increase long-term sustainable food production in return?' Better a few ecologically responsible fish species that do well than a menu with more beef, chicken and pork with their much greater impact on the environment, claimed Hilborn. Sustainability was a question of choosing the least bad solution. We have to eat after all. And you can't make an omelette without breaking eggs.

These kinds of argumentation found less and less favour amongst a growing group of fish and marine environment advocates. Could we afford the risk of certain ecosystems disappearing for good? Wasn't there the threat that we might be left with an empty sea, lacking in species and full of poisonous algae and jellyfish plagues? No one could predict with certainty, but the possibility of seas permanently impoverished on a large scale was no longer a theoretical exercise. How effective was the resistance and resilience of the global marine environment under pressure from the fisheries?

Pauly and Hilborn regularly attacked each other in public about a presumed lack of credibility in their data and calculations. This was not boring. The factional struggle was fought so fiercely that you gained the impression you had inadvertently landed in the science department of a free fight club. Pauly, who was an (unpaid) member of the board of the NGO Oceana, was accused by Hilborn of conflict of interests. Conversely, Hilborn was accused by Pauly of covering up his own research funding from the fishing industry. Greenpeace, which was frequently on the receiving end of Hilborn's vitriol, in turn branded the professor an 'overfishing-denier' working in the service of the 'fishing-industrial complex'. In 2016, the NGO submitted a complaint to his university for withholding information about fishing industry funding for his research amounting to millions of dollars, the suggestion being that his research would therefore contain conclusions that worked out much more favourably towards fisheries [119].

This was the new battlefield in which the fight over tuna and other fish was playing out. An arena that was determined by fierce scientific debates about the true state of fish stocks. With increasing frequency, scientific arguments were used to support or dismiss fisheries measures. All camps however agreed on one thing: fisheries measures had to have a scientific basis. In a globally ramified billion-dollar industry with great power concentrations in the value chains for fish and a fishing capacity that was only increasing, science became a weapon in this commercial mayhem. Here, science reached much further than mere academic application: without adequate information about fishing stocks it was impossible to design something that looked remotely like a fisheries policy. Without information, it would be impossible to ascertain which fish species were about to collapse and which measures should taken. This touched on an essential question: what was actually known about the stock of fish swimming the globe?

That brought up a fundamental question that was constantly being rekindled: how do we count the number of fish in the sea? The ancient Greeks knew the problem; Martín Sarmiento had struggled with it: unlike the creatures that inhabit land or sky, we cannot measure how many fish there are in the sea. Underwater counting is a laborious if not impossible process. On top of that, fish like tuna are migratory, so do not have the decency to stay in one place. But science had moved on over all those centuries. Using catch data over the years, models were developed to estimate the quantities of fish in the sea. According to some fisheries scientists it nonetheless remains guesswork. Catches are what they are: catches. When catches for a particular fish plummet, this might be because there are no more fish. But it is equally possible that fish have abandoned a specific fishing ground. Or that fluctuating water temperatures have put such pressure on the fish that they have decided to alter their migration routes. The latter was a recurring phenomenon for tuna in the Pacific, where 'El Niño' and 'La Niña' led to amendments in migration routes. The band of warm water around the equator that develops in the central and eastern part of the Pacific (El Niño) and the cooling (La Niña) are probably influenced by climate change and as a result of the associated changing pressure zones have a far-reaching impact on the weather and rainfall. But fishing fleets, too, amend or tailor their fishing grounds under the influence of fluctuations in warm and cold ocean water.

What in fact physically happens under water is difficult and extremely expensive to investigate. That's why the catch statistics are the only measure available to us. Serious catch statistics have only been around for a relatively short time. The biennial report of the State of World Fisheries and Aquaculture (SOFIA) is considered the standard for the global database for fish; this has been maintained by the Food and Agriculture Organisation of the United Nations (FAO) since the 1950s. In 2016, the FAO announced that more fish than ever was eaten: on average 20 kg per capita of the global population. Trade in fish had developed in value to one of the most important food commodities of the world, half of which originated in developing countries. This spectacular growth in consumption particularly over the past few decades was due to fish farms, by then responsible for half of all fish consumption. China had climbed to become the biggest fish producer in the world and the largest exporter of fish and fishery products. Number two and three on the list were Norway and Vietnam, with farmed fish playing a major role. The biggest world fish importer was the European Union, followed at some distance by the US and Japan [94].

What strikes us right away in the above table (Fig. 19.1) is that despite the growth in the fishing vessels' catch capacity, since the late 1980s wild fish capture has been hovering around 90 million tonnes. The number of fishing vessels amounts to 4.6 million, 75% of which are in Asiatic countries. The percentage of populations that were fished sustainably had decreased from 90% in 1974 to 68.6% in 2013. Almost a third of all fish stocks, 31.4%, were overfished! This means that more fish was caught than was considered advisable to keep stocks level according to biological standards.

These trends and figures were not immediately encouraging. But anyone who had looked around world fisheries also knew they could only give a limited view of the harsh reality. Firstly, data exists only for the 500 or so fish species that are fished commercially. Many of the fish a diver encounters under water—possibly ever more infrequently—constitute a black hole statistically. What's more, many countries offer inadequate or incomplete information about their fisheries, meaning large quantities of fish caught remain under the statistical radar. When no scientific observer is present during capture or even more during the landing of the catch in

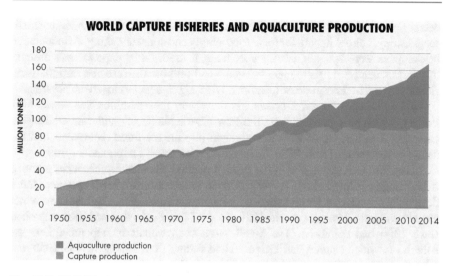

Fig. 19.1   FAO fisheries catch and production statistic: stable catch, growing production (By courtesy of FAO)

portessential information about the composition of the catch (size, age, sexually mature fish or not) tends be missing. And while within de European Union these data have been improved fundamentally, with millions of tunas counted for, many mayor fisheries still lack a systematic data gathering. There are more black holes in fish statistics: great quantities of illegal, unreported and unregulated (IUU) fish, the catches of many small-scale and amateur fishermen and the so-called bycatch fish, or discards, usually thrown overboard and not recorded anywhere. Contrary to what its name suggests, this so-called 'bycatch' quite often comprises most by far of the fish that ends up in the net. So, these are not small deviations from the official catch statistics, but fundamental quantities which are not recorded anywhere if no independent observers are present to monitor them at close quarters.

Daniel Pauly has made it his life's work to unravel the true state of world fisheries. As a neophyte scientist in 1975 he saw, in the Java Sea off Indonesia, how the bottom trawlers with scientists on board emptied their nets on deck. He was disconcerted to realise that the enormous mountain of crawling fish and molluscs would be impossible to chart using existing scientific methods. When counted, the snapper (red filament threadfin bream) which the fishermen were targeting turned out to make up about one per cent of all the fish that was caught. 'I realized then and there that for fisheries science to be able to deal with tropical biodiversity, it would require more than the application of the tricks and approximations that were fashionable in fisheries science at the time,' he wrote. Like a fish detective, Pauly went looking for the fish that had disappeared in the statistics. He developed his own methodology to 'reconstruct' the true status of the stocks. Using diverse sources within fisheries, he tried to frame models and time series of catches, taking into account practical data about fish that is normally left out of official records. Entire categories which had previously been ignored were now included, such as small-scale fishing, coastal fishing for local food supply and amateur fishing. Also included in estimates were bycatch and the specific gear used. Local experts were brought in—from scientists to fishermen/-women—and all possible research into the local situation was checked in order to estimate missing figures as closely as possible.

This approach took on a systematic quality in the ambitious Sea Around Us Project, which was supported by the Pew Charitable Trusts and was focused on gathering data that would give more insight into the maritime ecosystems and the health of the oceans [189, 190]. The project found several remarkable results.

Using the stock status plot of the FAO, a method whereby catches were grouped from rising to falling catches for 400 different fisheries since 1950, Pauly and Zeller developed a stock plot for all fisheries for which there was data. The results were not exactly cheering. During the mid-1990s 20% of fish stocks that had been exploited since 1950 had collapsed. The trends were not particularly hopeful either: the fisheries in which catches had increased had shrunk from almost 90% in 1950 to a paltry ten per cent. The overfished populations had expanded from zero to more than 20% of the fisheries.

The catch reconstructions that Pauly and his team had compiled would eventually result in several publications about various global fish populations [189, 190]. The articles had one consistent conclusion in common: actual fish catches had been many times greater than could be gathered from the FAO figures. The estimated catches for a fishing country such as Portugal, in 2014, were more than a third higher than the officially registered landings. In Panama, the difference was as great as 40%. In the Ocean of Senegal, amongst some of the richest West African fishing waters, the reconstruction suggests that, over a period of 40 years, an amount four times greater than the official FAO statistics was fished out. According to estimates by Pauly's colleague Dyhia Belhabib, over the past few years 300 million dollars' worth of illegally caught fish is alleged to have disappeared from the Senegalese waters alone [34].

Globally the reconstructed catches between 1950 and 2010 looked markedly different from what had been registered officially with the FAO. Instead of the FAO top catch of 86 million tonnes caught during the late 1990s, the suggested estimated catch here was in excess of 130 million tonnes. According to Pauly and his colleagues, the world had fished substantially more than had been assumed. This not only meant that large quantities of fish had disappeared in the illegal, black-market economy, a billion-dollar trade no one had a grip on. It also meant that far more fish was discarded as surplus or landed without being checked. A further logical conclusion was that the stable total catches shown in the FAO figures since the late 1990s had largely been manipulated to present a more rosy picture. In truth, a nose-dive in catches had been set in motion. Fishing was in freefall. If that were the case, then it was highly likely that this also applied to the fish populations themselves.

The expansion of research criteria for catches was a substantial step forward in the knowledge of fishing and fish stocks. Other researchers also made use of the data gathered by Pauly. People like Boris Worms, who cooperated with both Pauly and Hilborn in research. He concluded partly on the basis of the Sea Around Us project that the global catch of shark, for instance, was three to four times higher than the official FAO statistics [255].

The figures caused disquiet, but, of course, not to everyone. Pauly and his team's reconstruction methodology was soon binned by Ray Hilborn. It was, the researcher thought, subjective nonsense. 'They end up just using someone's opinion. I think they are just pissing in the wind,' he stated to the press. Hilborn stroke back with a publication which concluded that human fishing for forage fish does not have as great an impact on the food chain as thought, given that humans typically catch fish of much larger size than those typically hunted and eaten by non-human species. The study also decouples the link between the size of forage fish populations and the populations of species that predate on forage fish. "What we found is that there is essentially no relationship between how many forage fish there are in the ocean and how well predators do in terms of whether the populations increase or decrease" [136, 250].

What was the average person to make of that? Was the fish on the threshold of a Big Implosion if nothing changed? Or was it pissing in the wind? What in fact did we see happening to fish and to tuna in particular? And above all: what were we not seeing? Could a smart approach avert the decline in tuna catches? Bluefin tuna had its own story to tell, which I had been following closely during the previous years. It was a story worth telling, particularly for those who like crime and suspense novels.

Tuna Wild West

The first stop was in the Mediterranean, in Croatia. The World Wide Fund for Nature (WWF) Mediterranean fisheries programme's tuna specialists have no happy memories of their trip to Dubrovnik in November 2006. The seaside town in Croatia, sometimes called the 'Pearl of the Adriatic' because of its gorgeous location, was the setting of the annual meeting of the International Commission for the Conservation of Atlantic Tunas (ICCAT) that autumn. During the 1990s, over a period of just 10 years, Croatia had been one of the first countries in the Mediterranean to introduce tuna farms and so it made sense to hold the tuna meeting there.

The WWF was prominently present at the annual debate about quota and harvest rules for bluefin tuna, in this case for the 2007 fishing season. The Catalan marine biologist Dr. Sergi Tudela, who led the Mediterranean WWF programme, was a well-known figure in the world of big fish companies, fishermen, civil servants, politicians and NGOs that convened like a travelling circus to discuss the future of bluefin tuna. Tudela was considered an enthusiastic champion of bluefin tuna, and an advocate of solid measures to limit the catch so that the extraordinary fish was spared potential extinction. It had made him enemies as well as friends. Earlier in the year WWF had published a research report about bluefin tuna stock levels in the Mediterranean. It had been compiled by Spanish tuna expert Roberto Mielgo Bregazzi, who had been involved for years in the setting up of commercial tuna farms [165]. The conclusions were sombre. The catches of the four *almadrabas* in Spain had nose-dived by 80% in 3 years. On the black market, at least 40% more Atlantic bluefin tuna was traded than the official rules allowed. Now that the western Mediterranean had been practically fished bare, the purse seine fleet had shifted increasingly towards Libya and the Levantine Sea, home of the last big zones where bluefin tuna could breed freely.

Bluefin tuna had degenerated into a trade involving hundreds of millions of euros, while checks on ship owners, farm owners and corrupt Mediterranean regimes were more or less entirely absent. While large-scale fraud was being committed with the catch figures, the trade in 'black-market tuna' flourished. The ICCAT protection measures had been a kind of joke that no one took the slightest notice of for 10 years.

S. Adolf, *Tuna Wars*, https://doi.org/10.1007/978-3-030-20641-3_20

The report's harsh conclusion: if no drastic action was taken, the entire stocks of bluefin tuna would soon disappear.

In its publications, the WWF was usually rather cautious when presenting the state of affairs. Now, the scramble and deceit of tuna fishing was examined [166, 167]. Some countries barely took the trouble to conceal the fact that the catch records they reported to the ICCAT were far too low. In 2004, France, for instance, gave catch figures to the Statistical Office of the European Union (Eurostat) which were 25% higher than the figures reported to ICCAT. That figure in turn was 50% above the permitted quota.

Libya, the country then ruled by colonel Muammar Gaddafi's family dictatorship, home to some 50 modern purse seine boats, was actively involved in IUU catches. These were estimated to be almost three times as big as catches officially permitted by ICCAT. In 2004, together with Malta and Cyprus, Libya was responsible for just under 8000 tonnes of illegally caught bluefin tuna. Turkey fished considerable amounts of tuna in 2005 without the country having been allocated any quota in the first place. This did not stop the country from building a fleet of 240 purse seine ships.

The rest of the international list went as follows. The catches by the Italian purse seine fleet were well above the allocated quota. The Japanese had longline ships navigating off the Algerian coast which were estimated to haul three times as much tuna as was officially reported. Export data for tuna from Algeria was entirely absent. Trade figures between Spain and Tunisia showed that 530 tonnes of tuna seemed to be missing from the catch statistics. Korea did not report anything about catches made by its longliners still active in the Mediterranean. Host country Croatia, too, turned out to have fiddled on a large scale with official catch figures, and left unreported thousands of tonnes of tuna from its fattening farms, almost twice as much as had been reported to ICCAT. In short, the international policy to protect Atlantic bluefin tuna was a mess.

In the ICCAT meeting room in Dubrovnik the WWF report appeared not to have gone unnoticed. Someone had left flowers for the WWF delegation on the NGO's table. It was a flower arrangement of white lilies and chrysanthemums. A funeral wreath. The classic way the mafia warn you one last time. You would not survive it the next time. 'He sleeps with the fishes,' as they say so appropriately in *The Godfather*.

'Our report clearly touched a raw nerve in some participants in tuna fishing,' a WWF spokesperson replied wryly when I asked after the wreath. It had not been the only warning. Other contributors to the WWF report were sent an envelope with a bullet. Some journalists attending the Dubrovnik meeting received telephone threats. This set the tone: at the beginning of the twenty-first century bluefin tuna had grown into big business in which illegality, political fraud and organised crime played a not insignificant role. Troublesome big-mouthed snoopers were warned.

Management of the stocks was urgently needed. But how do you keep an eye on tuna that swims freely in the vast, borderless oceans? This took the form of five international organisations which came to be known as the Regional Fisheries Management Organisations (RFMOs), who were tasked with supervising the catch of different tuna species. They were supposed to become the answer to the problem that keeps Pauly's powerful fishing-industrial complex in check: five consultative bodies made up by the countries bordering the oceanic zones and countries whose long-distance fleets fished in these waters (Fig. 20.1). They were public organisations with a string of impressive names and abbreviations, hardly known with the broad public, operating in the shadow as a kind of intended international rescue for tuna. One we have already met: the International Commission for the Conservation of Atlantic Tunas (ICCAT). ICCAT was established in 1969 by 17 countries, against the background of a flourishing international canning industry. The aim was to manage and conserve tuna species and tuna-like species such as swordfish, in the Atlantic, the Mediterranean and the Gulf of Mexico. The thinking was to establish a system of total allowable fishing quotas (total allowable catch, TAC) that were divided over individual quotas for the various member countries. Sister organisations with a similar remit are the Inter-American Tropical Tuna Commission (IATTC, the first RFMO, established in 1949) for the eastern side of the Pacific, the Commission for the Conservation of the Southern Bluefin Tuna (CCSBT, 1994) which oversees the southern bluefin in the south of the Pacific and the Indian Ocean Tuna Commission (IOTC, 1996) that has the Indian Ocean under its protection. The youngest RFMO, but eventually most important in terms of catch was the Western and Central Pacific Fisheries Commission (WCPFC, 2004) that was established to conserve and manage tuna and other highly migratory fish stocks across the western and central areas of the Pacific Ocean. It commenced operations in late 2005 [195].

Although the international tuna organisations tend to operate outside the attention of the wider public, occasionally one of its committees hits the limelight. In 1995, the IATTC achieved success with the signing of the Panama Declaration: the 12 countries involved committed themselves to a legally binding agreement to protect dolphins while Japan opted out. This came after tuna purse seiner nets had been responsible for a massacre of dolphins in the eastern part of the Pacific. For the first time ever, a colossal bycatch of tuna fishing had been successfully clamped down on and fishermen were involved in a more sustainable form of nature management at sea.

ICCAT was the first RFMO I had to deal with in my quest for management policies for the salvation of the bluefin tuna. ICCAT had expanded into 46 member

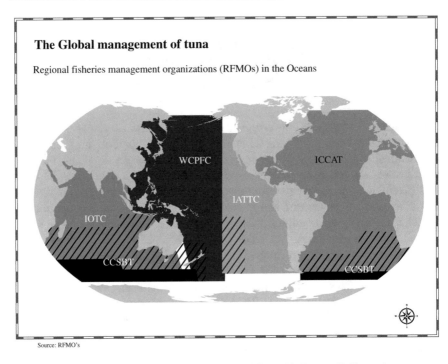

Fig. 20.1   The world according to the tuna RFMOs (Infographic Ramses Reijerman)

countries, including the US, Canada, Japan, Korea, Russia, the European Union and a range of South American and Central American countries. This impressive assembly has been gathering for decades to save tuna from extinction. ICCAT was an exceptional organisation. While the affiliated scientists warned in increasingly shrill tones about the impending disappearance of the bluefin tuna, ICCAT struggled from agreement to agreement intended to make bluefin tuna fishing sustainable. The organisation seemed more like a toy for the enormous financial interests of the fisheries sector than a powerful supervisor making a stand for sustainable tuna stocks. This took on such proportions that ICCAT widely went by the nickname of 'International Conspiracy to Catch All Tuna'. The joke gradually began to acquire an ever grimmer undertone.

We had agreed to meet at an appropriate location for spies: an anonymous café terrace at Atocha train station in Madrid. No suspicious eavesdroppers around, much distracting background noise that renders any attempt at recording impossible and plenty of escape routes to take to our heels should the need arise. Roberto Mielgo preferred to call himself a 'technical consultant'. But it's true, you could call him a tuna spy, he admitted and laughed.

Mielgo, then an energetic middle-aged man with a portly figure and an alert look behind his glasses—had just arrived from Barcelona by high-speed train and would soon have to go on to his next appointment. 'These are busy times,' he said and lit up a cigarette. Following the controversial report he had written in 2005 for the WWF, Mielgo was a much-consulted man. There was a great deal of interest in where bluefin tuna was coming from, who was catching it, which farms were fattening them and where the fish would end up eventually. And with his wide network of colleagues, plane spotters and informants, Mielgo knew the tuna routes. In the 1990s, an experienced diver, he had become involved in setting up tuna farms near Cartagena, where Ricardo Fuentes pioneered his first tuna farms. Mielgo helped to design cage nets and methods for transporting captured tuna, and to shoot and slaughter the animals. Working as an independent advisor in Spain and Croatia, he then helped to set up farms in Italy, Greece, Cyprus and Malta, and later also along the North African coast in Algeria, Tunisia, Libya and Turkey. Mielgo was not a conservationist, but what he saw began to fill him with increased loathing. In less than 10 years bluefin tuna fishing had degenerated rapidly. The proliferation in tuna farms, combined with pressure from the Japanese market to keep prices low was increasingly looking like common theft, with a powerful whiff of corruption and black-market trade growing ever more intense.

He therefore decided to deploy his knowledge to saving the fish. 'And with it, to save fishing. Because the exploitation of bluefin tuna has set in motion a process which is threatening the fishing tradition with ruin. That should be averted,' said Mielgo. So he now acted as advisor for NGOs, but governments and tuna farmers made use of his services as well. Everyone who wanted to know more about tuna's shadowy routes would end up with Mielgo sooner or later. Gathering tuna information was not without risk, Mielgo explained. World trade in tuna involved big interests, and no one enjoyed their tuna being tracked from the moment it was caught until it ended up on a plate. Transparency was unwelcome especially where it concerned illegal tuna. It was a treacherous world, including double agents who deployed fake information in order to intimidate. Tunisian dictator Ben Ali's tuna mafia threatened to kill Mielgo's workers. Mielgo himself also regularly received threatening phone calls. Sicily and Malta, where the black-market trade in tuna was flourishing, were places where you had to watch your step. Libya, where dictator Muammar Gaddafi was personally involved in the bluefin fisheries, also had a bad name. Mielgo was prepared. If something nasty was to happen to him, his lawyer had an envelope with photos and other incriminating material. Possible perpetrators were warned, in other words.

A great deal of money, little control. According to Mielgo this, in short, was the reason why the industrial catch of bluefin tuna had been derailed to such an extent. Tuna is ideal for the black market and criminality, whether sanctioned by power or not, Mielgo said and lit another cigarette. Loans for setting up farms and pre-financing the catch are ideal for laundering money. No wonder the Sicilian mafia was up to its ears in tuna.

In one of his WWF reports Mielgo [165–167] related how, in 2005, Libya and Tunisia substantially enlarged their exclusive offshore fishery zone. This included the zones home to the last large bluefin tuna breeding areas. Translated freely, the tuna deals were directly controlled by the Gaddafi clan and the powerful Tunisian Trabelsi business family, connected to 'president for life' Ben Ali. In 2004, Libya had stopped reporting its catches to ICCAT. According to Mielgo's analysis, Tunisia's reports were evidently far too low. He estimated that in 2005 both countries caught around 9000 tonnes of tuna, two and half times as much as their allocated quota.

The European Union, too, indirectly assisted illegal tuna fishing. Using European subsidies running into the millions, French, Italian and Spanish purse seine ships were completely overhauled from the end of the 1990s onwards. More efficient types of ships with better engines, equipped with state-of-the-art radar and sonar technology combined with chlorophyll and zooplankton meters, led to a fishing capacity completely out of proportion to the officially permitted quotas for bluefin tuna.

The outcome was that the European Union once again began to pay subsidies, this time to take ships out of service. This did not stop many French purse seine boats, having netted capacity adjustment subsidies, from simply carrying on fishing under a Libyan flag, often with the same crew and captain and for the same tuna farms.

Technology also served the tuna wars. Frozen bluefin tuna can be kept almost indefinitely. The people who control the supply of this commodity, have the key to a market worth many millions of euros. If you have plenty of cold stores at your disposal, you can build up supplies and drive up prices in times of scarcity. If you are big and powerful and can store on a massive scale, creating a shortage even has a monetary advantage.

When freezing technology turned tuna from a fish into a commodity, big Japanese players decided to reinforce their grip on all the links in the production and distribution value chain, from the moment tuna was caught until the end product was handed over to the consumer. Roberto Mielgo saw this process happen from close up. The Japanese trade giants in the tuna market—Mitsubishi, Mitsui, Sojtiz and Maruchu—surfaced more and more in the fisheries and farms, the production side of the value chain, in the Mediterranean. They sent their purchasers to buy tuna directly from the *almadrabas* in southern Spain. 'In the end, it's the Japanese companies who pull the strings in the tuna market,' Mielgo said. 'As powerful buyers they determine the market and the prices that are paid for tuna. There are no other big customers. Market power is concentrated.' Thus a traders' cartel was formed in which prices, market sharing and supply were deftly agreed.

It worked as follows: at the beginning of the season, when the first tuna swam into the Strait of Gibraltar, the headquarters in Tokyo told the big tuna farms what they might expect to get that year per kilo of frozen tuna. The local tuna farmers worked out what they in turn could pay to the fishermen per kilo. Based on this price, many fishermen asked the big tuna farms for a pre-financing payment. The tuna farms in turn needed guarantees from the big Japanese tuna buyers. Both the fishermen and the farmers were at the mercy of their pre-financers in Tokyo. That's where the price was set and the financing conditions fixed.

'The Japanese have control over the fish,' concluded Mielgo. 'They buy a luxury product for knock-down prices and hold sway over the tuna from our sea. We have sold out hook, line and sinker like a bunch of lackeys.'

It was not the first time that a limited group of companies in key positions in the value chain had bent things to their will. The world of tuna has often reeked of conspiracy and other collusion practices that did not mesh with the rules of free international trade. Another story you tended to hear was that the big Japanese trade houses built up strategic bluefin tuna reserves in their cold stores. While stocks in the sea were heading for biological collapse, in Asia, a Fort Knox of bluefin tuna was being stockpiled. And just as the United States has its strategic gold reserve to guarantee the value of the dollar, Japan has its reserve of frozen tuna in order to be able to drive up the price of bluefin. It's a classic way to corner the market: the party that has purchased enough of a particular commodity is able to manipulate the price. Roberto Mielgo estimated the Japanese national strategic tuna reserves to be some 20,000 tonnes of bluefin tuna. He suspected the cold stores beyond the Japanese border—in China, Vietnam, Thailand and the Philippines—held similar quantities, meaning that the true strategic reserves would be double that. Mielgo repeated his accusations in the film documentary *The End of the Line*. The accusatory finger was pointed above all at the Japanese trade conglomerate Mitsubishi, which at that point managed 35–40% of trade in Atlantic bluefin tuna.

It was a terrifying image: an enormous stockpile of tuna, at a temperature of −70 °C, waiting patiently for the right price to be thawed and turned into sushi or sashimi. The theory was consistently and forcefully refuted by the Japanese tuna houses. It was pointed out that in time the costs of freezing might rise sharply. But those who believed in a tuna Fort Knox went a step further: a significant part of the strategic reserve was lodged in the grey market, they believed: the big cold stores being erected outside Japan were allegedly used for channelling black-market tuna from the Mediterranean. It was suggested that by fiddling with the catch certificates, the illegally caught tuna, after an intermediary stop, was channelled back into the value chain as official tuna and crossed the border into Japan. Rather like in tuna farms in Malta, a kind of 'tuna laundering' was taking place, but in frozen form.

Whether there was a true secret stockpile of bluefin tuna or not, it was considered a real possibility by many players in the market. Mitsubishi did not deny keeping frozen bluefin tuna on a large scale, either: this was normal practice in a market characterised by an extremely seasonal harvest. The company did however deny manipulating the market, which would be difficult to prove in any case.

Even without its frozen mega stockpile the global tuna value chain was an example of an industry in which everything that could go wrong did go wrong. Its technology, trade power and fraud paved the way for depletion of the existing populations. A case of systematic overfishing existed despite, or rather because of, the relatively low prices commanded by the Japanese sashimi market. Despite existing overcapacity, the fleet of Mediterranean purse seine ships was enlarged systematically. Out of nowhere, Turkey managed by 2008 to create a fleet comprising 200 purse seine ships; enough to account for 15,000 tonnes of bluefin tuna, which ICCAT scientists believed should constitute the maximum total sustainable catch up to that point for all countries.

The scaling-up of the fleet paved the way for a new dynamic which only increased the pressure to catch even more. New, modern purse seine ships needed greater catches per ship to cover their costs. The breakeven point for the entire French fleet—that is the catch volume at which costs are covered—rose within a few years from 72 to 120 tonnes of bluefin tuna, including European Union subsidies.

In this way, the market had drifted far away from the original *almadraba* artisanal fishing. 'No one fishes out of love for the trade any longer,' Roberto Mielgo lamented. 'It's become a ruthless run on hauling out the last bluefin tuna. The fishing variant of an arms race. Everyone tries to get one up on the other, hoping not to fish behind the net. But in the end the fish cuts off his nose to spite his face.' As far as he was concerned there was only one way left to keep Atlantic bluefin tuna from extinction: a total ban on fishing in the Mediterranean and the Atlantic. Actions enforced by military means. 'Deploy the NATO fleet to track down the offenders. That will teach them.'

Tokyo's Tuna Fort

21

It was time to have a closer look at the 'fishing industrial complex' at work in the bluefin tuna trade. Tuna was still snugly tied up to money and power, as it had been for centuries. It was only the main players that had become bigger and more global. The trail of bluefin tuna now stretched from the Mediterranean to Japan. That's where power over tuna resided, and so that's where there might be an answer to the question of how the catch could be made more sustainable.

With large cranes on its roof, the skyscraper of the Mitsubishi Corporation's headquarters looks like a big, grey, unapproachable giant. A giant fort casting its shadow: on one side the moat around the Imperial Palace's park, and on the other Marunouchi Naka-dori, Tokyo's Fifth Avenue. A symbolically chosen location between closed imperial power and the swankiest shopping street in Tokyo where the wealthy business elite shop in Tiffany, Bottega Veneta and Armani after work. The headquarters' colossal lobby oozes organised power. Cars driven by chauffeurs in white gloves come and go; smiling uniformed hostesses rush towards visitors in the enormous glazed atrium. Mitsubishi Corporation is Japan's largest business conglomerate. An industrial and financial global behemoth in which Japan's biggest bank and trading house are united. Mitsubishi Corporation produces and trades a wide range of products and services: not only cars, ships and engines, but also electrical systems, transport equipment and lifts. And more: oil, coal, petrochemical products, liquefied gas, chemical products, synthetic fibres and metals. Even food and clothing can be obtained from the Mitsubishi empire. And bluefin tuna, of course.

With its enormous market share, Mitsubishi Corporation ranks as the bluefin tuna powerhouse. It's estimated that half of all annual Japanese tuna consumption comes through this conglomerate. Mitsubishi has its own tuna fishing fleet, holds shares in tuna farms and buys 'wild' tuna from longline and purse seine fleets as well as from traditional *almadraba* fishermen. The conglomerate supplies its tuna to large retailers in Japan. And Mitsubishi Corporation has at its disposal an unknown stockpile of bluefin tuna in its cold stores.

© Springer Nature Switzerland AG 2019
S. Adolf, *Tuna Wars*, https://doi.org/10.1007/978-3-030-20641-3_21

The Japanese conglomerate is a recurring target for environmental movements that devote themselves to saving bluefin tuna from extinction. In times in which 'sustainable enterprise' forms an inevitable part of the image of companies that take themselves seriously, this is an uncomfortable issue. And Mitsubishi Corporation takes itself seriously, as my Tokyo visit proved.

Once the hostesses in their skirt suits and gloves had announced my arrival, I was escorted a few floors higher to a meeting room. Immediately striking in the corridors were the posters with photographs of coral and texts about the importance of sustainability. The official company philosophy was neatly encapsulated in three principles. *Shoji Komei* stands for conducting business with fairness and transparency, I was told. *Ritsugyo Boeki* is about striving for trade expansion to a global scale. And *Shoki Hoko* refers to the company's responsibility towards society. Trade expansion, but with fairness and responsibility. Not an easy job, finding the right balance between these principles. Especially when dealing with a complicated story like the bluefin tuna. The Mitsubishi Corporation principles: 'Strive to enrich society, both materially and spiritually, while contributing towards the preservation of the global environment.' Number two on the list of basic principles, immediately below the guideline around respecting human rights, we find the rights of tuna: 'Ensure that our business is conducted in an environmentally sustainable manner.'

In the meeting room, a small platoon appeared to have gathered to explain the ins and outs of Mitsubishi's involvement in trade. Three people from the tuna division, a spokeswoman for Mitsubishi Japan, a spokeswoman for the London office, two members from an external British public relations office and an interpreter awaited me, seated around a table. That's not the kind of welcoming committee an interested writer expects to see every day. Mitsubishi did not only take itself seriously, but me too, it seemed.

'We are just a simple link in the tuna trade supply chain' Akihiko Soga, manager of the tuna team in the sea products unit of the Fresh Food Products Division said. In disciplined staccato Japanese, he explained the role of his company. Mitsubishi Corporation was but a modest link in a long value chain, he wanted me to know. In 1971, the subsidiary company Toyo Reizo began the trade in tuna, which was largely caught by longline ships at the time. The tuna farms, which were launched during the 1990s off the coast of Australia and in the Mediterranean, were a big breakthrough for trade.

During the initial years of the twenty-first century the amount of tuna supplied by the farms began to stall. The production trend was downward now, Soga said. Mitsubishi Corporation did not publicise any details about the amount of tuna that was being traded annually. But the company controlled around 40% of bluefin tuna imports into Japan. According to Soga, at that time the Mitsubishi Corporation participated in just two farm companies: one since 1999 in Spain and another since 2002 in Croatia. Supermarkets are the most important customers. This is the 'less expensive part of the market'. In addition to this, Mitsubishi Corporation was active at Tsukiji, where the more expensive, more exclusive fish of the highest quality is sold.

Guaranteeing supply to clients required a sizeable stock of bluefin tuna, Soga admitted. He was not allowed to say anything about the size of this stockpile, but all those rumours about a large, speculative amount of bluefin tuna, they were made up, Soga wanted me to know. All around the table people nodded in approval. 'As you know tuna has a yearly trade cycle. The catch of tuna takes place from May until July, the tuna is fattened over 3–6 months. Between October and the end of year the fish is slaughtered, frozen and shipped to Japan. Tuna arrives in February and March, and will be sold over the rest of the year,' Soga said. 'It is a very simple stock situation.'

Mitsubishi Corporation was no more than a nexus in a complex world and was not able to change much in the thorny situation surrounding the Atlantic bluefin tuna population. That was the work of politicians and international organisations, not Mitsubishi's. 'To maintain the stock of bluefin tuna we do believe that it is necessary that organisations like ICCAT take measures based on the scientific recommendations for the total allowable catch and the period to catch the fish,' the tuna trader argued in a practised manner. 'And we think that all participants of the agreement have to respect those rules. If ICCAT decides to reduce the quota drastically or even introduce a moratorium on tuna fishing during the spawning season, then Mitsubishi won't raise any objections.' On the contrary, said Soga. 'We would support these measures. As long as the measures are transparent, effective and enforceable, of course.' The assembled company in the meeting room gave their tuna spokesman satisfied looks.

The Mitsubishi Corporation tuna principles were announced in a press release at the end of 2008 [170]:

'If it becomes apparent that the sourcing of bluefin tuna cannot be carried out in a way that enables the tuna industry to operate on a sustainable basis—protecting both the fish stocks and the fishermen and their families whose livelihood depends on it—we will reassess our involvement in this business.'

If science, if the competitors, if fisheries, if ICCAT. . . . It was a formulation that kept many options open.

What made it all such tough going, I soon discovered in Tokyo, was that Mitsubishi was by no means alone in its attitude of considered passivity regarding bluefin tuna. A few kilometres down the road in the capital, the business and diplomatic district Akasaka was home to Mitsubishi's competitor in the bluefin tuna market, Sojitz Corporation. Sojitz's headquarters were manifestly of more modest proportions than Mitsubishi's. But this, too, was another purveyor of oil, gas and coal, iron and steel, electronics, cosmetics, plastics and chemical fertilisers, engines, aeronautic engineering, food and textiles. It goes without saying that Sojitz also trades in tuna, tuna products, as well as shrimp, herring and mackerel. Again, an HQ with the official company slogan plastered everywhere: 'New way, new value'. Here, too, much effort is devoted to sending out a positive company image as I read the official company statements.

'Now that it has become such a pressing (. . .) issue, Sojitz considers environmental protection to be one of its most important management challenges. Striving to bequeath an Earth that can provide abundantly for the next generation, Sojitz Group is doing its utmost (. . .) to realise sustainable growth whereby economic development and environmental preservation coexist.'

Of course, Sojitz has stockpiles of bluefin tuna, acknowledged manager Masami Ueda of Unit 1, Marketing of Sea Products, Food and Retail Department in the Retail and Lifestyle Business Division. But this had nothing to do with speculation and money laundering. As a wholesaler, you had to keep a decent reserve of tuna, that's all. There was no need for that with bigeye and yellowfin tuna: longline fishermen in Chinese ports supplied plenty every day. Two months' worth of bluefin reserve was enough for Sojitz. He estimated the maximum bluefin tuna reserves to be around 30,000 tonnes. And it goes without saying that Sojitz does not deal in black-market tuna.

Sojitz recognised that the situation in which bluefin tuna had ended up was worrying, but the notion of a moratorium during the breeding season, as advocated by the NGOs and increasing numbers of scientists, was clearly going too far for Ueda. 'Illegally disrespecting the quota declared by ICCAT should be the target of action,' he said. 'Maybe we can find reasonable quotas, combined with strict control. There is no sense for the farms, the fishermen and the trade if illegal fishing just continues. That is the main problem we have,' Ueda said.

No, bluefin tuna would not have an easy future, the tuna trader had to concede. The profits in the trade had been small over the past few years. It was a difficult period of low prices, too much supply and fierce competition. The market for sushi was growing in popularity globally. China had become an economic giant. 'We have the know-how in the market for frozen tuna. That creates possibilities in the future,' Ueda believed optimistically.

But suppose that the pessimistic scenarios are proved right and bad luck would have it that the bluefin tuna does not survive in a way that is still attractive to fisheries? Then there won't be much left to trade, I suggested. Tuna manager Ueda needed a moment to think about this. 'If the stock of wild tuna is decreasing, then the farmed tuna has to take over the market,' he said finally. 'With salmon that has also been very successful, so why shouldn't it work with tuna? In five, six years the so-called Kinki technology will be much further developed. Then we will have 100% farm reproduced tuna of good quality.'

I had heard that before: the secret hope of farmed tuna. All things considered it was true: why no farmed bluefin tuna? There was farmed salmon, after all, and this had expanded into a successful international product with an enormous market. There were farmed shrimps too. And when it comes to bluefin, we are not talking about fattening farms, but real fish farming from egg to fish. The Japanese government was working closely with universities and private companies on developing methods to spawn and propagate tuna in captivity. But farmed bluefin tuna had not moved beyond the laborious experimental stage. For the time being it would not be offering true salvation.

So, it wasn't a bad idea to call in on the Fisheries Agency of Japan, which defines and executes the country's fisheries policy. The agency is housed on the premises of the powerful Japanese Ministry of Agriculture, Forestry and Fisheries, a hulking concrete monster such as you might have encountered in Eastern Europe before the Berlin Wall came down. It was, appropriately enough, within walking distance of the Tsukiji market in Tokyo. I had an appointment with Masanori Miyahara, who as chief advisor at the Fisheries Agency, had included himself in ICCAT negotiations about the Mediterranean tuna quotas during years, growing into a veteran participant of the commission.

Miyahara seemed conscious of Japan's image as the bad guy in the world of bluefin tuna. 'You have to understand that our fisheries are involved in a long crisis,' he explained patiently. For years the fishing fleet had been in a state of reorganisation and restructuring. And that was tough in a sector that played such a major role in the country traditionally. 'We have to rethink our fisheries in a structural way. Maybe we have to look at the capacity of our distant water fleet, look at the efficiency with fewer boats and more overseas ports instead of Japanese harbours as operating bases.' Fish had always been a national interest in Japan; the Fisheries Agency was a powerful public body. With an annual consumption of around 450,000 tonnes of tuna in the first decade of the century, Japan was by far the largest tuna market in the world. In fact bigeye and yellowfin tuna, followed by albacore, were the most important species where consumption volumes were concerned. With some 45,000 tonnes, bluefin tuna claimed but a small part of the market. However, it was conversely just about the entire haul of worldwide bluefin tuna that finds its way to the consumer via Japan: 70–80% of international trade flowed via Japan [127].

What measures did the Fisheries Agency think it could introduce to avert the bluefin tuna population's disappearance through overfishing? 'Our position always has been clear and simple: management of the stock has to be based in respecting proper scientific advice,' Miyahara told me. 'And for good advice scientists always need proper and sufficient data about the tuna stocks.' Unfortunately, that's often lacking, he wanted me to know. I did already know. 'Very unfortunately,' according to chief advisor Miyahara. 'We lack reliable data, indeed. And without these data management will be difficult. So the priority should be to find a solution for the problem of data collection.'

There wasn't much to argue against in this: without figures it was difficult to make good arrangements. But it also looked like a fine excuse of convenience: as long as there are no reliable figures, it is impossible to formulate policy. With the result that management aimed at maintaining the population would come to nothing. Meanwhile time was beginning to run out. Japanese bluefin tuna catches in the Pacific had been terrible for years. Those in the Mediterranean were showing a steep downward curve as well. What if reliable data only became available when the last bluefin tuna had been fished out of the seas? Then we would know for sure that the population had been wiped out, but it would also be too late to do anything about it. To be on the safe side, might it not be a good idea to ban fishing in breeding grounds? Or to introduce a ban on fishing young bluefin tuna that had not yet been able to spawn? 'Of course you might think of that,' the Japanese negotiator acknowledged, 'but the moment of spawning is ideal to fish with the purse seiner. You have a lot of tuna concentrated on the surface water. That is an opportunity not to be missed.'

What's more, the tuna value chain was complex, the tuna official advanced. Policy had to be drawn up with a range of different stakeholders: fishing fleets, dealers, the processing industry and all the governments of the countries who were members of the ICCAT. The Mediterranean catch was moreover an issue that concerned Europe more than Japan, Miyahara said. Europe supplied the major part of the purse seine fleet in the Mediterranean. And the European partners did not seem all that inclined to tackle their own fisheries, the Japanese negotiator pointed out astutely. 'I think there is still a lot of negotiation for us to do with some European countries.'

That did not sound particularly hopeful. But Japan was doing its best, Miyahara insisted. All imported bluefin tuna had to have a catch certificate showing its origin. This system had been introduced in 2008. If it worked well, then all illegally caught tuna would disappear from the Japanese market. The big Japanese trade conglomerates were not keen on negative publicity around bluefin tuna either. 'We strongly advise our big companies to listen to the complaints of the international NGOs.' Was this working? 'All big companies are aiming to get a bigger market share,' Miyahara stated in an apologetic tone. 'There is strong competition going on.'

The management of a sustainable tuna fisheries would not be easy, that much was clear.

My search for the power behind tuna in Japan had not increased my optimism. If you penetrate more deeply into the globally diffused value chain of bluefin tuna you increasingly realise the complexity with which the interests are interlinked in a wide network of interested parties. This is what makes policy to ensure that this migratory fish is caught sustainably so laborious and complicated. Few fish are caught and traded in such a labyrinthine chain of fishermen, dealers, processing companies, importers, brands and retail channels as tuna. Bluefin tuna is an extreme version of this. Then there are the other interested stakeholders around the tuna chain: the governments, NGOs worrying about sustainability, international trade organisations and regional management organisations. A chaotic mass of often contradictory interests of political power, employment, and money, of course, a great deal of money [127].

Still in Tokyo, I took the opportunity to visit Professor Makoto Peter Miyake, probably the most experienced amongst tuna science veterans with a deep knowledge of the world of tuna policies.

For more than three decades, since the start of its foundation, Miyake had been assistant executive secretary of ICCAT. The doyen of tuna scientists (1933) had retired, but at that point in time was more active than ever. He had researched bluefin tuna for more than half a century and was now an advisor of the Japan Tuna Fisheries Cooperative Association; we met at their headquarters in Tokyo's the business quarter. You'd go to any tuna conference in the world and Miyake would be a striking presence with his sturdy physique and calm, natural authority. What did he think of the alarming issues around bluefin tuna and how the world was struggling to find a way to manage the fisheries and trade in a sustainable direction?

Miyake did not deny that the industrialisation of bluefin tuna fisheries had created a monster of global proportions. It was therefore not an easy job to install effective tuna management policies. First of all information and transparency was vital. Miyake described how the trade in tuna was struggling to achieve a transparent supply chain due to inadequate checks on the fish's origin. He clearly had his doubts about the new labels stuck on the packaging of sushi and sashimi in Japanese supermarkets stating the tuna's country of origin and the sea in which it had been caught. 'In Taiwan and China they now have to report that the tuna has been caught in the Indian Ocean, that kind of information,' Miyake chuckled.

The system was supposed to give more control and oversight of the catches, but he believed it was totally unsound. An unknown number of smaller boats in the Indian and Pacific Oceans fished on bluefin tuna and sold it fresh on the local market. No one knew the quantities involved. The checks on bluefin tuna catches in the Mediterranean were even less trustworthy. Take Malta: 'A very remarkable country,' Professor Miyake said. 'First they catch their bluefin tuna quota and sell it to Japan as wild tuna. Then they buy the tuna of the other European purse seiner fleets and put them in the tuna farms to fatten. But we have little or no information about what size the tuna was caught or in what amount. We only know something after they are raised in the cages. So the cage is a black box: we know what comes out, but not what goes in,' the tuna professor explained.

According to him, the statistics were largely unreliable. Miyake estimated that in 2007 40,000–45,000 m of Atlantic bluefin tuna had been caught. That is less than the 61,000 m estimated by the ICCAT scientific committee itself, but still considerably more than the 29,000 m total allowable catch (TAC).

Back in Japan, most tuna fishermen went bankrupt in the meantime, that much was clear. This was the case primarily for smaller companies who, after having fished their own waters clean, had not been able to make the step to worldwide expansion. During the 1980s there were more than 1000 Japanese longliners fishing for tuna. By 2010, only 200 were left. High labour costs and oil prices made survival for the small ship owners impossible. The low tuna prices in the supermarkets and tough competition did the rest. In 2008, the price for bluefin tuna had been hovering around 3500 yen (24 euros in 2008)/kg for years, a lot of money, but all things considered, obscenely low for a fish whose survival was at stake. The Japanese do not want to spend much money on their tuna, it was as simple as that. And while bluefin tuna's existence was threatened, unknown reserves were kept in cold stores.

Miyake believed there was really only one solution: tighter action with a temporary fishing ban. 'Management through quotas has its limitations. As a scientist I say: install a closure during the spawning areas from May until July. Monitoring that closure is relatively easy. Arrest the purse seine fishers that are caught in the act. But of course, if the European countries are not able to realise these kinds of measures things become difficult to manage.'

But surprisingly, despite all the difficulties of managing the stocks, Miyake was not entirely pessimistic about the future. By nature, tuna had more resilience than many people thought, the scientist argued. As far as he was concerned, tuna was not yet a lost cause. 'In the global market of bluefin tuna almost anything can happen,' he smiled.

That might be true, I thought when I left the tuna professor. But things got a little fishy around the bluefin as far as I could judge. It was time to look for the European security forces that could save the fish from disaster.

The Tuna Police

22

Control and enforcement of strict harvesting rules for bluefin tuna. How would this work in a sea without borders? And what, in fact, was Europe doing on this front? The EU Common Fisheries Policy (CFP) is a story in itself. A largely untold story, in fact, about a painstakingly built cathedral of management policies to recover the stocks of different species that took decades of complicated negotiations between the member states and their fishing interests. This only began to pay dividends with respect to sustainable fishing from 2010 on, with remarkable results in terms of improved stocks in the common waters [137, 191]. But within European policy tuna always had been a different story with an external stage outside the common seas. It was European distant water fleets, especially those registered in Spain and France, that fished tuna in the oceans around the world. Bluefin tuna fishing had become a bit of an exception, happening for the most part in the Mediterranean. This sea was an intricate patchwork of different claims on fishing grounds with monitoring systems of extremely varying quality. The EU was the most important force in terms of industrial interest within ICCAT. But this management organisation, which was in charge, seemed to have the greatest difficulty in taking steps to protect bluefin tuna. Could it play a role when it came to enforcement of compliance?

To find out more about that, I travelled to Vigo, the port in the northern Spanish region of Galicia. Galicia is home to the majority of Spain's present tuna canning industries, and with it the largest tuna processing industry in Europe. Vigo is the most important discharge port for tuna caught throughout the world by the state-of-the-art Spanish fleet of gigantic purse seine ships, largely skipjack, yellowfin and albacore tuna which disappear into cans as a bulk product or are presented as steaks on the supermarket shelves.

Here, in the office of the European Fisheries Control Agency (EFCA), housed in a grand building on the port's central avenue, I was shown how the tuna fleets in the Mediterranean and Atlantic waters were monitored. It was somewhat reminiscent of a James Bond film control centre: the wide computer screen in the office showed a chart of the Mediterranean, with dots and lines steering a course across it in different colours. Harry Koster, at the time the Dutch director of the European Fisheries

© Springer Nature Switzerland AG 2019
S. Adolf, *Tuna Wars*, https://doi.org/10.1007/978-3-030-20641-3_22

Control Agency, explained to me how it worked. 'Blue for the French, red for the Spanish and green for the Italian purse seine ships. These are the movements of the European bluefin tuna fleet, displayed within the Vessel Monitoring System (VMS), which picks up signals from special boxes fixed to the boats via a GPS system,' Koster told me. A tracking system has thus been set up which shows where the boats are and in what patterns they are sailing. In the thick of the short, hectic fishing season for bluefin tuna, the line patterns converge in areas where tuna is spawning. Here the fish comes to the surface and thus becomes easy prey for the ships' seine nets. Koster zoomed in on the coasts off North Africa, Sicily and the Balearic Islands for me. 'Look, this ship is drifting, which means it has tuna in its nets,' Koster surmised as he tapped one finger on the screen. 'They are waiting for a tug to transfer it to the fish farm.'

On the screen chart, we scanned the coast towards the east. A blue line ended just off the coast of Egypt. Koster was now studying the screen intently. 'This is an illegal transfer of undeclared tuna from a French ship to a reefer, a refrigerated transport ship.' An informant had told him of a rumour that a cargo of illegal tuna was being transferred at sea off Port Said. 'And now we can see it online, cruising for 3 h at low speed just outside the port. Then you know the score.'

In September 2006 Koster had been appointed director of the newly established agency, as a kind of European Eliot Ness of tuna. From his office in Vigo he coordinated the battle against organised tuna crime: fiddled catch certificates, the deployment of prohibited reconnaissance planes and trade in black-market tuna. By tracking the movement patterns and speed of the ships on the screen, his team was able to monitor the activities of the fleets from Vigo. Long straight lines from the port: en route to the fishing ground. Spirals: looking for bluefin tuna. Bobbing in the same place: fishing. Long straight lines from the port followed by a sudden treading of water: fishing guided by an illegal plane.

A European variant of a coastguard tasked with monitoring fishing restrictions: this was something new. Since it concerned bluefin tuna it was what you call an important challenge. When Koster swapped his job as head of the control and inspection unit of the Directorate-General for Maritime Affairs and Fisheries at the European Commission in Brussels for the agency, quite a few of his colleagues thought he was mad. With a budget of 5 million euros and hardly any resources, the monitoring of upwards of 1000 ships—purse seine, longliners, tugs and freezer vessels—seemed a hopeless task. After all, the agency would never be able to become a powerful coastguard along American lines. Statutory rules, including those for catch restrictions, were enforced at national level, so this left Europe empty-handed. Direct inspection is a mandate belonging to the member state, in other words. The agency did not have its own ships. 'They laughed at me when I came up with a list of resources we wanted to use for control,' Koster remembered.

These colleagues—and a growing section of the tuna fleet in the Mediterranean— did not laugh for long. In its short existence, it succeeded in building up a solid reputation among fisheries, interest groups and NGOs.

By deftly joining the forces of all monitoring resources in the individual member states at a stroke, Koster had an unexpectedly large control mechanism at his disposal. For his tuna campaign in 2008 he had a fleet of 53 patrol vessels, 16 planes and a force of nearly 150 inspectors available. Linked data files, with which the various member states were able to follow their fishing vessels, delivered a reasonably complete overview of fleet movements in the Mediterranean. On the agency's instruction, spot checks were carried out. Member states with insufficient numbers of inspectors were given assistance at their own request. In Malta, it was Spanish and Italian inspectors who monitored the landing of tuna for the tuna farms. 'We check the guidelines and send observers on national inspections. We are the inspectors of the inspectors,' Koster said.

Tuna was by far the most important activity for the agency, which had been set up as part of one of the European fisheries policy reforms. The EU fisheries policy was known as an extremely labyrinthine minefield of member states' interests and as an extraordinarily troublesome and complicated dossier. The expectations of the fisheries police were not particularly high, that much was true. A cynical joke went around Brussels that everyone in the fishing industry relished the idea. Half of the fishermen were pleased that, as a result of tighter controls, their colleagues would no longer be able to ignore quotas and catch restrictions scot-free. The other half was pleased because they believed that an agency under the umbrella of the European Union would be the best guarantee of efficient controls coming to nothing. To everyone's surprise, things worked out differently: it was the latter group that got the short end, while the former was rightfully satisfied.

The agency's first campaign involved cod fishing in the North Sea. In 2002 this was banned, as cod was heavily overfished. A catastrophe such as the one that befell the cod population off Newfoundland had to be avoided. Strict quotas were laid down within the EU and with Norway, and these needed to be adhered to. At first the advent of the agency elicited few reactions from the fishermen. 'They would have a go and continue fishing anyway,' Koster told me. But that soon came to an end. The efforts of the member state control teams, an aeroplane and some steep fines soon put paid to the infringements. This catch management would develop into one of the success stories of the CFP. Thanks to European policy and enforcement, cod had returned to abundant levels in less than a decade. Consequently, the agency's activities were expanded to limiting cod catches in the Baltic and Greenland halibut off the coast of Canada.

But bluefin tuna was a different story. Matters there had really got 'a little out of hand', as Koster put it mildly. The Mediterranean turned into the tuna Wild West: an area where no one would take much notice of law or order. Here the law of the jungle held sway. The problem was 'much more serious than with cod', Koster believed.

When it came to the technicalities of controls, tuna fishing was not easy to manage. Large shoals of bluefin tuna were fished and hauled out of the sea, fattened, shot and transported. Catch certificates used to check origin and quota passed by different people in various countries at various points in the supply chain. Tugs transporting cages with living tuna swapped stocks under the radar. The banned light aircraft used to trace tuna shoals were able to fly over the spawning grounds untroubled, dozens at a time. Right at the end of the official season, catches were registered in tuna farms which did not hit the nets until much later. Fiddling with the regulations was the rule rather than the exception. Tuna had become a hotbed of tinkering, bribery and falsification.

The conclusions of the first year of tuna controls, which Koster sent to the Fisheries Commission of the European Parliament at the end of 2008, were there for everyone to see. Despite talks with the fishermen in the European Commission and the member states, 'It is not a priority for the majority of participants in fisheries to comply with the ICCAT's legal requirements,' Koster concluded wryly in his report. There was 'considerable' tampering with catch documents and de-activating of VMS satellite receivers on the ships, according to the report. The illegal catches in the race for bluefin tuna were getting so out of hand that the European Commission, on the agency's recommendation, closed the fishing season on 16 June, 2 weeks before the official end. 'That was just in time,' Koster said. His agency preferred to catch perpetrators in the act. Member states and ICCAT satellite detection systems, which are able to follow all of a ship's movements via an on-board transmitter, should be able to supply their data more quickly and in a more standardised way. One idea mooted was an automatic alarm system that starts to bleep as soon as ships begin to make suspect manoeuvres. Satellite photos of ships are also taken these days. Every afternoon, a schedule based on actual data has to be issued to the patrol vessels and the inspectors on site.

But more than anything else the control on the tuna supply chain, what is termed the chain of custody, had to be improved so that the consumer could count on the certificated tuna really being legally caught tuna. The checks on tuna chains, and not just bluefin tuna, would expand into a global challenge which would determine not only tuna but the entire world of fish. The year 2009 would be the breakthrough for control on bluefin tuna, Koster thought. Not only was the credibility of his agency at stake, but also the future of bluefin tuna, and with it the reputation of the common fisheries policy of the European Union.

Tuna professor Makoto Peter Miyake was right. Virtually anything seemed possible in the world of bluefin tuna at the end of the first decade of the new millennium. Even if the dominant possibilities were primarily negative. That same European Commission that had closed the fishing season in the summer of 2008 as a precautionary measure when fraud was running riot, appeared a few months later to be engendering another tuna policy failure. In November 2008, the annual ICCAT meeting was held to set catch restrictions. This time the meeting journeyed to Marrakech in Morocco. 'Marrakech' would go down in history as ICCAT's biggest disaster so far. On the eve of Marrakech, WWF and Greenpeace were still optimistically disposed, hoping that robust protective measures would be introduced for tuna. For months, the rumour had been buzzing around the corridors that there might even be a moratorium.

It worked out quite differently. At the meeting ICCAT decided to fix the total allowable catch for 2009 at 22,000 tonnes. In doing so, they turned a deaf ear to pressing recommendations by their own committee of scientists to restrict the catch to 15,000 tonnes. The experts believed this to be the responsible maximum amount for maintaining sustainable bluefin tuna stocks. But under pressure from the EU representatives and above all from fisheries interests from member states such as Spain, ICCAT took an entirely different decision.

This was incomprehensible. Everybody—governments, NGOs and industry alike—always defended the idea that decisions on sustainable management of tuna fisheries should be firmly based on scientific research. But when it came to it, ICCAT turned a deaf ear to the recommendations of its own scientific committee. An understanding of how these things could happen requires a look into the dynamics of politics. France, which for years had made a mockery of the checks on quotas and undersized fish, felt it had been snubbed by the EU. In 2006, it had been given a smarting rap on the knuckles with a fine of almost 58 million euros for insufficient checks on undersized hake during the 1990s. In 2007, France decided to monitor its bluefin tuna quota strictly. It turned out to have been amply exceeded, which landed France a 3 year punishment in the form of a quota reduction. But in that same year

Italy, too, had generously exceeded its quota and refused to acknowledge guilt. This was even more striking because a year previously, a spokesperson for the Italian ministry of fisheries told a British newspaper that its country followed a zero tolerance policy and applied 'strict rules' in prosecuting illegal fishing that endangered bluefin tuna. A special investigative report by EU accountants in 2007 revealed what these solemn pledges of Italian checks and 'zero tolerance' amounted to in practice: little to nothing. In Italy, there were no reliable catch statistics; the local port authorities knew nothing about tuna quantities landed and there was no centralised system for inspection reports. According to the WWF report on Italian fish markets [167], local mafia clans held sway, producing catch figures and trading bluefin tuna that bore no resemblance to the truth. Some fish markets in regions representing the traditional heart of the trade sold not a single kilo of tuna, according to the official figures. In 2007, the ocean conservation organisation Oceana uncovered more than 80 fishing boats using driftnets, a fishing method which had been banned in the Mediterranean since 2002. The Italians simply got away with it [178]. No wonder it angered the French, who had incurred a fine.

So the French decided to take revenge on the Italians. In order to satisfy its own tuna fishermen, France used Marrakech to push through a higher total catch. The moment was ideal: at the time, France held the presidency of the European Union. As an experienced player in power diplomacy France was able to leave its stamp on the EU tuna dossier. The French argument was this: the agreed catch was far above the ICCAT scientific recommendations, but with stricter control measures the damage would, on balance, be easily cancelled out.

All member states agreed, and afterwards an EU spokesperson declared himself 'extremely satisfied with the agreed consensus'. Environmental organisations reacted furiously to ICCAT's decision [117, 260]. Sebastian Losada, Greenpeace's tuna specialist: 'The game is over. ICCAT has missed its last chance to save the bluefin tuna from stock collapse. These past 7 days have demonstrated that ICCAT is a farce—it has run a stock under its management into the ground and is not even prepared to face the consequences.' Sergi Tudela from the generally diplomatic WWF concluded, 'This is not a decision, it's is a disgrace which leaves WWF little choice [but] to look elsewhere to save the fishery from itself.' Not one single tuna scientist had a good word to say about ignoring the ICCAT scientific committee's recommendations. Even the tuna traders cried shame over it, although it remained to be seen whether they were in fact shedding crocodile tears. Mitsubishi Corporation declared itself officially to be 'extremely disappointed'. The trading giant announced that, together with the other big Japanese businesses, it had urged Japanese Fisheries Agency chief advisor Masanori Miyahara to adopt the scientists' recommendations. 'Mitsubishi Corporation believes that the Mediterranean bluefin tuna is being overfished and, without effective regulation and conservation management being set and observed, this situation will worsen,' the official statement concluded.

This did not stop Mitsubishi from continuing to trade in bluefin tuna. The press release explained why. 'Going forward, if the global community's consensus is that (...) it is necessary to implement regulations that exceed the current scientific committee recommendations, or to require a full moratorium on the fishery or even to completely ban the commercial trade of BFT, Mitsubishi will fully support this consensus if it is implemented as an international rule that is fair, effective and enforceable' [171].

Without consensus, no policy. And without policy, no cooperation in sustainable tuna fishing and, in the end, even less consensus. Mitsubishi, and the Japanese businesses with their cold stores full of frozen tuna, were able to wash their hands of it. It was nothing to do with them, they argued. The problem lay in Europe.

To make matters even worse, there were stricter controls, which, according to the French, would compensate for the higher quota in the end. Shortly after Marrakech, this was reiterated by the European Fisheries Commissioner Joe Borg. The man who was formally responsible for the Marrakech fiasco declared control and enforcement of catch restrictions the cornerstone of the common fisheries policy.

But in Malta, notably the home base of Joe Borg, tuna farms had a particularly bad reputation. Reports estimated that a consignment of tuna, possibly illegally caught, worth 25 million euros, had allegedly been manoeuvred onto the legal market via Malta.

Laundering black-market tuna, how does that work? According to Malta's official catch figures, at the end of 2007 a total of 1350 tonnes of caught tuna was alleged to be present at the farm and intended for 'harvest' in 2008. At New Year, that quantity was still in the nets of four fattening farms. A calculation by tuna spy Roberto Mielgo showed that the cage net capacity of the four farms involved was far too small to hold the amount of tuna declared. In some cases, even less than half of the declared tuna fitted in the nets. What's more, the catches were attributed to boats that Mielgo believed were not equipped for tuna fishing. Because no one counted the exact number of tuna that swam into the nets, the so-called fattened tuna could be bumped up in kilos and the illegally caught tuna was laundered.

The case around Malta is illustrative of the complex problems involved in effective control. The indications are clear, but the submitted calculations are not conclusive evidence. And there was always an expert somewhere who was prepared to state that there might be far more fish in the nets. In particular in Malta.

Caged Tuna

23

The tuna cages are a vortex, a maelstrom of giants. In the deep blue water the shoal of bluefin circles slowly around an invisible axis somewhere in the middle of the cage of circular nets. There must be many dozens of them, in a deep whirlpool of fish that reaches to the bottom of the sea. They inch forward, their tailfins swishing majestically through the water. The stately calm disappears as soon as the engine of the boat with fresh fish comes within earshot. Once the sardines, herring and cuttlefish are thrown over the raised edge that keeps the nets of the circular cage afloat, the surface of the water transforms into a whirling mass of dorsal fins, splashing foam and gaping jaws into which the fish disappears. Feeding time at the tuna farm. Tuna is always hungry.

Once they've been inside the cages for a few months, the excitement wears off, say the men who shovel the fish into the cages. The tuna become fatter, slower and wiser. Experience has taught them that there is plenty of food, and so they glide along and allow the slowly falling fish to drop into their jaws.

Slaughtering season begins in October or November, depending on demand in the market in Tokyo. Divers descend to kill the fish with a well-aimed shot from a speargun. On other farms the fish are killed using a special gun with an explosive bullet. This has been developed as a quick and efficient way to slaughter fish. The tuna dies before it realises what's happening. Less stress in other words, something only to be welcomed from an animal welfare point of view. It's also good for sales: no adrenaline is released in the fish to change the meat's colour. The tuna is hoisted out of the water, cleaned and frozen for transportation. The best specimens are packed in ice and flown to Tokyo's Narita airport and from there transported as quickly as possible to Tsukiji fish market. Over the past few years a growing proportion of fresh tuna has been destined for the European home market, particularly for restaurants.

The striking fact remains that most tuna farms have been set up off the coasts where the Phoenician and Carthaginian trade colonies founded their beach heads as a bit of living ancient history. Malta is such a spot where time has stood still for

S. Adolf, *Tuna Wars*, https://doi.org/10.1007/978-3-030-20641-3_23

2500 years as far as tuna is concerned. Ricardo Fuentes and Sons' tuna empire is situated off the dry, desert-like coast around Cartagena, the new Carthage of General Hannibal, whose family established the first large tuna trading power on the southern Spanish coast. The name of the hilltop fort of El Atalaya recalls the old lookouts where *almadraba* guards alerted fishermen when tuna came into view. These days, the main tourist attractions are the ancient Roman theatre, a few towers and old forts.

The tuna farm—or rather, fattening centre—struggled with a subpar image. There were frequent press reports about bluefin tuna slaughtering methods. At one point Greenpeace had thrown floating white crosses into the cage net waters to turn the farm literally into a graveyard for bluefin tuna. Tuna scientists, too, pointed an accusatory finger at the farms as a link in a business chain that facilitates overfishing of bluefin tuna. This leaves its mark. Ricardo Fuentes, who died in 2015 aged 80, abhorred publicity. His son Paco, who leads the empire at present, is not particularly keen on it either. 'It creates a lot of misunderstanding,' he told me at an ICCAT conference.

In 2008 Ricardo Fuentes Group director David Martínez was prepared to talk to outsiders. 'The situation around tuna is difficult,' he acknowledged to me. 'Because it was war on the tuna market: every man for himself, and the tuna for all of us. I know we always get the blame,' the director said. Yet Martínez did not think tuna farms were to blame for the rapid decline in stocks. Things were more complicated. The 1990s had been a golden age for tuna. 'A lot of money was earned in farmed tuna,' the director admitted. 'Now it's increasingly difficult to keep your head above water.' Producers struggle with falling prices as a result of competition. It was around the start of the new millennium that the market situation for farmed tuna began to shift. Because the yen was on a continuing downward slope from 2002, the margins became ever tighter. The Japanese consumer who buys farmed tuna is not prepared to pay more for his fish, and the feed does not become any cheaper.

The problem with the tuna farm, Martínez said, was that the industry was becoming a victim of its own success. Cheap tuna has created an enormous market, which has to contend with consumers prepared to pay only a relatively low price for their sashimi and sushi. Under increasing costs, production can only be kept profitable by upscaling. And that means even bigger catches on even bigger tuna farms. The illegal catches in the Mediterranean disrupt the market even further. New tuna farms continued to be granted licences. And at the core of the issue was the overcapacity of purse seine ships fishing for tuna in the Mediterranean. 'That should be cut,' Martínez said. 'But no country wants to be the one to take the first step.' The number of tuna-fishing ships continued to grow apace. 'It's got a bit out of hand with tuna,' Martínez said. It was not the first time I had heard this.

Even the expensive, traditional *almadrabas* had become more profitable than fattening on the farm, he told me. 'Wild' tuna can command much higher prices on the market. That's why in 2001 Fuentes bought himself into the Barbate *almadraba* and why there were 16 *almadraba* licences that had yet to be awarded in Morocco. Four of those were now being used, all near Larache. The ancient Phoenicians' Lixus was thus being restored.

Diversification is the magic word. Since 2004 the Fuentes Group has begun to cultivate other sea fish: seabream and seabass. Just in case the market for tuna ran into lasting difficulties.

Martínez did not think that a total ban on bluefin tuna fishing would be a good idea, for that matter. Understandably, as a few years' moratorium would be the kiss of death for his farming business. A drastic reduction in the Mediterranean tuna fleet was a solution in which he saw more future. As far as he was concerned this could be shrunk by half its existing size. If all countries then kept to their fishing quotas, the sector would take a significant step forward. A shame that everyone tried to carry on fishing regardless. But that was also understandable. 'What's the point in stopping catching bluefin tuna when your colleagues carry on unabated?'

Martínez hoped that European research projects to grow bluefin tuna from the egg in captivity would yield results. Tuna Graso, the tuna farm Fuentes owned together with the Japanese Mitsui also takes part in the research. Fully cultivated, from roe obtained in captivity, just like with seabream or seabass. 'This must be the future.'

David Martínez kept his word. In 2014, the Fuentes Group launched its first cultivated Atlantic bluefin tuna weighing 25–30 kg onto the market. The two fish arrived in the Netherlands on ice in a polystyrene box, fully intact. The recipient was Jan Van As, a fish dealer in Amsterdam known for his interest in sustainable fish. From there the cultivated tuna landed on the table of an exclusive restaurant in Amsterdam. It was meant as a stunt, but the Fuentes Group expected the sales of young cultivated bluefin tuna to take off from 2018 and to become profitable.

Indeed, why not: farming tuna, from roe to adult specimen, could become the tuna industry's big hope. Growing fish in captivity is far from new. China has been culturing carp since as far back as 3500 BCE. Europe has been cultivating trout on a large scale since the nineteenth century. Modern fish farms sprung up in the 1960s after Norway developed salmon farms in its fjords. At the beginning of the twenty-first century, pangasius farming expanded into an important export sector in Vietnam. The UN Food and Agriculture Organisation, the FAO, underlined this trend in a report from 2006: most fishing grounds have reached their maximum capacity or are overfished. The global growth figures for aquaculture, on the other hand, are undergoing an impressive expansion. From a million tonnes of fish in 1950, production in the farms grew to almost 60 million tonnes in 2004. The FAO estimates that at the beginning of the twenty-first century a little under half of all fish consumption was coming from farms. China even provides as much as two thirds of its fish for consumption from farms. At the current global population growth, production over the next 20 years will have to increase by 40 million tonnes if fish consumption per global head of population is to remain constant. Carp in central

Europe, oysters in Asia, wolf fish in the United States, salmon in Canada, shrimp in Latin America and the Caribbean: farms are spreading like never before. Catching 'wild' fish is not an option for the future in order to meet growing demand. So farms are the solution.

Yet farmed fish had its own problems. Feed and pollution of the marine environment were cause for concern. Take salmon. While in 1980 just three percent of global salmon production came from farms, 20 years later this had risen to 65%. The concentration of great quantities of fish in fixed nets led to a build-up of excreta, food waste and pesticides. A report by the University of Victoria estimated that, at the end of the 1990s, the waste products in the salmon farms of British Columbia corresponded with the sewage production of a city of half a million inhabitants. Great swathes of the Latin American coasts are taken up with large-scale shrimp farms constituting a concentrated source of pollution for the vulnerable coastal environment. Packed closely together, fish are extra-vulnerable to parasites and diseases. This is countered by mixing antibiotics in with the food, the residue of which ends up in the human body, leading to potentially antibiotic-resistant bacteria.

In addition, in order to save costs, feed of dubious origin is dabbled with. Some food specialists believe this explains the high concentrations of harmful substances that can be found in farmed salmon. Plenty of healthy omega-3 oils for your cardiovascular system, but mixed in with carcinogenic PCBs and other organic chlorine compounds: not a pleasant thought. Since appearances count, farmed salmon is fed colouring agents to compensate for the loss of pink colour when the feed lacks shrimp. Usually harmless, but not exactly a key requirement.

Then there is a genetic problem in the coastal waters surrounding salmon farms. Farmed fish can obviously escape and so affect their wild cousins by weakening them genetically or by infecting them with rapidly spreading diseases. And we are not talking about small numbers here. Off the North American Pacific coast alone, half a million farmed salmon escaped during the 1990s, often as a result of storms that wrecked cages. In September 2017, a spectacular escape off the coast of Washington State followed after more than 160,000 Atlantic farmed salmon mixed with their native Pacific cousins once the cages had been opened. Research suggests that, genetically speaking, 40% of the 'wild' salmon in the North Atlantic originate from farmed salmon. The original fish is thus decimated by a species with entirely new characteristics, the consequences of which are not yet fully known.

In the big global markets, in which consumers are more interested in low prices than in a sustainable piece of fish, many producers are looking for loopholes in the law for cheaper production. Many European salmon farmers emigrated to British Columbia after they faced tighter regulation in Europe to combat pollution. A country like Chile could set up enormous salmon production and, using an aggressive pricing policy, force the competition elsewhere out of the market.

Yet many innovative things are happening in fish farming. Substitutes for fishmeal or bait are busily being sought, with food based on algae in particular looking very promising. Other problems are being tackled with more widely enforced, stricter regulation. The development of fish vaccines has pushed back the use of antibiotics. What's more, many initiatives for more sustainable production come from the industry itself. That's a hopeful development. Such as the application of certification by the Aquaculture Stewardship Council (ASC)—sister organisation of the Marine Stewardship Council (MSC)—an independent organisation that uses its label to strive for a sustainable approach to fish farming and a measurable reduction in the burden on the ecosystem. Or take the Global Salmon Initiative (GSI), which represents companies that make up 70% of the global production of farmed salmon. In order to make fish farming more sustainable, the GSI aims to have all its farmed salmon ASC certified as early as 2020.

Tuna had a reputation as an impossible fish to cultivate. The farmed tuna that came into being in the 1990s had not been grown from spawn, but had been caught and fattened. There should be another way around this, believed Professor Hidema Kumai, head of fisheries research at Kinki University in Higashiosaka. In 1970, he became obsessed by the curious idea of growing tuna from spawn obtained in captivity. For well over 30 years his university pulled out all the stops to hatch bluefin tuna in captivity. The person who succeeded in fertilising all those millions of tuna eggs per tuna and in rearing the larvae in a protected environment would have the future of bluefin tuna in their hands.

Initially Kumai's experiments were not particularly successful. The tuna larvae died en masse. Young tuna tried to break free from the nets by swimming into them at full speed, breaking their necks in the process. Bigger specimens ate the smaller ones. The tuna farm was an unproductive killing field.

The biggest problem the experiment ran into was the artificial prompting of tuna to spawn. Bluefin tuna only appeared to be able to produce ripe eggs under very special circumstances and at particular temperatures. It took until 2002 for the Japanese scientist to be successful: using the right food and water temperature the first tuna were reared. Two years later, in 2004, the first adult specimen appeared on the market. The Japanese Maruha Group, a big player in the tuna market, became the fish's main distributor.

Kindai, the fish was christened, after Kinki University's new name. The name change was a rare concession to the international community from a Japanese institute, once the realisation had dawned that in English Kinki Tuna resembled 'kinky tuna', inviting off-colour jokes. Kindai seemed more neutral.

Kindai became all the rage amongst tuna specialists. It could be an important step towards the first fully 'sustainable' bluefin tuna, leaving the existing population of 'wild tuna' undisturbed and guaranteeing the continuation of the species. If it worked, you would be able to eat as much of this tuna as you wished without losing your sustainable high ground. The first *Kindai* found its way from Japan to the United States as an exclusive snack in the restaurants of leading hotels in New York and San Francisco. The prices are comparable to those of 'wild' tuna from the *almadraba*. In Japan itself the acceptance of farmed tuna was as yet mixed. Some sushi chefs were alleged to have turned their nose up at the fish's 'rancid' taste, other culinary masters conversely praised the *Kindai's* excellent quality, saying it could not be distinguished from wild tuna.

Anyhow, *Kindai* set in motion a global race in farmed tuna. Hagen Stehr's Stehr Group from Port Lincoln, one of the biggest fish farms in Australia, began its own project for farming southern bluefin tuna with support from the Australian government [81]. But Stehr called time on the project after just a few years when development costs almost plunged the entire company into bankruptcy. A team of European researchers led by the Spanish tuna expert Fernando de la Gánderia, supported by European Union subsidies and the assistance of the Fuentes Group's Tuna Graso farms, has been working on obtaining fertilised tuna eggs and developing a new feeding regime of vegetarian extracts and algae. This has eventually yielded results.

Farmed tuna caught on. The Japanese tuna firm Sojitz joined forces with Kinki University for their *Kindai* project. As the superpower amongst the bluefin tuna dealers, the Mitsubishi Corporation likewise threw itself into researching *Kindai*, working with partners both in Japan and overseas on the farm project.

In the race to produce farmed tuna, Japan and the Fuentes Group appeared to be ahead of their direct competition. In 2017, after 10 years of research, seafood giant Nippon Suisan—better known as Nissui and occupying second place behind Maruha Nichiro—introduced its first significant quantity of tuna farmed fully in captivity (from spawn to adult) into supermarkets and restaurants. Around 10,000 specimens, weighing on average 50 kg each, were launched onto the market. Maruha Nichiro promptly promised ten thousand 60 kg bluefin tuna for the year 2018. Engine manufacturer Kaneko Sangyo, which dealt in tuna, followed suit with an announcement of 100 tonne in 2019. In total, 1700 tonne of fully farmed Japanese bluefin tuna was anticipated in 2019. South Korea, too, was attracting a great deal of attention with farmed tuna. This was beginning to look promising. According to the Japanese Fisheries Agency the demand for tuna was picking up due to the pressure of the problems with the wild catch which Japanese fishing was wrestling with. Now that the Chinese market has also discovered sushi, and China's newly rich want to eat sushi made exclusively with bluefin tuna, the demand will only increase over the next few years.

In the Mediterranean, farming was supported by a couple of EU programmes (SELFDOTT and TRANSDOTT), which, together with private initiatives from 2015 onwards showed positive results. Farms in Spain, Turkey, Italy, Greece and Egypt were founded but in many cases lack of success led to their closure. The companies involved, such as the Fuentes Group or the Turkish Kılıç Seafood, were not particularly forthcoming with information about their fingerling bluefin tuna farming: because Mediterranean bluefin tuna was reputed to be an endangered fish amongst the wider public, and tuna farms did not have a particularly good image at the best of times, there was a reluctance to start talking about fish farming. According to estimates by the Dutch tuna expert Jonah van Beijnen [33] from 2011 on, a total production reaching 47,500 young fingerling bluefin tuna would have been generated by 2015. But the following year production collapsed when the farming projects were halted. It continued to be difficult to grow bluefin tuna in captivity and Europe was clearly lagging behind Japan. In 2011 in Japan a billion fertilised eggs were produced, while the TRANSDOTT programme in 2013 yielded 40 million fertilised eggs. Enormous mortality as a result of disease, collisions with tank walls, cannibalism and lack of the right food left far too little to set in motion a commercially interesting production process.

While Europe was muddling along, almost 60 bluefin tuna farms were active in Japan by 2016. It was a clear example of a 'developmental state': the government, in partnership with universities and businesses, had for decades financed the development of a new, sustainable tuna industry. Behind it was an innovative long-term vision to invest in the future and to overcome the numerous problems around farmed tuna. In wild bluefin tuna fishing, Japan had been far from playing a leading role and was responsible for decimating worldwide bluefin tuna fishing. But when it came to the commercial development of farmed tuna, a new sector with a clear sustainable character, Japan could be regarded as setting an example.

Pioneer Kinki University made an international name for itself as a knowledge centre for bluefin tuna farming. In 2016, after 15 years of its project, Kindai saw an annual number of around 3000 farmed tuna swim in its basins. Reason to come up with an extremely innovative consumer product: a special 'lip scrub' based on the protein obtained from the skin of farmed tuna. The scrub, which looked like a kind of soft sweet, moisturised lips and was sold in grapefruit and berry flavours. Lip beauty did come at a price, however: every 100 g of lip scrub required 40 kg of bluefin tuna. But it had been made from sustainable, farmed bluefin. A product European farms were not even able to dream of at that point.

* * *

There was little to argue against. Farmed tuna was more sustainable from the point of view of wild population management. It was closed circle aquaculture. And although the production capacity of farmed tuna was as yet on the modest side, in future tuna farms might even be able to bring down the demand for fishing in the wild. But from the standpoint of nutritional efficiency there was a lot to say against the sustainability of farmed tuna. Because farming is an attack on existing biomass. Big fish eat small fish. A salmon eats three kilos of fish to gain one kilo in weight; every kilo of cod equals five kilos of small fish. But every kilo of weight gain in a bluefin tuna requires some 20 kg of feeding fish. This makes tuna the Hummer amongst fish: in terms of food energy it is an extremely inefficient glutton. It could be compared to eating meat: the protein in a steak takes up an enormous amount of scarce resources. You do not have to have read Jonathan Safran Foer's *Eating Animals* [100] to realise that eating farmed tuna entails comparable moral issues. It also constitutes the central argument in *The Perfect Protein* [220] by Andy Sharpless, CEO of Oceana, one of the best informed NGOs where protection of the marine environment is concerned. One of Sharpless' meta-management recommendations to provide the world with fish in the future, is to avoid eating large predatory fish such as tuna. Better to have a kilo of small fish than a kilo of tuna which necessitated 20 kg of small fish. That's another condition for sustainable tuna: fattening or growing tuna should not be allowed to wreak havoc on other fish species that serve as food. Artificially obtained proteins, especially those derived from plants should offer a solution here. Japanese firm Nissui has taken steps in this direction with food pellets specifically designed for tuna, which resemble small sausages, made of a mixture of fish and plant material which in terms of taste and texture have been specifically developed for bluefin tuna. The fish appear to go wild about these sausages.

Maybe all our farmed tuna will grow up on a diet of pressed algae-extract pellets and small feeding fish will no longer be part of the picture. But that leaves us having to grapple with another sustainability problem: the circular nets and tanks in which the farmed tuna have to swim if forced to grow to full size. I am of course no tuna, but if something had become clear to me in my quest for this fish, it was that the idea of a living torpedo, built to roam freely over thousands of kilometres of ocean, now raised in a net measuring a few dozen metres across, in which it can only swim around in circles, is not a pleasant one. It must be exceedingly boring for a tuna. It is the marine variant of the wild animal cage in an old-fashioned zoo, with tigers and lions pacing up and down in their cramped space. Or like a battery cage for chickens. Pig farming. Crated calves. One of the major causes of death amongst farmed tuna is that adults during spawning, as well as the larvae and small tuna, crash against the walls of the tanks in which they are grown [33]. This is often attributed to stress and flight behaviour, which may well be the case, because the tuna is a nervous fish that gets spooked by unexpected movements or sounds in such an artificial environment. Anyhow, the sustainability of tuna is not only related to maintaining the wild population, but also to the welfare and health of tuna in captivity.

Tuna Welfare

24

Animal welfare is an elastic concept. Slippery ice. It undoubtedly involves a great deal of anthropocentric interpretation. A tuna is an animal, a fish at that. And yet I had looked into the eyes of a living tuna too often not to recognise that more was going on in that big head than rudimentary thoughts about eating, swimming and spawning. A thought, of course, one which you would often detect in swooning whale and dolphin conservationists, who believed they had discovered a notion of understanding when looking into the eyes of their charges instead of into their big-teethed jaws. (My theory: we attribute excessive self-awareness to all animals with a large head and big eyes. There is some perceptible movement in the eye that makes us wish to believe that there is such a thing as mutual recognition or a dawning empathy; which does not necessarily have to be the case. Somebody should investigate this). But wasn't the danger just as much in the denial of any feeling or self-awareness in animals? The Dutch primatologist and behavioural biologist Frans de Waal had written some fine books about that, one of which was, as it happens, called *The Ape and the Sushi Master* [249]. But a primate was no fish, of course.

Ethologist Jonathan Balcombe is one of the few writers who has tried to figure out what is going on inside the heads of fish. His book *What a Fish knows* is subtitled *The Inner Life of Our Underwater Cousins* [30] and this comes quite close to covering the core of his reflections. Our knowledge about the inner life of fish has increased exponentially because we are gaining ever more insight into what is taking place under water. And guess what? Our image of rather primitive beings, whose peanut brains without memory are only able to process the most basic impulses, is based on a gross underestimation. Fish have consciousness, are social, know how to use tools and to communicate. They are even able to manipulate things, according to Balcombe. Hearing, seeing, tasting, smelling: as far as their senses are concerned fish are more than our match. Some fish can be summoned by whistling when they know they are getting food. Aquarium fish bash their tails against the glass to demand attention, others make a smacking noise when they want food. They recognise music. Herrings communicate by farting. They can smell each other. They spit out foul-tasting food ('Yuck!'). And, perhaps the most important aspect for caged tuna:

© Springer Nature Switzerland AG 2019
S. Adolf, *Tuna Wars*, https://doi.org/10.1007/978-3-030-20641-3_24

fish have a clear sense of the space they are in. They navigate using the sun, geomagnetic sense, the smell of the water's composition and possibly visual landmarks. This is especially the case in migratory fish of which tuna is a prime example. They also know when they are in a school of fellow tuna. They make eye contact with humans and can be curious. If you ever meet a stone bass in his cave under water, you will probably know all about it. Some fish like being stroked. They display behaviour that suggests they experience pleasure. Like the southern bluefin tuna in the Pacific who can tumble and turn on the water's surface in the sun for hours in something that looks suspiciously enough like sunbathing. Conversely it is also conceivable that fish experience pain and fear. Experiments with injected vinegar and painkillers in trout point in that direction. Although a topic of scientific debate, it is best to be on the safe side and allow for the possibility that fish, especially large fish like tuna, enjoy a degree of self-awareness.

This all confronts us with a potentially moral problem, now that fish appear to be significantly more than a kind of swimming vegetable. I no longer eat stone bass, for instance, after several encounters with this fish. It left its cave especially for me, swam towards me and inspected me with benevolent interest. That is fatal for my appetite. Even eating octopus causes difficulty, after a juvenile could not stop embracing and playing my fingertips during a dive. I admit, these are sentimental considerations. But what should we do if farmed tuna turns out to be a roaring success? It must be a dreary hell for these creatures to have to swim the same circles over and over again in their prison. Even if they don't know any better, they are genetically preprogramed to cover enormous distances, find their way, hunt their prey and sunbathe. Then there are all manner of plagues and diseases—Balcombe gives a bleak description of sea lice, painful for the fish—that descend on farmed fish in massive numbers. He compares eating tuna to eating a tiger. Both large predators that are threatened with extinction. Eating a farmed tuna would be like eating a caged tiger. Not something to be recommended.

In contrast to tigers, there are no protests where farmed tuna is concerned. Tuna's tragedy is its low level of cuddliness. The Dutch author Rudy Kousbroek coined a special term for this, *aaibaarheidsfactor* (literally, strokeability factor) [156]. This is the extent to which an animal arouses affection or fondness. Some animals invite stroking, others do not at all. Kousbroek did not eat pork because he had his own pet pig during his childhood in Indonesia. The tiger Hobbes from the comic strip *Calvin and Hobbes* could be spotted as a soft cuddly toy in many a child's bed. But the first toddler with a soft cuddly tuna has yet to be found. The tuna's face does not help either. It is rather inscrutable. 'It is of no use for frowning or smiling; if the fish could do these things, it would receive a great deal more sympathy than it does,' Brian Curtis writes in his book *The Life Story of Fish* [69]. At best, you might be able to tell something from his big eyes. It is in any case not an expressionless glassy gaze: the tuna sees you and examines you. Even so, a tuna is obviously not a seal pup with big, pretty eyes. Brigitte Bardot clutching a tuna: it does not work. No eternal smile on his lips like a dolphin. No amusing waddle on ice like a penguin. A tuna does not cry and does not make grunting noises like a whale. Instead it silently looks around and flaps its rigid tailfin every so often to move forward.

And yet, change is in the air. The sustainability debate has also given the animal welfare discussion new impetus. In sport and leisure fisheries debate is quite common on how to kill fish in a responsible, painless way. The first experiments are being initiated in professional fisheries. The issue is developing mainstream in animal welfare movements and fishery NGOs. It seems to be just a matter of time or tuna welfare will be another of the criteria that have to be complied for serious certification standards like MSC. The emergence of the Aquaculture Stewardship Council (ASC) label for farmed fish will mean inevitably that we become more empathetic towards caged fish that swims around in boring circular nets. (The problem is that caged bluefin until now is mainly a wild caught fish caged for fattening purposes and for that reason not apt for ASC certification). The introduction of new killing techniques for bluefin tuna causing instant death are important steps forward, despite the fact that the issue at stake here is the quality of the meat and not tuna welfare. But matters are shifting. Important too: children toys do really educate empathy with fish. Ever since clownfish Nemo, soft toy shark, and even Spongebob has found its way to children's bedrooms. Now all we have to do is introduce the first cuddly tuna.

How to Kill a Tuna

When it comes to decent treatment of tuna—maintaining the quality of the meat goes hand in hand with a quick and pain-free death—the Japanese have introduced a number of important techniques. *Ikijime* is a technique that causes the fish's brain to shut down rapidly, resulting in instant brain death. The advantage of this is that the fish is no longer conscious and is spared a protracted death by asphyxiation. The technique basically consists of striking a spike into the tuna's brains, which are located roughly diagonally above the eyes. Then the central nerve is pulled out of the fish's spine to prevent spasms in its body. Grooves scored behind the pelvic fins meanwhile ensure that the fish bleeds dry.

All these techniques are aimed at sparing the fish as much stress as possible. A key point is preventing muscle spasms, that lead to lower quality of the meat. This is caused by waste products in the body, such as lactic acid, well-known from muscular pain in human bodies. Lactic acid in turn leads to bacterial changes that acidify muscular tissue and give the meat a brownish tint and bitter taste. Without muscular spasms, this does not happen. Bleeding the fish dry also contributes to improved quality. Although *Ikijime* was devised primarily for the fish's culinary quality, it is also the most humane way to kill them. No stress; instead a quick, pain-free death. In Japan, the technique is used on a wide scale, also with other, smaller fish species. Other countries are increasingly following suit. Amongst industrial fisheries, it is expected that in the near future *Ikijime* will be used on an ever wider scale.

Tiger of the Sea

25

Africa had its rhinos and elephants, Asia its tigers and Europe its bluefin tuna. Three unique, endangered species which, at the beginning of the century were pursued by merciless poachers bent on quick profit. With one important difference: the bluefin tuna was being fished from under our noses with little ado. A collective of research journalists estimated that 4 billion dollars' worth of illegally fished bluefin tuna disappeared from Mediterranean waters between 1998 and 2007 [254]. During the top years of the big tuna pillage, half of bluefin tuna arrived on the global market illegally. Yet the tide of attention for the fate of bluefin tuna was turning. It may not have reached the level of that for seals or whales, but tuna had suddenly advanced to the vanguard of concern for the decline of the marine environment.

This provisional peak of interest culminated in an extraordinary attempt to save the Atlantic tuna from extinction from an unusual corner: the principality of Monaco. Primarily known as a tax haven for affluent tennis players and an annual Formula I race in its streets, the tiny principality did not have a fishing industry of any significance. But head of state Prince Albert II of Monaco appeared to be impressed by the efforts of the WWF to protect bluefin tuna. With its monumental Oceanographic Institute of Monaco—set up by his grandfather Albert I and with Jacques Cousteau as its director for more than 30 years—Monaco boasted a serious tradition of protecting the Mediterranean Sea. Albert decided actively to mobilise his foundation for sustainability projects on behalf of the endangered bluefin tuna, working jointly with the WWF. Monaco was proclaimed the first 'bluefin tuna-free' nation in the world. In 2009 the fish was deleted from the menus of all restaurants. Albert went a step further: Monaco took the lead in the action to get bluefin tuna on the UN list of endangered species in 2010. Once on this list, included in the so-called Appendix no 1 of the Convention on International Trade in Endangered Species (CITES), international trade would be banned for 2 years. This meant that bluefin tuna would have a similar status to the rhino and tiger. Albert was quickly able to get the most important NGOs behind the initiative. But the project was given true impetus when countries like the Netherlands, the UK and Sweden rallied around the proposal and managed to persuade the EU to support the project.

© Springer Nature Switzerland AG 2019
S. Adolf, *Tuna Wars*, https://doi.org/10.1007/978-3-030-20641-3_25

The idea was simple: once all international trade in bluefin tuna had been banned, massive exploitation of expensive, fresh tuna would soon be over. An important part of the trade chain from the Mediterranean towards Japan would after all be blocked off, which in fact came close to a moratorium. It was clearly a creative application of thinking in terms of global value chains about the tuna trade. In March 2010, the proposal was due to be discussed during the CITES annual world conference in Doha, the capital of Qatar. During the months leading up to 'Doha' an expectant excitement grabbed hold of tuna conservationists. One European country after another stated that they were in favour of protecting bluefin tuna. Even France appeared to consider seriously making the fish a protected species. What had not succeeded in all those years under ICCAT management suddenly seemed to grow new wings.

But just like ICCAT, CITES with it's 183 member states (including the EU countries) was an unwieldy international behemoth. Moreover, big tuna fishing nations had decided to throw their full weight behind steering the CITES conference to their advantage. Japan sent an extra-heavy delegation to Doha which, plied with bags of money and clever diplomacy, went to work on the affiliated developing countries. This worked brilliantly. It couldn't quite be labelled bribery, but in exchange for aid projects in the countries concerned, Japan obtained increasing support to vote against the Monaco initiative.

The definitive blow to the proposal came in the end from an unexpected corner: Libya. It was the delegate from this country who called for shooting down the proposal on the eve of the vote. Following his Libyan leader Gaddafi, the man whose family was deeply involved in massive illegal tuna fisheries, he treated everyone to a fine spectacle. In a screaming voice, the delegate accused the 'imperial nations' of refusing to let Libya have its fair share of the lucrative bluefin tuna trade. It all worked out fine: when the proposal to get bluefin tuna included in CITES Appendix 1 was put to an early vote, it was rejected by a clear majority.

Those who had been following the battle around bluefin tuna all those years had got used to the cynicism of power. That it was Libya of all countries which put the kibosh on the CITES initiative may not have been surprising, but nonetheless could be called a new low. The comic performance of the delegation leader masked a serious business: the criminal agenda of bluefin tuna stock plundering and fraud. Here was a regime which had totally ignored the catch restrictions and had enriched itself to the tune of tens of millions of euros by illegal tuna fishing and smuggling practices. Together with his colleague dictator Ben Ali from Tunisia's Trabelsi clan, Gaddafi was known as the biggest tuna plunderer amongst the Mediterranean heads of state. Although affiliated to ICCAT, Libya and Tunisia disregarded licences and catch figures on a grand scale. Just as when, more than 2000 years previously Hannibal's Barca family managed to earn a fortune from bluefin tuna, so too did the Gaddafi's and Trabelsi's.

Gaddafi beat them all. In 2005, Libya's 'brotherly leader' unilaterally decided to expand the Libyan fishing water zone to 65 miles from its coast. The bluefin tuna's entire spawning area in the Gulf of Sirte, just to the north of the Gaddafi clan's base of Sirte, were included in Libya's EEZ. From then on, fleets from Italy, Spain and

France were only allowed to fish there if they paid a hefty licence fee. Gaddafi threatened to take the toughest action against any transgressors. The Libyan patrol vessel Maradona (named after the Argentinean footballer whom Gaddafi admired so much) would spring into action immediately. 'The Maradona is coming' became a much-heard threat on maritime radios. In order to avoid problems, many boats decided henceforth to fish under the Libyan flag, which in practice meant that they would form part of the dictator's fleet. Many tuna boats originally based in the southern French port of Sète ended up in Gaddafi's hands. As there was no overview of the Libyan fleet (Gaddafi did not allow ICCAT inspectors on board) these boats did not need to take much notice of the official fishing quotas. In his WWF tuna report (*The Plunder*, 2006) [165] Roberto Mielgo estimated that in 2005 alone, almost 3600 tonnes of illegal and unreported bluefin tuna had been fished out of Libyan waters, almost double the official ICCAT quota for Libya. Gaddafi's second son Saif al-Islam was the family member who personally kept the accounts, worth millions, for the illegal trade in tuna. His London School of Economics (LSE) PhD degree—awarded following Gaddafi International Charity and Development Foundation donations to the LSE—proved to be useful after all.

Tuna biologists regard the Gulf of Sirte as the most important place for bluefin tuna to gather and spawn. And lying in wait for the fish every year were the biggest poachers in the Mediterranean. It did not stop Libya's continued membership of ICCAT. This had to do with bigger geostrategic interests: Europe did not want to have any problems with the dictator, who threatened, at the drop of a hat, to help illegal immigrants from his country to reach the European continent. So a blind eye was turned to Libyan tuna fraud: less tuna was preferable to more immigrants. So illegal tuna from Libya was fattened on tuna farms off the Libyan coast, but also in Malta, Tunisia, Spain and Croatia. It was an unassailable stronghold in the battle against bluefin tuna fraud. Only when the dictators disappeared would light glimmer on the horizon again.

After the failure to declare it a protected species under CITES rules, ICCAT was again left as the last hope to save the bluefin tuna. After the deception of the 2008 meeting in Marrakech, prospects were not really developing in a favourable direction. Having previously cold-shouldered their own scientists, ICCAT seemed now to take notice of their recommendations. But that did not reveal a particularly cautious attitude. The scientists recommended that to achieve 'at least' 60% probability of recovering Mediterranean bluefin tuna stocks by 2022, the quota should be set 'between zero and the current 13,500 tonnes'. So, of course, an annual catch of 13,500 tonnes was decided upon—more than double the amount of 6000 tonnes that was considered reasonable by WWF.

To give an impression: 13,000 tonnes is 52,000 large tuna with an average weight of 250 kg each. Not much perhaps in absolute terms, but still considerably more than was being argued for by tuna scientists and NGOs. And then: was there any point to the ICCAT quotas if compliance and catch monitoring left so much room for fraud and deception? Even considering the fact that the data gathering was still in a deplorable state according to the scientists and compliance with the catch certificates was only 43% of the total catch? With the establishment of the European Agency in

Vigo—which also carried out surveillance on behalf of ICCAT—the first steps had been made towards improvement, that much was true. But plenty of juicy reports continued to circulate about illegal catches, registration fraud and other large-scale fishy business. A report by the European Agency showed that a group of Panamanian tugs working for a Spanish tuna farm had simply vanished off the chart after its satellite signal had been turned off. No one knew where they had got to or where they were tugging their load of tuna. Something similar happened to the Croatian fleet of 16 purse seine ships. The greatly expanded Turkish tuna fleet, for its part, was guilty of fiddling with catch certificates. Sanctions for violations, often a local affair, hardly got a look in. No investigation and prosecution of fraudulent officials took place. Estimates suggest that international trade in Mediterranean bluefin tuna was more than double what should be available according to the official fishing quotas.

Large-scale cheating was still going on at tuna farms, not least in Malta, where a complete closure of the tuna farms on the island was suggested by WWF. Obviously this idea did not receive much approval within ICCAT. Neither did a ban on large-scale industrial purse seine tuna fishing or the introduction of a fishing ban on spawning grounds, for example in the Libyan Gulf of Sirte and off the Spanish Balearic. All possible measures were besieged by the industrial fleets plying the Mediterranean waters from east to west. Countries such as Turkey, Libya or Tunisia alone had so many ships that their fishing capacity was many times greater than the allocated quotas. All these fishermen had to scout out the tuna to cover their costs and pay their backers, staying one step ahead of the others in order to catch the last bluefin tuna.

So no wonder that at the end of the first decade the atmosphere amongst the tuna experts was pretty gloomy when it came to Mediterranean bluefin. (Fig. 25.1) How did we come to this point? Marine biologist and WWF tuna specialist Sergi Tudela blamed human shortcomings, he told me when we met with a group of tuna addicts in 2008, during the Seafood Summit in Barcelona. The Seafood Summit is an event in which the global fishing industry, together with scientists, NGOs and policy makers, tackles the issue of fish caught sustainably. For Tudela, bluefin tuna had come to represent the inability to manage fish populations in the oceans sustainably. 'The fate of tuna stands for the fate of global fisheries policy,' he said. If we don't succeed in saving such an enormous fish that fires the imagination, what are the chances of all the other species which are nominated for extinction? The French veteran tuna scientist Alain Fonteneau cited biology as the biggest problem. All things considered, was it not the fault of bluefin tuna itself?, the professor joked. 'Tuna is essentially a very stupid fish. Over thousands of years it has spawned at exactly the same spots in the sea. And of course all the fishermen have to do is wait there to catch it.' American food anthropologist Ted Bestor, who had explored every corner of Tokyo's Tsukiji fish market and had written a book about it, called it an important sociological problem: the Japanese saw their sushi with bluefin tuna as an acquired right, a tradition which they would not easily allow others to take away. If you touched sushi, you touched Japan. The fact that in Japan ever fewer consumers could distinguish between the different kinds of tuna they are eating is irrelevant.

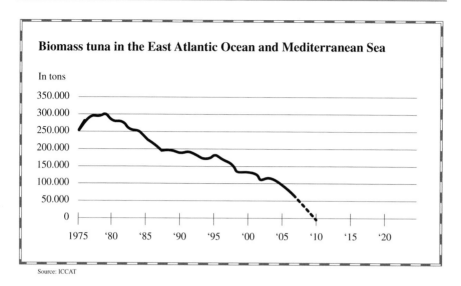

Fig. 25.1 Estimates of bluefin tuna stock at the start of the first decade of the twenty-first century. The end of tuna history in the Mediterranean? (Infograph Ramses Reijerman)

Bestor: 'It is not so much important what you eat, but what you think you eat.' Bluefin tuna was also lifestyle, entertainment and image, after all.

'I feel a little ashamed,' the professor Peter Makoto Miyake responded. There was an ironic undertone in his voice, but he clearly felt uneasy about this criticism about his fellow countrymen's behaviour. The large-scale global and borderless economy, that was the issue, he thought. The fisheries industry and the tuna farms. With their constant supply of tuna, they had created a market that, like a hungry beast, had inflated the demand for bluefin tuna tremendously. That in turn had produced an enormous overcapacity of European fishing fleets. If this was not vigorously axed, bluefin tuna's death warrant was signed. Paid for by European subsidies, far too many tuna ships had been built. And now there were these European subsidies to reduce the fleet. In the meantime, the same ships continued to fish regardless, often for the same tuna farms, with the same middlemen and the same consumers. The only thing that changed was the flag the ship was sailing under.

Thunnus thynnus found itself in no less than a perfect storm, it seemed. The book *The Perfect Storm* [152] described a real-life disaster involving the crew of the fishing boat Andrea Gail, based in the tuna port of Gloucester, when it hit upon a devastating storm off the coast of New England, caused by an accumulation of exceptional circumstances that led to a monstrous hurricane with extreme waves. Tuna's perfect storm was more of a maelstrom of economic upscaling, competing interests amounting to millions in a global economy, borderless corruption, indifference, political horse trading amongst policy makers and a technological tour de force used to fish dry a complete species with extreme efficiency. Just as in the book, a happy ending seemed almost certainly ruled out.

Part IV
The Tuna Challenge

The great challenge of the twenty-first century is to raise people everywhere to a decent standard of living while preserving as much of the rest of life as possible.—Edward O. Wilson

Tuna Deus ex Machina

26

In the era of the Phoenicians, when large-scale tuna fisheries started to conquer human civilisation, the Greeks were the great storytellers of the Mediterranean. One famous technique in the Greek theatre was the deus ex machina (ἀπὸ μηχανῆς θεός): when the tragedy was unfolding in a rather predictable and thus boring way, the plot could suddenly change totally unexpectedly, sometimes against all logic, bringing a new dynamic to the storyline and refreshing the attention of the audience. A god might enter the stage to change the course of the story by some divine intervention. In modern language it was a 'god from the machine'. (Aeschylus, writer of the *The Persians* where the invaders were slaughtered as if they were tuna, was famous for his use of deus ex machina. In the end it is all connected to tuna).

The year 2012 was the deus ex machina in our bluefin tuna tragedy. Something totally unexpected changed everything, as if the gods—in this case probably Poseidon—had intervened to avert a predictable tuna tragedy. Poseidon had somehow swung it so that there was still tuna swimming in the Atlantic and the Mediterranean, despite the fact that some of the statistics (Fig. 25.1) predicted ground zero for the populations. What's more, if you were to believe the *almadraba* fishermen's stories on the southern Spanish coast, substantially more rather than less tuna was entering their nets. A year later, orcas were back in their traditional bluefin tuna hunting grounds. The year after that, the *almadraba* nets filled up in record time with big, fat bluefin tuna. Visits to the *almadraba* fishermen during the fishing season in April and May gradually began to lose their funereal character. In 2017, Diego Crespo, director of the *almadraba* at Zahara de los Atunes and chairman of the Spanish *almadraba* organisation OPP, was pleased to tell me that 'the nets are full' in a record time. In the port of Barbate, the Reina Cristina, Tokyo Seafoods' freezer vessel, was ready and waiting to freeze its tuna bought from the *almadrabas*. By tradition, the surrounding cities were celebrating their annual tuna festivals, which seemed to improve year on year. In Zahara de los Atunes, on the Ruta del Atún, visitors were able to eat artfully crafted tuna tapas for a whole week. You could have a picture taken of yourself with a dancing tuna, there was tuna chinaware, tuna art

© Springer Nature Switzerland AG 2019
S. Adolf, *Tuna Wars*, https://doi.org/10.1007/978-3-030-20641-3_26

and tuna jerseys. From a forgotten and almost extinct fish, bluefin tuna had mushroomed into a local tourist attraction, not only giving work to the *almadraba* fishermen but also to thousands of chefs, waiters, hotel staff, taxi drivers and owners of rental accommodation alike, in this region plagued by unemployment.

Crespo showed me his brand new tugboat in the port from which a man-sized bluefin tuna was being hoisted. In the cold store, a team of 50 butchers were standing by with chainsaws, hooks and enormous, razor-sharp knives. The head and tail were sawn off, the meat cut from the spine into four large fillets, and the best chunks of tuna were stripped off the fish's head. Within a few minutes the tuna had been completely dissected and portioned up, ready to be frozen in the cold store on stainless steel shelves.

Changes had been introduced, Crespo mentioned. Three of the four *almadrabas* had joined forces in the Gadira trading company, which had taken the sale and marketing of tuna and tuna products into their own hands. The only *almadraba* not to participate was that of the Fuentes Group. The fishermen had thus penetrated deeper into the tuna trade chain as processors and dealers. An ever greater part was destined for the European market instead of Japan. This did not mean that the Japanese had lost their influence: on the contrary, their inspectors were still there sampling their wares in the port's cold store. Following the Japanese example, the treatment of tuna had only become more sophisticated. 'Look: *Ikijime*,' Crespo said and pointed out the latest Japanese killing technique they had adopted, whereby the fish was rendered braindead as quickly as possible by knocking out the central brains with a spike. This is inserted into the spine via the tuna's head to remove the central nerves and counteract the dead tuna's spasms. Following this almost surgical operation, the tuna, stress-free and selected by weight, was placed on ice for transport to the port.

To prevent the release of stress hormones, which turns the meat brown and affects its taste, the *almadraba* had switched permanently to methods causing the tuna as little anxiety or harm as possible. The bloody theatre of lifting the net horizontally between the boats, the so-called *levantá*, whereby dozens of giant tuna splash up fountains of foaming seawater with their tails, the showers of fish, were definitively past history. These days each individual tuna was killed in the nets by a diver. Before this, the fish had been neatly selected on a weight of 160–180 kg; those were the tuna with the ideal amount of fat, Crespo explained. The diver shot them using a harpoon with a small amount of explosives which led to instant death. Then very quick lateral cuts were made so that the tuna could bleed dry. It was certainly less spectacular to watch, but that was the price of progress—both for the tuna and for the consumer.

Most important: there was plenty of tuna, Crespo said. He hoped that ICCAT, on the basis of its latest scientific committee reports, would increase the quota to 30,000 tonnes during its annual meeting in Marrakech later that year. This was a significant leap: 5 years previously, at the nadir of the tuna crisis the quota for eastern bluefin tuna had been reduced to less than 13,000 tonnes. This would make the amount of tuna that could be caught two and a half times as big. The western stocks, which were being fished by Canada and the United States, lagged far behind this. From 1750 tonnes at the worst point, the quotas had been raised to 2000 tonnes. Poor

Wicked Tuna: the situation on the American east coast remained dire. But the bluefin tuna in the east of the Atlantic and the Mediterranean seemed to have risen like a phoenix from its ashes. How could this be?

Following my multi-year quest for bluefin tuna, one thing was certain: the more I knew about this extraordinary fish, the more I was aware of what I did not know. It's a cliché, but even clichés are true sometimes. And I was not the only person to have woken up to that fact. Historically, there had been other instances of alarming plunges in bluefin tuna catches. The dukes of Medina Sidonia worried about it centuries ago. But the tuna population always seemed to recover, even now, after the perfect storm of industrial overfishing, shameless plundering and poaching and a world market that liked to have bluefin tuna in its sushi. What was the cause? It obviously had to do with how much was being fished, but also with the biological characteristics of the fish, and with the situation in its environment, with the amount of food available, perhaps even with climate change and the changes in temperature and sea currents. Since Von Humboldt, we have also known that we form part of an ecological system that could be an intricate network of various influences. In this network, besides biology and physics, entirely different issues such as the economy, politics and social developments played a part. And above all; power. For thousands of years the giant tuna had been closely bound up in power and geopolitical interests. At the beginning of the twenty-first century the situation was no different.

So it could not be a coincidence that there were links between the Arab Spring and the bluefin tuna spring. The uprisings in North African countries came out of the blue, something of a deus exmachina, in fact. In December 2010, a desperate street vendor called Mohamed Bouazizi set fire to himself in the backwater of Sidi Bouzid in Tunisia, after the police had confiscated the wares on his cart. Bouazizi did not have any money to bribe the officials to leave him alone. Self-immolation had happened before, but this time it was the spark to set alight growing North African dissatisfaction with its authoritarian, dictatorial leaders. The popular anger around the death of Bouazizi, who passed away on 4 January 2011, got totally out of hand.

To everyone's amazement, President Ben Ali, who had ruled the country for 23 years as a cruel and corrupt dictator, had to flee the country with his family 10 days later. Bouazizi became a posthumous hero, not only in his own country, but also in neighbouring ones. All over North Africa and the Arab world, numerous men set fire to themselves in protest against their poverty-stricken existences and against corrupt power. In Egypt, President Mubarak resigned amidst great unrest and massive uprisings in Cairo. In Libya, civil war broke out between several tribes. Muammar Gaddafi, since 1969 the uncontested leader of the country, had to flee Tripoli. In October, virtually the entire world witnessed online the mobile phone footage of the lynching of the dictator, who had tried in vain to hide in a sewer pipe after a surprise attack on his convoy. Gaddafi was killed in Sirte, the bastion of his family clan, with its port, which had been the centre of the local tuna catch for centuries, if not millennia.

The uprisings did not only finish off the harsh regimes that had held sway over the region for decades. The popular revolt proved to make quick work of precisely the biggest and most corrupt tuna mafias in North Africa. In Tunisia, the tuna trade had been controlled by Mourad Trabelsi, dictator Ben Ali's brother-in-law. Trabelsi, who had personally been in charge of the tuna fisheries, was arrested and given a 10-year prison sentence for widespread fraud and corruption in the fishing industry. That meant the end of his tuna empire. In Libya, the death of Muammar Gaddafi ushered in a similar end to the tuna mafia. His son Saif el-Islam, who managed the tuna affairs, survived the battle, but in the end was found in the desert and arrested. The Libyan tuna fleet was grounded in the port of Sète. At the request of EU Fisheries Commissioner Maria Damanaki, ICCAT issued a fishing ban. Libya had always been a bottomless pit of tuna fraud, but now that the Libyan authorities no longer appeared capable of submitting a plan for the coming season, the fishing waters were officially declared a no-go area. Tunisia crippled, Libya at war, NATO naval ships patrolling Libyan waters: tuna fishing amidst its crucial spawning and breeding ground largely came to a halt. Finally a more powerful de facto moratorium than had ever been conceivable took effect, complete with NATO frigates as guards. Nonetheless, mysteries and question marks remained around the illegal bluefin tuna catch in Libya. A few months after Saif el-Islam's arrest and the death of his father, the BBC [32] published a leaked map of the Gulf of Sirte showing the VMS signals left on the tuna fishing boats' monitor system during the catch season. The Maradona was nowhere to be seen.

This was a mystery, because the Libyan fishing boats officially had not left the port of Sète, and the European Fisheries Control Agency had hermetically sealed the Libyan waters to every tuna vessel that wanted to leave it, so it said. Was it a secret military operation that had been registered unintentionally? Or had the pursuit of quick profit won out in the end, and had some ships, despite the war, wormed their way into the Libyan waters after all? Whatever the case may be, it remained a mystery: the chart with satellite signals ended up in a drawer. The catch had been illegal, but did not amount to very much compared to the usual plundering which had taken place for years.

In any event, the Arab Spring had turned into a bluefin tuna spring. We will never know to what degree. But the disappearance of two zones which had been home to a significant share of IUU tuna undeniably contributed to its recovery. The quantity of tuna fished largely out of Libyan waters as IUU alone, according to the WWF report, amounted to more than 10,000 tonnes in 2005. This was a substantial quantity when measured against the total catch in the Mediterranean. If anything had fared well as a result of Gaddafi's death, it was tuna.

Of course there were further powers benefitting bluefin tuna. The sudden recovery of stocks coincided with a surge in effectiveness from both the EU fisheries policies and the ICCAT measures. After years of political wrangling and extremely complex negotiations, the EU Common Fisheries Policy (CFP) began to bear fruit with respect to more sustainable fishing. The politics of quota distribution and above all its enforcement had gradually acquired flesh and bones. This was especially the case for internal European fishing. Externally, the checks in the battle against IUU fishing, through controls of catch certificates, for instance, had acquired more substance. The Mediterranean, with its complicated fishing grounds and interests, had always been a headache dossier for the EU. But the catch certificate with information about bluefin tuna, validated by ICCAT, became mandatory from 2010 for all import and export as part of the EU anti-IUU measures. The European Fisheries Control Agency in Vigo in the meantime had been given more muscle than had been expected. What's more, the EFCA operated with a mandate on behalf of ICCAT, which elevated its efficacy to a higher level. Nonetheless, reports about fraud and fiddling with bluefin tuna catches continued to be leaked. The mystery of the boats spotted in the Gulf of Sirte during the Libyan civil war was just one example of this. But the jamboree of deception which held sway from the end of the 1990s, was no longer in evidence. Moreover, ICCAT, and with it the EU and member states such as Spain, France and Italy, after years of proverbial negotiating failure (ICCAT as the International Conspiracy to Catch All Tuna) finally made a U-turn towards a more decisive, effective policy. The effects of this should not be underestimated: ICCAT not only turned out to be capable of saving bluefin tuna stocks from collapse, but the organisation once considered a laughing stock and example of how not to do it across the globe, had promoted itself to the most resolute of Regional Fisheries Management Organisations (RFMO) and had become an example to ailing tuna organisations elsewhere around the oceans. That was quite something for an organisation which, after the Marrakech fiasco in 2008 had been described as 'economically and biologically bankrupt' as regards sustainable fisheries management and an 'international disgrace'. The shift came with a radical rescue plan for eastern Atlantic bluefin tuna which was adopted at the end of 2012. This included reducing the Total Allowable Catch (TAC, the total quota) to 13,400 tonnes per year from 2014, in keeping with scientific advice. Perhaps even more important: a minimum catch weight of 30 kg had been set earlier, which corresponds to the size of a mature tuna. This set a basic guarantee for the amount of spawning tuna that could reproduce. Strict enforcement and control formed the tail end of the approach. The NGOs, which had been tearing their hair out for years at ICCAT's failures, were barely able to suppress their amazement. Following decades of overfishing and mismanagement, the package of measures gave bluefin tuna 'a fighting chance', according to Susan Lieberman, director of international policy of the Pew Charitable Trusts.

Responsible management of an international borderless migratory fish such as bluefin tuna proved to be possible. And amazingly enough it was the case with bluefin tuna. If you had been following tuna you knew that there had been a power shift in sustainability policy which, like a deeper undercurrent, had effected change. It had been the 'fishing-industrial complex'—the powerful fishing industry with its long-distance fleets—which, through their governments and the European Union, had left the greatest mark on the policy. In the case of bluefin tuna, it was primarily large-scale purse seine fishing fleets and fattening farms which had the final say in the global value chains. In these chains, large trading firms, especially from Japan, ruled the roost, controlling marketing and production financing. But along the way, other stakeholders such as NGOs and environmental groups acquired ever more weight in thinking and decision-making. Publicity had given tuna a face amongst consumers and voting citizens. Sometimes literally, as in the WWF campaign with the sketched image of a panda mask strapped onto the head of a swimming tuna. Also other NGOs, from Greenpeace and Sea Shepherd to Oceana and Pew, had likewise put bluefin tuna on the map. The attempt to get the bluefin tuna on the CITES list had failed, but it had achieved part of its aim. Suddenly, tuna was included in the category of endangered animals like the rhino, gorilla or tiger. All over the world, tuna was beginning to take a place in the collective sustainability consciousness. The culmination in this respect was the offical ratification of 'World Tuna Day' during a UN session in New York in 2017, due to be celebrated annually on 2 May. The initiative originally arose from a joint action between the tuna-rich island state of Papua New Guinea in the Pacific and a sustainability-conscious Dutch tuna dealer, as representatives of a sustainable tuna production chain (see Flipper Wars).

More and more people realised that something had to happen regarding bluefin tuna, and perhaps all tuna. This struck back in governmental and political decision-making: more politicians and administrators had to take into consideration that their voters were consumers who valued sustainable tuna fishing. Analogous to this were effects in the market. A shift occurred in the balance of power in this chain. In the nascent canning industry, just as with the Phoenicians and later with the Italians and Americans, this power had been largely in the hands of dealers and brands. Now, power was shifting increasingly to the end consumer and to the retail market in which the tuna was bought. The value chain of fishermen, dealers and retailers was increasingly reminded of its own responsibility towards the sustainability of its products. If you failed, you might incur reputational damage. This could translate to lower company results, which is something companies hate. Consumers in Japan were at best lukewarm about a sustainable catch. But elsewhere in the global markets, things were rumbling. In the UK, the big retailers feared Greenpeace protest actions on their doorsteps against canned tuna which was not particularly sustainable. On the continent, supermarket chains in northern Europe did not want to take any risks and started their own policy of sustainable purchasing. In the United States, many big retailers committed themselves to similar projects to ensure that their tuna and other fish products were sustainable within a few years. It turned out that such promises were not always so easy to fulfil, but the trend had been set. In a demand-driven market for canned tuna and tuna steaks, the consumer—or the NGO which used consumers to exert pressure—was able to exercise considerable power over the rest of the chain.

ICCAT and the EU proved that international cooperation within regional organisations was able to come up with something that was beginning to look like a sustainable management plan for fisheries based on scientific advice. A plan, moreover, that in the short space of a few years had contributed to the recovery of fish stocks which appeared to have been irreversibly in free fall. It was a practical example of an approach Oceana's Andy Sharpless and Suzannah Evans had outlined in Sharpless' *The Perfect Protein*. The principle: working together with the countries that have the majority of global catches within their economic waters, a management plan can be devised that can lead sustainably to far higher production. Substance-wise, the measures were really a question of common sense. If, years after ICCAT introduced the management plan, you spoke to the veterans amongst the tuna experts, you would notice that there was widespread agreement about one particular reason behind the success of the measures. 'It is very probable that quota manage-ment for the bluefin tuna has been an important contribution to the improvement of the stock,' Japanese tuna researcher Makoto Peter Miyake said. 'And of course also the minimum size limit of 30 kg has contributed to the comeback of the bluefin tuna,' his French colleague Alain Fonteneau added. Of major importance in increasing the catch limit from 10 to 30 kg—from 4-year-old tuna, just over the brink of sexual maturity—is that by this method the annual cohorts of reproductive tuna had been dramatically reinforced. This measure came into force in 2007 following consent of the European Commission. From Spain, José Cort confirms the effect of the mea-sure. According to counts he carried out amongst the *almadraba* fishermen in the Strait of Gibraltar and bait boat fishing in the Bay of Biscay, it is estimated that this contributed to some 840,000 additional young sexually mature bluefin tuna every year. This means that during the first 10 years of the twenty-first century, many millions of bluefin tuna were able to reproduce.

All these measures would have been worthless if it had not been for better compliance and surveillance. Significant progress was made here with the European Fisheries Control Agency and the catch certificates. Traceability of the fish, the idea behind the certificates, was progressing far from perfectly. It appeared that every member state had its own methods of surveillance and its own sanctions on the fish trade, as environmental lawyer group Client Earth discovered [75]. Almost 90% of the fleets of Spain, France, Italy and the UK proved to be monitored insufficiently, especially when it came to catch capacity as recorded in the boat logs. An NGO coalition of WWF, Oceana, Pew and the Environmental Justice Foundation (EJF) [80] for their part ascertained that the paper-based system of catch certification left rather a lot of room for tinkering. Digital traceability of certificates would not be available from the EU until 2019, replacing forgeable stamped paperwork. Bluefin tuna capture was one of the EU's priorities as far as controls and traceability was concerned. But this too did not follow a very smooth course: when bluefin tuna was landed, port inspectors in Italy and France did not use the extensive checklist, for the simple reason that there was no firm requirement to do so.

It was far from ideal, therefore. If you were going around the Spanish south coast during the *almadraba* season in April and May, you would notice that the era of large-scale tampering carried on apace. But not when the fish was caught or landed in the port of Barbate, the landing port for the *almadrabas*. There, everything was monitored properly by the inspectors, with scientists from the Spanish Oceanic Institute standing by as well at the quayside to inspect the hauled tuna. It was in other parts of the production chain that shady dealings continued. If you thought you were being served *almadraba* bluefin tuna in restaurants, you might well be disappointed. A restaurant owner: 'If you order *almadraba* bluefin tuna at El Campero, you know this is what you will get on your plate. But for the others it is a simple calculation: look at the amount of bluefin tuna caught in the *almadraba* and compare this to the quantities you find on restaurant menus. It's simply not possible!' The impression was that much of the bluefin tuna came from fattening farms. Or maybe it was not even bluefin tuna at all, but yellowfin. Most people were not able to taste the difference anyway. And who was going to ask for a catch certificate while sitting down to a meal in a restaurant?

Despite all the improvements, illegal trade in tuna carried on. A restaurant owner in southern Spain: 'What were those Italian refrigerator lorries doing at the tuna processing companies here in the village? Illegal tuna trade!' A wholesaler for restaurants in northern Europe: 'What did you expect? If I want, I can get bluefin tuna without papers at the drop of a hat. They simply ask you: do you want it with or without papers? That stuff comes from Greece. Or Malta. Or Turkey. The further east you go in the Mediterranean, the easier it is.' Inspection was still falling short, in other words: too few controls in the ports and during transit across Europe. A significant development was an outbreak in 2018 of histamine in fresh tuna along the coast of Almería, in the middle of the bluefin fishing season. Histamine, an organic bacterially-generated toxin which can develop when tuna has not been cooled sufficiently, had nothing to do with the species of tuna being served, but everything to do with inspection of the fish, which evidently had not taken place. That same year a new outburst of tuna scandals hit the European market. The *Malta Independent on line* reported that 5000 kilo of illegal bluefin was smuggled into Malta to be exported to other EU markets like Germany, The Netherlands, UK and France [239]. That is 5000 kilo a week. The illegal bluefin tuna racket was later, rather cautiously estimated at more than 20 million euro a year. It proved that Malta was still something of a pirate nest, with a high concentration of experienced black tuna farms. Hell broke out later that year, just a few weeks before the yearly ICCAT meeting in Croatia. The united European police force Europol started 'Operation Tarantelo', an anti-mob like operation against an international bluefin smuggling ring with the arrest of 79 people in Spain [88, 240, 261]. Over 80,000 illicit bluefin tuna was seized in addition to half a million euro in cash and seven luxury cars. The size of the illegal market was estimated a yearly 2.5 million kilo of bluefin tuna, twice the size of the official market. The tuna was caught in Italian and Maltese waters and smuggled onwards, where it landed without any documentation or custom control.

So the system was, to put it mildly, still far from watertight. All things considered, it would have been a miracle if the IUU bluefin tuna catches in the Mediterranean had effectively stopped. But the total madness of the uncontrolled tuna rush, with its massive, large-scale fraud, replete with authoritarian regimes and organised crime groups, had appeared to be over. It proved to be otherwise. But looking at it from the bright side: at least the tuna mob was now caught in the act. It was a hopeful start of a sustainable management policy that was beginning to bear fruit.

Unfortunately, the fraud had an overwhelming dark site. Not only the continued existence of a flourishing black and criminal market as such was a depressing fact. It also pointed out? another monster problem that came up from the deep. With an illegal fishery still that big, the data gathering of the scientists—that was supposed to be the main pillar of the governance and the bluefin fisheries management policies—obviously had some very serious flaws. It seemed to underline, for Atlantic bluefin at least, the Daniel Pauly findings of fundamental higher catches in fisheries than reported [189].

That was a serious problem if you came to think of it in management terms. In particular considering the bluefin tuna: the lack of trustworthy data might well have been one of the main reasons of the remarkable recovery of the species after the crisis in the second half of the first decade of the century. That was at least the opinion of senior tuna scientist like Alain Fonteneau, who had dedicated a life studying the species. In the period of deep bluefin crisis 2005/2008 the biomass of adult bluefin was in fact widely underestimated, mainly because of very poor data used in the models then, Fonteneau explained to me. The diagnosis was that the bluefin stock was vanishing, when all the data we have in hindsight show that the stock was deeply reduced, but much less than estimated by scientists. Errors that were seldom discussed between the tuna scientists of ICCAT. In fact, the very fast recovery of the bluefin tuna can be explained by these technical errors, Fonteneau told me. 'All the today stock assessment analysis are estimating that the biomass in the 2000–2008 period was at least two or three times more important. In fact they were totally wrong, and widely underestimating the BFT biomass.'

Fortunately, the management policy was supported by broader, ongoing scientific research into bluefin tuna. We do know more and more about bluefin tuna now. Over more than a decade, the countries of the European Union have invested tens of millions of euros in scientific investigations in the species. By tagging the tuna, the breeding zones and migration patterns are charted, which can further support future policy. Growth, movements, stock structure, catch at size and catch at age, spawning: never before did we know that much about the bluefin. Studies revealed that a larger group of bluefin tuna than had been assumed thus far not only spawns in the Mediterranean, but also stays there, using it as its more or less fixed habitat. The idea going back thousands of years that most tuna only come to the Mediterranean to spawn, spending the rest of the year eating to their heart's content in the Atlantic, seemed to fit together in a more nuanced and complex way. And although bluefin tuna from off the American coast and their family members on the European side meet each other in the ocean, few tuna appear to swim from the western side to the eastern side to spawn. The separation between the two families could also explain why the American stocks are barely able to recover and remain on the small side. Later sexual maturity and other food situations could also play a role.

So there we are. Bluefin tuna, the Tiger of the Oceans, had been saved from extinction, but it had been a messy and confusing affair, a multi-layered wicked problem, as always with this wicked tuna. It was too early to raise the flag permanently, so much was clear, because the power games and the large financial interests around tuna were complex and fickle. Fraud and illegal fisheries seemed to be endemic. There was still a great deal that could be improved around the protection of the fish: take for example the idea of closing the tuna breeding grounds to fishing within Marine Protected Areas, as had been mooted by WWF and Greenpeace around the Balearics and off Libya. A glance at the ICCAT boat register alone showed that the capacity of the purse seiner fleet in the Mediterranean was still many times bigger than the bluefin tuna stocks justified. This exerted great economic pressure on the catch. Now that more tuna was available, many new fishermen saw favourable momentum for entering the market, as in the case of the Canaries pole and line and Turkish purse seiner fishermen. The latter did not have a particularly good reputation where compliance with the ICCAT rules were concerned. The fishermen, who had made great sacrifices even in *almadraba* fishing because of the drastic catch restrictions to save tuna from collapse, observed with disquiet that the improvement of the stocks might pass them by to some extent. Then there was the major issue of the value chain: the eventual demand for more sustainably caught fish may help save tuna from extinction, but there was still a great deal of doubt as to whether the tuna which ultimately landed on the consumer's plate was, according to available information, sustainably caught. The transparency of the different stages within the trade, from boat to port or farm, from processing in the fish factory, the cold store, and transport, up to wholesalers and the restaurants, was a complex series of transactions in which the tuna's traceability was still far from guaranteed.

Nevertheless, bluefin tuna was swimming around again in numbers nobody had expected just a decade earlier. What mattered now was to maintain the stock levels in a sustainable way. Bluefin tuna was already an icon because of its fisheries culture dating back thousands of years. In a similar way, this fish could develop into a practical and valuable benchmark of management that brought a sustainable catch within reach, not only for bluefin, but also for other tuna species. Bluefin tuna was but a small portion of the tuna that was eaten worldwide, after all. During the second half of the twentieth century, millions of tonnes of other tuna species had joined the global consumption pattern every year. The challenge for the new century now was even bigger: sustainability for all tuna. In the next chapter we dive into all those other tunas that you have probably found on your plate without knowing what species they were, or whether you were supposed to feel guilty eating them. And we have a closer look at how the global management of tuna works and the related conflicts of interest that are sometimes referred to as tuna wars.

Twenty-first Century Tuna Wars

27

From the start of the second millennium BCE, finding tuna was always a journey without borders. After the Phoenicians introduced the tuna industry from the eastern Mediterranean, it had gone on to conquer Western civilisation. From the twentieth century on, the tuna value chain had spread out globally. If you wanted to know more about the tuna on your plate at the start of the third millennium you had to travel to the Pacific and Asia. Meet Huynh Phi Minh [44, 203]. This 50 year old fisherman from the Phuoc Dong commune in the Vietnamese Nha Trang City witnessed something in the spring of 2017 which he will remember for the rest of his life. Huynh Phi Minh was sailing the South China Seas when the crew of his small vessel hooked a bluefin tuna that was so big and strong that his battle with the animal would qualify for a bout in *Wicked Tuna*. It took eight hours before they were able to hoist the creature on board. An epic fight on the edge of the Western Pacific, in the middle of the territorial waters of the Paracel Islands. The Pacific bluefin tuna weighed exactly 307 kg. This was worthy of mention for several reasons. Never before had a Vietnamese fisherman caught such a big tuna. Nominated by the Vietnam Association of Tuna, Huynh Phi Minh received an official certificate that stated this remarkable record. Not only that, it was a Pacific bluefin, the species that was overfished so heavily that it was extremely hard to catch. Tuna scientists even believed this fish was on the verge of being wiped out in the Pacific Ocean. But for the moment that wasn't Huynh Phi Minh's greatest concern. He calculated that he was well over 55 million Vietnamese Dong (2200 euros) richer with his haul on board. That was a lot of money in Vietnam; the fisherman could be happy. Although it was not very likely that the Japanese trader, who bought the fish as soon as it was landed in the port of Nha Trang, did pay the highest price for this catch. An eight-hour battle is likely to have produced a quantity of stress hormones that would have greatly impaired the quality of the tuna's flesh. Not a fish for the premium sushi and sashimi on the Tokyo market maybe, but a good catch for a Vietnamese fisherman.

In 2017 the Pacific bluefin that Huynh Phi Minh caught, found itself in a similar downward spiral to that of its bluefin brother in the eastern Atlantic Ocean and the

S. Adolf, *Tuna Wars*, https://doi.org/10.1007/978-3-030-20641-3_27

Mediterranean 10 years previously. Decades of persistent overfishing, especially by Japanese fishing fleets, had decimated bluefin tuna in the Pacific. Sustainable fisheries management by the fleets and coastal countries involved was very slow in getting off the ground. No wonder, in an area that ranked high as an intersection of economic, political and strategic conflicts between superpowers. Armed intervention and violence were always ready to surface. The Pacific bluefin tuna was stuck in a similar situation to the African elephant, the lion, the tiger and the rhino: endangered species in conflict areas that are difficult to control.

If 3500 years of tuna history makes one thing clear, it is that where tuna is captured, human conflict and issues around governance are never far away. (Part of the explanation is that today canned tuna demand often follows civil strife as an affordable shelf stable form of protein). This law prevailed, independent of time and place. It is probably related to the major interests behind the availability of commodities, in this case healthy and relatively cheap protein from the sea that could be shelved. Conserved food is power? and closely interrelated with conflict. What had changed was the nature of the conflict. During the twenty-first century it was primarily the scale and complex interconnectedness of the power between the actors in tuna's global value chain—the coastal states and flag states, fishermen, major economic conglomerates financing the fisheries, traders, big tuna brands, retailers and consumers. Now the entire chain for its part could not escape getting involved in geostrategic developments in the seas frequented by tuna. Tensions and political unrest could lead to a deterioration in the functioning of the state and the market, and made joint action in the interests of tuna fisheries harder to effectuate. Policy and governance of sustainable fishing of the different tuna stocks globally thus became an increasingly complicated process, a battle involving ever more parties under ever more varying circumstances (Fig. 27.1). Thus, the battle for access and production of a healthy seafood protein fundamental for food security became closely interrelated with commercial and geostrategic interests. During the twenty-first century 'wicked tuna' as it was coined in the popular television series had become a 'wicked problem' [205], the fish version of a Russian doll, in which a new problem would present itself hidden within each apparent solution.

This was a battle, moreover, which revolved around a new concept: sustainability. As the first person to put sustainable capture on the agenda in the eighteenth century, the Spanish monk Martín Sarmiento would never have been able to divine such a thing. Three centuries later, the governing of sustainable tuna fisheries had advanced from being a relatively unimportant, marginal phenomenon to a mainstream issue. Sustainability grew rapidly into a policy theme in public and private regulation. This was the case with fish in general, as explains Simon Bush, professor and Chair of the Environmental Policy Group at the Dutch Wageningen University whom I got to know through my research work. He happened to be as obsessed with sustainable fisheries as I became. He explained to me that it was not until the turn of the millennium that social scientists like himself started to notice that something interesting was happening on a global scale around the so-called 'governance' of fisheries and aquaculture [180]. All kind of new governance arrangements, with a new group of stakeholders and various partnerships, started to pop up with the

Fig. 27.1   Challenges of the twenty-first century: the Wave of Sustainable Tuna Fishing. (Woodblock print by Hokusai by courtesy of the Rijksmuseum Amsterdam)

purpose of improving the fisheries in a more sustainable way. And, as we will see, this was in particular the case with tuna, with its special characteristics of a global fishery and a global value chain that made it the figurehead in these developments. Not only did sustainability become a marketing tool that began to steer global trade in tuna. Sustainable fisheries, which were supposed to guarantee the future of tuna, led to a new power dynamic in the supply chain and all associated economic, political and social interests. In concrete terms, within a few decades the sustainability of captured fish had become a factor in the choice around production and purchase of tuna. Sustainability penetrated into daily power relations between states and powerful companies, and between relatively new players in the world of tuna, such as environmental organisations and other NGOs representing a range of interests in fisheries and fish consumption. These were the new stakeholders, as the industry jargon would have it, a jargon that was spoken in a colourful, global array of tuna conferences, certification assemblies, meetings between regional tuna organisations, scientists and governments. These gatherings resembled a global sect meeting in its temples dedicated to tuna, where high priests cast spells in an attempt to balance a world-encompassing sustainability policy for tuna fishing and growing tuna consumption. This sustainable tuna cult encompassed a select circle of initiates, largely hidden from the wider public's interest.

The 'wicked problem' of tuna capture had to do with globalisation and the diversification which the tuna trade had undergone during the last century of its long and lively history. The international battle around tuna had once started in the long distant past in the Mediterranean, where powerful families that controlled fishing, processing and trade became embroiled in conflicts such as the Punic Wars. During the Middle Ages, new rulers emerged on the Spanish south coast, who financed themselves with tuna monopolies. Large scale bluefin tuna fisheries lost much of its importance during the eighteenth century as a result of dwindling catches, but the advent of canned tuna engendered a new development in the tuna industry. Power shifted from fishermen to traders and to a new processing industry that conquered the market. Tuna itself also changed: fisheries discovered different species within the tuna family which were suitable for the canning industry. Capture was spread out over the oceans. Tuna became a borderless, global industry. After WWII, this expansion took another exponential flight, especially after the introduction of new purse seine gears and freezing techniques giving distant water fleets a virtually unlimited range [31]. The Japanese fleet, once admitted to rebuild massively their fleet of big longliners, conquered the Pacific, the Indian Ocean and the Atlantic Ocean in the mid and late fifties. They were followed by the Taiwanese and Korean longliners in the sixties. On the other site of the earth, from the mid-fifties on Spanish and French fishermen ventured onwards to the West African coasts with their pole and line fisheries, later converted into purse seiners. The French purse seiners explored the Indian Ocean first in 1981, followed by the Spanish 2 years later. A massive invasion of both European fleets followed in 1984. The United States and Japan's tuna boats pillaged the Pacific Ocean in search of new tuna for a growing canned fish market and the Japanese demand for *katsuobushi*. The American can pole and line fishermen had come across immense tuna stocks when their boats had been deployed in the Pacific by the navy during the war. The fishing industry, based around San Diego first started to fish in the Eastern Pacific, but moved away almost completely in the 1980's to the Western and Central parts of the Pacific.

The boom in sushi and sashimi market gave tuna a further global impulse from the eighties on. The market was widened from the canning industry to fresh and frozen products allowing longer distances and global distribution. Initially sushi had been the poor man's sashimi and rice and vinegar used to mask the spoilage. It became increasingly popular in Japan and in global demand today. The Japanese fishing fleet switched to providing high-value tuna intended for raw consumption, canned trade and tuna flakes. Improved freezing techniques at ultra-low temperatures broke down the last barriers to commercial interests, preserving tuna over long periods allowing Tuna, itself a migrant across all the oceans, was now turning into an international fish that had truly conquered the world. It became a commodity that supplied a worldwide industry as a raw material, involving economic and geostrategic interests you would never guess from the simple can on the supermarket shelf. No fish had such long and complex global value chains from net to plate, measured in thousands of kilometres and the interests of an army of stakeholders. A tuna could have been caught in the Solomon Islands Exclusive Economic Zone (EEZ) in the Western Pacific by a Korean fishing boat, transferred to a carrier sailing under a Panamanian flag for an Italian-American trader, canned by a Thai processor near Bangkok, then transported in a container to the port of Rotterdam, where it would end up in a Swiss supermarket via a German broker. Tuna had been a biological and historical icon, but it now had become the fish that represented the borderless global economy like no other.

The absence of borders and the question as to who really owned the fish, formed an important new layer in the 'wicked tuna problem' that demanded an answer. Since the beginning of the seventeenth century the legal concept of high seas had not changed. The basic principle of the free sea had been formulated by the Dutch jurist and scientist Hugo de Groot, more widely known by his Latin name Grotius. In 1609, Grotius, a great Renaissance scholar, wrote his book *Mare Liberum, The Freedom of the Seas* [120]. It was primarily meant as a legal tool serving the Dutch Republic, which in the seventeenth century dominated the oceans with its large merchant fleet, frequently seizing adversaries' ships with their merchandise in the process. Grotius formulated a starting premise for trade: the ocean was free for all countries, and no one could invoke any rights as to what could be found in it. This arrangement no longer sufficed in an era when fisheries were extracting raw materials from the world's oceans. A new principle was long overdue, to stave off conflict on the increasingly crowded waters. Thus in 1982 the UN Convention on the Law of the Sea (UNCLOS) was drafted [183], coming into force 12 years later. Perhaps the most important adjustment was the global introduction of the 200-mile (370 km) zone measured from the coast in which nations were able to claim their Exclusive Economic Zone (EEZ). Coastal states in Latin America already introduced the 200-miles zone in the fifties, the Pacific Island states implemented EEZ's at the start of the eighties [129]. The 200-miles zone was considerably larger than the 12-miles national waters, where the states could exercise absolute sovereignty. The coastal nation was able to assert its EEZ property rights above and below water, including what swam through this water and was fished out if it.

Everyone agreed that a new approach to the organisation of the oceans was vital. Nonetheless, the introduction of the EEZ created an entirely new battle, with overlapping claims on several parts of the sea. All of a sudden, fisheries had to take into consideration certain countries' rights in sections of the sea where they had been fishing for decades without any interference. This applied in particular to tuna, which was caught internationally by the 'distant water fishing fleets' (DWF), rapidly growing in capacity. Fleets consisted of ever larger and more efficient boats from countries such as the United States, Spain or Japan, generally built with sizable state subsidies from the flag state. These modern tuna fleets became the new large-scale fisheries that cast their nets and lines in seas which had suddenly become the economic property of the coastal states, many of them developing countries. These drastic changes in the composition of economic interest and the realisation that tuna for many was their only economic opportunity, happened at the same time that the need for greater sustainability in fishing and the battle against illegal and unregulated fisheries began to take shape. The new borders and the new objectives for fisheries were the factors which would decide what the oceans would look like over the coming centuries. But it also created the necessity for new forms of global governance of fish stocks. After the UN Conference of Environment and Development in 1992 underlined the importance of a fisheries management in coastal areas, the FAO developed the Code of Conduct of Responsible Fisheries. As a result a range of regional fisheries management organisations (RFMOs) between states was established [180]. The existing tuna RFMOs like IATTC and ICCAT had been the frontrunners in this new public framework of governance.

The solution of new borders at sea promptly created a further deep layer in the 'wicked tuna problem': sea battles in the shape of an economic power struggle around sustainability. Large distant water fleets from former colonial states and fast-emerging new economic powers could not avoid interacting with coastal states, often former colonies and low-income countries, which wanted to protect their fish against illegal fishing and exhaustion through overfishing. The resulting economic battle intermingled with an ecological, geostrategic and social power struggle in which the interests of countries in different continents clashed. This battle is sometimes referred to as a hybrid warfare [74]. From this perspective an estimated figure of 36 billion dollar of worldwide fisheries is related with global security implications like transnational crime, terrorism, geostrategic threats, slavery and illegal immigration. In the case of tuna and its distant water fishing fleets we can see an increasing battle for power in which borders between private industrial interests and public authorities blur. These miscellaneous conflicts sometimes are settled in the name of sustainability. The new battle grounds were negotiation tables at packed regional fisheries management organisations, trade commissions in dogged trade disputes concerning hundreds of millions of dollars, courts finding leading tuna executives guilty of criminal conspiracy against tuna buyers, diplomatic quarrels between countries in negotiations about tariffs, trade embargoes, import restrictions and, where necessary, at sea with raw maritime violence. NATO and South Korean warships had to be deployed in order to protect tuna fishermen off the east coast of Africa against pirates. Crews were taken hostage for ransom [15], South Korean coastguard patrol boats rammed Chinese poachers' fishing boats and sank them. The Indonesian government, led by the energetic Minister Susi Pudjiastuti, ordered boats from Vietnam, the Philippines and Malaysia which had been caught fishing illegally to be blown up with explosives in open sea. In just one August day in the summer of 2018 a total of 125 mainly foreign fishing vessels were sunk after they were found fishing illegally, bringing the total to almost 500 vessels destroyed in less than 4 years [21]. There was raw violence at sea: Vietnamese fishermen suspected of illegal fishing where killed by the Philippine navy. Plain murder was no exception: independent observers on ships monitoring tuna capture were killed as unwelcome spies and dumped overboard if they threatened to report a catch that breached the rules. Tuna mania increasingly resulted in conflicts, involving legal disputes, diplomacy and even the use of weapons.

Lethal Victims in Tuna battles
Keith Davis was last seen on 10 September 2015 [16]. The 41-year-old biologist disappeared from his cabin on the Victoria No. 168, a Panamanian-flagged reefer (cargo ship), while tuna was being transferred on deck. When a

(continued)

crew member went to his cabin later that day to have his papers signed, Davis had disappeared without a trace. It was a calm day, with light winds and smooth seas. The likelihood that he was knocked overboard due to heavy weather was negligible. Later that evening, a search was mounted with several fishing boats in the area, but to no avail.

Davis had dedicated himself to improving the sea environment and to sustainable tuna fisheries. As an independent observer he worked for the IATTC (Inter-American Tropical Tuna Commission) to monitor the transfer at sea of caught tuna for transport to the landing ports.

In a detailed email to friends a year earlier he had included an attachment with a shocking video on YouTube. It featured four helpless men clinging on to the wreckage of a dinghy and shot dead by a group of fishermen speaking Chinese, Thai and Vietnamese. Others believed the men being shot to be Somali pirates from a Taiwanese tuna boat, following their failed attempt to hijack it. 'One way or the other, the video depicts murder,' Davis wrote in his mail, which was later published in an article in the Canadian online magazine Hakai. He felt that readers should '. . . know that there is other awful stuff that happens out there that goes unpublished' [245].

This butchery at sea was extensively investigated by Ian Urbina for The New York Times in a series called 'The Outlaw Ocean' [248]. Due to lack of cooperation by the authorities of Taiwan and the Seychelles nothing could be ascertained, but it can be assumed with some likelihood that the shooting took place in the Indian Ocean, off the Somali coast, from a tuna longliner flying a Taiwanese flag. The victims may have been local fishermen whose boat had been sunk. This fits in with the bad reputation many tuna longliners have: ships that are able to evade virtually all control because they hardly ever call in at any port. Aside from illegal fishing, longliners can be a prime setting for flagrant violations of human rights, including slavery and killings.

A year later Davis himself disappeared at sea, just as tuna was being transferred from a tuna longliner onto the reefer, or transport ship, that he was working for. Independent on-board observers are extremely vulnerable; Davis knew this better than anyone. In 2013 he had helped draw up an International Observer Bill of Rights. How could observers be protected when they saw things out in the open sea that would not pass muster? Such as illegal fishing practices, or worse, when violence and other crimes were committed? They were an easy, unprotected target in their lonely position on board a ship in the middle of nowhere. And now it appeared he had become the victim of his own integrity. 'Keith was not going to compromise his morals for anyone,' said his good friend Alfred 'Bubba' Cook, WWF Tuna Programme Manager in the Pacific.

Independent observers are the frontline soldiers in the battle against Illegal, Unreported and Unregulated (IUU) tuna fishing. In some coastal nations and regional management organisations their presence has been made mandatory

(continued)

for all purse seine boats fishing in their waters. This is the case with the tuna-rich Pacific Island countries that form the Parties to the Nauru Agreement (PNA) where the role of observers is just to record (and not to act as policemen), and there is 100% coverage of all purse seine fishing. Further 100% of all transhipments are also monitored in port. Tuna fisheries that take their certification seriously use their services. The pressure on observers is enormous: a great deal of money is at stake in fisheries; occasionally looking the other way can make a big financial difference. There is a constant threat of crews attempting to change the observer's mind using violence. Between 2010 and 2016 at least five observers disappeared from fishing boats in the Pacific. The true number of victims is probably very much higher. Governor Allan Bird of the East Sepik province of Papua New Guinea reported at the beginning of 2018 that during the previous 5 years 18 of his fellow countrymen working as observers had disappeared while on duty. According to the minister responsible, the number was lower [22]. But no one had been prosecuted for the disappearances, let alone convicted. A much greater share of a total of 2500 observers operating globally experienced violence and threats. This too seldom reaches public attention. The observers lacked much of a protection. There are only few formal obligations to report deaths or persons missing from the ships. In the case of vessels that sail under a 'flag of convenience' (registration in countries with hardly any supervision) obligations are often non-existent. Bubba Cook: 'The bigger issue is that nobody in the industry is held accountable when things like this happen on these boats. Big fishing companies stand above the law, either through an non-transparant supply chain that hides "beneficial ownership" of the catch and boats, or through blatant organised crime and corruption. That is the real tragedy: that there is no justice.'

After Davis' disappearance, the Victoria No. 168 remained at sea for a further 10 days. The only thing the FBI agents awaiting the ship in port were able to establish was that the boat had been cleaned and given a new lick of paint here and there. Three years after Davis had disappeared, the investigation by the authorities had produced nothing. It was feared the case would be closed. This despite the fact that enough was known to justify charges for suspected murder. The Victoria No 168 on which Davis had been stationed sailed under a Panamanian flag, was in Japanese hands, exploited by Chinese fisheries with a Burmese-Taiwanese crew. Keith had sent photos of large tuna without head or fins to a biologist friend, and complained that this made it unclear whether he was dealing with bigeye or protected bluefin tuna. There was a whiff of unlawful and corrupt trade in illegal tuna. The crew may have got wind of his suspicion; it may have cost him his life.

The lack of protection of observers forms part of a broader and systematic violation of human rights around tuna capture. On land too, merciless murders take place. In August 2015, Gerlie Menchie Alpajora was shot dead in her bed

(continued)

at home in the Philippine region of Bicol, where she was sleeping with her children, aged two and four [12]. Her husband was away fishing for tuna. As secretary of the Sagnay Tuna Fishers Association, set up by WWF, she was known to have exposed illegal fishing of baby yellowfin tuna by the big ships that plundered the coastal waters with fine-mesh purse seines, Fish Aggregating Devices (FADs) and dynamite fishing. She knew that corrupt local politicians were behind such practices, but refused to keep her mouth shut. She paid for it with her life.

The region of Bicol, an island group known for its small-scale handline fishing for large yellowfin tuna, was often the stage for killings related to illegal tuna fisheries. Not long before, the chair of the local municipal fisheries management had been killed. Everyone in the small community had an idea of who was behind the murders. No one dared to cooperate with the investigation. Equally horrifying was the murder of the Frenchman Jean Marc Messina (54), who was killed together with his wife and their four-year-old son, whose bodies were found at the end of January 2016 [54]. They had been shot dead in their car next to an arterial road near Narra, a town not far from Puerto Princesa on the southern Philippine island of Palawan. Friends suspected him to have been lured into a trap by fake police officers. This Foreign Legion ex-commando and Afghanistan veteran was known as 'Rambo', after shooting tuna poachers who were blowing up reefs with dynamite on the island where he lived with his family. Messina was the fifth person in 10 years to have been killed in this region because of his efforts against illegal fishing.

Violation of human rights became part of the tuna sustainability agenda from 2010 on. Besides killings, the poor working conditions of the tuna boat crews, especially those on longliners, who spend large amounts of time at sea, became an issue. The working conditions of immigrants in Thai tuna factories were raised by several articles in the international press. A growing number of people called for social and human rights to be part of the standards for sustainable tuna. The MSC certifiers decided to investigate how this could be included in their assessments.

Thankfully Vietnamese fisherman Huynh Phi Minh remained unharmed when he hauled in his giant tuna. A quick glance at the chart showed that this was not to be taken for granted, since he was fishing in an area where many nations fought each other over claims on specific sections of the sea (Fig. 27.2). Mr. Minh was quoted saying that his crew caught the fish in Hoàng Sa waters, the Vietnamese traditional fishing grounds, under Vietnam's sovereignty. This was actually a matter of perspective: two Vietnamese fishermen were not as lucky as Mr. Minh; they were killed by the Philippine navy a few months later, not that far from the lucky bluefin catch of Mr. Minh. According to the Philippines, they had been caught red-handed fishing illegally in the Philippine EEZ. When they tried to escape the naval ship with their boat, warning shots were fired, killing the fishermen. On board, the marines found a

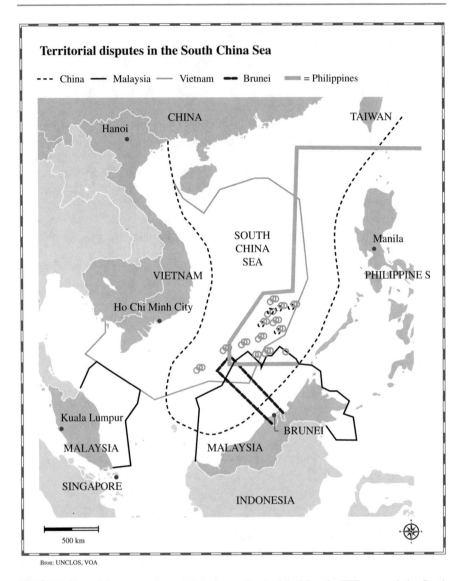

Fig. 27.2 Tuna fishing tuna in troubled waters. Territorial claims in EEZs around the South
Chinese Sea (Infograph Ramses Reijerman)

few yellowfin tuna and a dried squid. Five further crew members who survived the
shootings were arrested [17].

Hybrid warfare always lay in wait in this part of the world. The Paracel archipelago of some 130 coral islands is called Xisha in Chinese and Hoàng Sa in Vietnamese. The island group had long been a stake in a dogged battle in which China,
Vietnam and Taiwan in turn contested ownership. It includes islands such as
Discovery Reef, Money Island and Triton Island, exotic names referring to a colonial

past, piracy and strife. As in all tuna regions, it is a mythical stretch of ocean, full of threatening references. Near Discovery Reef is the 300 m deep Dragon Hole. Fishermen call the dark blue hole the 'eye of the South China Sea'. Sun Wukong, the Monkey King from sixteenth century Chinese mythology, is said to have found his golden staff here. The sea chart around the Paracel islands displays a colourful mosaic of overlapping EEZ claims (see Fig. 29.2) from various countries. China had occupied and populated the largest islands after WWII with some 1000 inhabitants. In 1974, during the Vietnam war, the area became the scene of a sea battle between ships of the Chinese and South-Vietnamese marine [253]. China won the battle of the Paracel Islands, and further escalations with the United States ally South Vietnam were avoided with the release of several American soldiers by China. After the reunification, the new Socialist Republic of Vietnam renewed its claim on the Paracel Islands against their Chinese communist brothers, making it a problem of national importance.

The presence of fish, not least tuna, had given the power struggle new impetus and now the island group was claimed once again by Vietnam. The Philippine fleets lay in wait for fishing in its waters. Everyone thought they would be able to assert their rights here, with China and its fast-growing fishing fleets as the most powerful player in the region. Here, survival of the fittest reigned, and IUU fishing was the rule rather than the exception. The Paracel Island: remember the name. It might well that the next big war will start here.

What made things even worse was that the different claims made it impossible to define the western boundary of the Pacific under the convention of the regional fishery management organisation (WCPFC). The convention only defines the convention area as the waters of the Pacific with defined eastern and southern boundaries only. Of all places, it was in this barely manageable ocean region that one of the biggest bluefin tuna was hauled up from a population threatened with extinction. It all sounded pretty hopeless from a management perspective. Almost as hopeless as it had sounded for the species in the Mediterranean during the first decade of the twenty-first century.

Wicked Tuna Problems

28

By 2010 the bluefin tuna situation in the Mediterranean seemed hopeless. I wrote a sombre book about the large-scale 'wicked problems' of the bluefin tuna catch and trade in the Mediterranean. It seemed merely a question of time before the last bluefin tuna would be fished out of the waters by a gang of poachers, sold by local despots and mobsters, and shipped to the Tsukiji market in Tokyo into the hands of a powerful tuna industry. Sustainable management seemed a lost cause. Despite the book's gloomy message, I received many positive reactions from readers. Tuna had touched the hearts of an ever-expanding audience. NGOs made tuna one of the spearheads in their campaigns. Who would ever have thought this of a cold fish with a cuteness factor even lower than a shark? A growing group of consumers solemnly resolved only to eat sustainable tuna from now. Or they stopped eating tuna altogether. In short, tuna had suddenly become incredibly cool. Some people were so infected by the tuna bug that they took action themselves. One of these was Dutch violin maker Wietse van der Werf, one of the founders of The Black Fish Foundation: a citizens' initiative whereby concerned tuna lovers disguised as tourists photographed the illegal driftnets and undersized tuna in Italian fishing ports as evidence. Something akin to a liberation movement for bluefin tuna sprang up: Black Fish managed to release hundreds of undersized, immature tuna from a fattening farm off the coast of Croatia [235]. Sea Shepherd, Greenpeace's radical offshoot led by Captain Paul Watson, had even managed to free 800 undersized, illegally caught tuna from a purse seine off the Libyan coast in which they were being transported to a farm in Malta [236]. This was not without risk. Muammar Gaddafi, Libya's 'brotherly leader', must have been furious about a small fortune's worth of 'his' tuna having been cut free. The liberation campaigns, the pressure from the NGOs as well as Prince Albert of Monaco's efforts for an international trade embargo on bluefin tuna: all this contributed to bluefin tuna suddenly becoming an item on the agenda of governments, the EU and tuna management organisation ICCAT. For the first time, management measures to save the fish from extinction received serious, firm impetus. And, to the genuine surprise of many, it actually appeared to work.

© Springer Nature Switzerland AG 2019

S. Adolf, *Tuna Wars*, https://doi.org/10.1007/978-3-030-20641-3_28

The recovery of the population of eastern Atlantic bluefin tuna, which was becoming increasingly evident from 2013, could enter history as the important turning point of a paradigm in fisheries policy. Such a statement is always dangerous; where history is concerned you never know what the future will bring. But I would happily take the gamble here: the importance of saving eastern Atlantic bluefin tuna extends much further than the restoration of a population as such. Against all odds, it seemed that certain measures for sustainable fisheries management at least could contribute to the result. Even if there were different causes for the improvement—from dying tuna-dictators to failing data and underestimation of the stock and its natural recovery potential—this was an unprecedentedly positive message amidst the doom and gloom which usually dominated the debate about sustainable fisheries and ocean management. A fish had been saved here, the management of which was a 'wicked problem' in itself. A biologically and ecologically extremely complex issue, moreover belonging to one of the most intricate, large-scale and barely transparent global value chains imaginable for a fish. There was hope: if this had worked for Atlantic bluefin tuna, there was, in principle, a good chance it could work out for Pacific bluefin tuna too. And why not for all the other tuna species as well? For all fish, in fact, lessons could be learned from the restoration of bluefin tuna, our oldest industrial super fish. Contrary to all appearances, it was never too late to initiate a truly sustainable fisheries policy.

This positive message was worth quite a lot, you might think. So it was time to have a look at the actual situation in the management of global tuna fisheries. To start with ICCAT.

When I descended the plane's steps in Marrakech on a glorious November morning in 2017, there was little evidence of a festive mood in tuna land. ICCAT, the local management organisation for Atlantic tuna, had decided to have its plenary meeting once again in the royal city on the dry plain at the foot of the High Atlas Mountains. The air in Marrakech, with its palaces, palmeries and the beautiful Djemaa el-Fna square, was filled more with the scent of dates and mutton than of tuna. ICCAT, for years the tuna world's laughing stock as the International Conspiracy to Catch All Tuna, had in the end helped to save the Atlantic bluefin tuna from extinction. This was something to be proud of, something to broadcast as an inspiring example to the rest of the world. But if there was a festive mood amongst the 700 or so delegates assembled here for a week in the big meeting room of the conference centre, they were very good at hiding it. Even the European Union, firmly present with its DG Mare directorate, remained remarkably reticent about putting itself in the spotlight as the saviour of bluefin tuna. 'You know, we're not all that used to display our success,' Stefaan Depypere told me. As Director for International Affairs of DG Mare and former ICCAT chairman he was one of the most influential delegates. Next to him stood the Japanese Masanori Miyahara, another old hand from the tuna circuit who, 10 years previously in Tokyo, as director of the National Research and Development Agency, had explained to me that the management of bluefin tuna simply is a complex business everywhere and at all times. Miyahara—a man with a dry sense of humour—looked back with satisfaction on having survived all the ICCAT meetings he had chaired, reflecting that he would be able to retire more or less unscathed.

But there was not much delight about the recovery of bluefin tuna on any of the other familiar faces from the international ICCAT meeting circus. A sultry atmosphere hung in the meeting rooms, the kind that prefaces the turmoil of battle rather than the mirth of a party about to start. By far the weightiest delegation comprised the EU, but representatives from Japan, the United States and the African and Latin American countries were also very much in evidence at the Atlantic tuna summit. Arranged over three rows behind the chairs for national delegations sat fisheries representatives. This setting left no doubts about the presence and influence of the industry. Now that there seemed to be plenty of tuna swimming around again, everyone had a large wish list. The meeting table was flanked by rows of tables crammed with NGO observers, scientists, traders, sector organisations and other stakeholders from the tuna industry. The bluefin tuna may have been saved from extinction, but now the true battle commenced between all those who came to claim their share in the recovered bounty. The *almadraba* fishermen, for instance. 'We've made big sacrifices with the tight tuna quotas we've stuck to for years,' said Diego Crespo of the Zahara de los Atunes *almadraba*, representing the united *almadrabas* from the south of Spain. 'Now that things are once more going well, we want our share in the fisheries again.' Crespo was clearly at loggerheads with Rafael Centenera, the prominently present fisheries sub-director general of the Spanish Ministry of Agriculture, Fisheries and Food. Centenera supported higher quotas, in particular for the large-scale Spanish purse seine fleet and made no secret of his preference. This fleet supplied the fattening farms, such as those owned by the Fuentes Group, which was in attendance with a sizeable delegation led by Paco Fuentes (Ricardo's son). Together, the Spaniards formed a front against all kinds of newcomers amongst the fishing countries, such as Turkey, Albania and Algeria, who had sniffed new opportunities. 'In the end there'll be nothing left for us to divide,' Centenera's loud complaints echoed through the corridors.

Or take Norway, which no one had previously really taken into account where bluefin tuna fishing was concerned. Now a hush descended over the hall when the delegation leader addressed her fellow conference attendees in a freeze-dried tone. In a well-argued speech she reminded everyone that during the 1950s, 15,000 tonnes of bluefin tuna had been caught off the Norwegian coast. Now multiple thousands of bluefin tuna had returned well into the Arctic Circle and Norway was being fobbed off with 0.23% of the quota. A laughable quantity which was disproportionate to the catches of yesteryear. While the countries in the Mediterranean had always fiddled with their catch restrictions and had submitted large claims, it seemed as if Norway was being excessively punished, yet the country itself had suspended its catch. 'Apparently the more you overfish, the more you are entitled to catch,' was her conclusion. It was even worse: bluefin tuna made its way specially to the wide Norwegian fjords to gorge on the Norwegian pelagic fish in attendance. 'Do we have to watch how the bluefin comes and eats away the fish in our seas? And as a reward we are only allowed to catch 0.23% of the tuna. Is this fair?' No, it was not fair, that was beyond doubt. Iceland wrestled with a similar problem, the one-man delegation explained. While it was already difficult enough to decide whether the only fishing boat suitable for the catch available at that moment should be sent out to catch bluefin tuna or cod, an entire Japanese boat squadron was waiting just outside the 200 mile zone to scoop up the bluefin tuna as soon as they had filled their stomachs and swum out of the Icelandic EEZ again. That could hardly be called fair either.

While the government officials of the many country delegations squabbled about their share of the pie, with the fisheries industry delegates' eyes prying from behind their back, entirely different concerns prevailed in the NGOs' camp. These primarily concerned the fear that, under collective pressure from the fisheries, all caution was now being thrown overboard and the catch quota for bluefin tuna was being set far too high. There was plenty of reason for this fear. Before the ICCAT meetings, the figure of 36,000 tonnes had been circulating as the Total Allowable Catch (TAC) to aim for. This was roughly three times higher than had been allowed only a few years previously. The NGOs had always stressed that, in setting the quotas, the recommendations of the ICCAT scientific committee's tuna expert should be followed as much as possible. Science and not commercial interests would be the deciding factor when it came to the question of how much tuna could permissibly be caught. 'But we don't feel confident about it,' said Paulus Tak, the Belgian tuna specialist of the Pew Charitable Trusts. According to rumours, radically different scenarios circulated between the scientists about the recovery of bluefin tuna and the associated recommended catch.

This anxiety turned out to be justified. At the conference the ICCAT scientific committee presented a stock assessment of bluefin tuna which was so full of mysterious contradictions that it raised 'more questions than answers', according to a letter in the authoritative *Science Magazine* [60]. Obviously no one wanted to call into question the independence of the tuna scientists. But the strongly improved data in the analysis model (basically still the much used Virtual Population Analysis model), resulted in an advice that was so vague that nobody was able to make real sense of it. The state of bluefin tuna was much improved no doubt, but no scientist was able to say with confidence by how much. The estimates ranged from a population that had barely progressed to a total recovery of the original stocks in the eastern Atlantic and Mediterranean. 'The level of increase remains difficult to quantify,' stated the explanatory note to the scientific results. There were different scenarios about how the population might develop over the coming years and they all appeared to conclude that the stocks would diminish, regardless of how much was caught. Scientifically speaking these scenarios looked not very conclusive, let alone that they were very helpful in the design of any practical management of a sustainable fisheries. ICCAT and science seemed to maintain a difficult relationship. Scientists like Alain Fontenau, who was a longtime ICCAT collaborator, were convinced that the scientific committee had widely underestimated the size of the bluefin stock during the 2005–2009 crisis as a result of poor data in their models. These serious errors were seldom discussed. Now, after years of investing many millions of euros in better data, the committee came with such vague recommendations, that everyone could take from it what suited them best. And, of course, everybody did.

After 10 days of negotiations (aside from bluefin tuna, other Atlantic tuna species were discussed, as was the mako shark) a decision was made. Over the next 3 years, the TAC for Eastern Atlantic bluefin tuna would be raised from 28,000 tonnes to 32,000 tonnes and ultimately to 36,000 tonnes in 2020. Regarding the western Atlantic bluefin tuna stock, the 'wicked tuna' off the US coasts, it was admitted in so many words that the 20-year recovery plan ending in 2018 had failed miserably. The annual quota of 2000 tonnes was hiked up, despite the fact that the population had barely recovered, if at all. ICCAT bid farewell to a recovery plan here, without a recovery ever having taken place. It was at odds with its own fundamental principles of sustainable management. But it remained largely unnoticed in the noise about the new, much higher catch quota of the eastern stocks.

In the end, no one was happy with the outcome. The fishermen were dissatisfied, as they wanted to catch the full 36,000 tonnes from the first year onwards. They complained about how the quotas were divided between established operators and newcomers. The NGOs were disappointed that quotas would rise so much in such a short space of time. This was far from a policy of care and caution, they thought. Pew tuna-specialist Paulus Tak: 'This year was an enormous step backwards for sustainable tuna fisheries and shark conservation. The quotas adopted for the western and eastern stocks of Atlantic bluefin tuna go far above precautionary levels.' In the stock assessment, scientists had not been able to demonstrate a sustainable recovery of the Atlantic bluefin tuna. Quotas had been agreed which almost certainly would lead to a decrease in the population. 'This is an extremely concerning reversal of course for ICCAT, which actively jeopardises the future of a valuable and iconic species,' argued Tak.

Marrakech 2017 was no celebration, more a painful reminder that whilst bluefin tuna had successfully been saved from extinction, its sustainable management had once again become a 'wicked problem'.

The good news was that, contrary to all expectations, the stock had recovered. Maybe sustainable fisheries management had contributed to this result, although the scientific modelling had not been very conclusive and were still reason for serious concern. The bad news was that the industrial tuna complex was constantly ready to pounce, and might be able to exert strong influence in its quest for more tuna in order to meet the growing global demand. In terms of catch volumes, bluefin tuna had become a small niche in a tuna industry which, during the second half of the twentieth century, encompassed many tuna species roaming all of the world's oceans [45]. Many consumers understood that something was threatening tuna, but soon lost track. There were many more types of tuna than bluefin alone, after all. The fish you found on the market, in a can or on your plate in a restaurant actually was hardly ever bluefin tuna.

For most, the tuna family turned out to be confusingly big and diverse. Even if it had been clear what kind of tuna you had bought—on cans the description tended to be vague and on fish stalls or in restaurants it was usually absent—it was a hopeless task for the average consumer to find out the sustainability status of the tuna in question. At best they would recognise the label with the blue fish from the Marine Stewardship Council (MSC), a certification programme which ranked as the gold standard for sustainably caught fish. But now that sustainability had developed into an increasingly powerful marketing tool, cans or freezer bags of tuna tended to be plastered in an array of other labels all suggesting the fish was caught sustainably in one way or another: Dolphin Safe, Dolphin Friendly, FAD-free, pole and line, Friend of the Sea. It caused puzzlement. When asked, a significant part of American tuna consumers actually thought Dolphin Safe was a guarantee that there was no dolphin-meat in the cans of tuna. Sustainable tuna had become a confusing business, with sustainability claims that where lacking credibility. Some people decided to take the safe option and stop eating tuna altogether.

As a start to clarify the confusing issues of whether or not we might continue catching and eating tuna, the first step was to chart the fish. Bluefin tuna had been the historic pioneer in the large-scale industrial fisheries, but other tuna had followed suit and had risen swiftly during the twentieth century. This had developed into a global market running into many billions of dollars. Because of its high price, bluefin tuna continued to be of importance, but its cousins had long overtaken it in catch volume. As canned tuna, bluefin had virtually disappeared. The tuna found in cans or sold as frozen steaks was usually a different species. If you wanted to know more about the 'wicked problem' of sustainability and the new 'hybrid warfare' waged around it, getting to know more about the tuna family and their relations became unavoidable.

The Tuna Family

People eating canned tuna probably don't realise that strictly speaking they are not eating tuna at all. Most canned tuna is **skipjack tuna** after all, and skipjack (Fig. 29.1) is a member of the Scombridae family, to which next to the real tuna's as Atlantic Bluefin tuna (*Thunnus thynnus*) the mackerel and Atlantic bonito (*Sarda sarda*) also belong [23]. So skipjack is the tuna family's bastard cousin; the upstart who, because of its economic significance to the industrial fisheries, was welcomed warmly into tuna aristocracy. Besides being the most important ingredient for canned tuna, skipjack is also used in Japan for the smoked, dried and shaved *katsuobushi*, a key ingredient for fish soup and many other typically Japanese dishes [127]. The Americans know it, once canned, as 'light-meat' tuna [81]. Skipjack, that is found in tropical and subtropical waters, does not grow very big, to around 1.2 kg and 40 cm length at maturity or exceptionally 1 m in length. Horizontal stripes on its belly make it easily recognisable. Like the rest of its family, skipjack is a cosmopolitan migratory fish that moves around in shoals of thousands of fellow fish to hunt for food jointly. This makes it ideal for large-scale purse seine capture as well as for pole and line. In both cases floating rafts or FADs can be used, leading to large amounts of unsustainable bycatch. Skipjack, which can live up to a reported maximum of 12 years, had hitherto been considered a robust stock because of the speed with which they procreated. The fish is sexually mature within a year. They are sometimes called the cockroaches of the sea, although this comparison is seldom made in professional circles, presumably because of the negative connotations as far as hygiene and palatability are concerned.

Despite its alleged fast reproduction, the skipjack population at some places has been reduced to al level where the fisheries is no longer economically interesting. Thought still not overfished or subject to overfishing, the catch has lost much of its industrial exploration in the Atlantic. These days the Indian Ocean and especially the Pacific hold the larger stocks, with by far the largest population of skipjack in the Western and Central Pacific. Perhaps half of the total global stocks of skipjack are

© Springer Nature Switzerland AG 2019
S. Adolf, *Tuna Wars*, https://doi.org/10.1007/978-3-030-20641-3_29

Fig. 29.1 Skipjack (*Katsuwonus pelamis*, Esp. listado, Fr. bonite rayé, Chin. 鲣鱼, Jap. カツオ, Ind. cakalang) (by courtesy of NOAA/Fishwatch)

located in the immense sea area belonging to the eight island countries of the Parties to the Nauru Agreement (PNA), established in 1982: Papua New Guinea, Kiribati, the Federated States of Micronesia, the Solomon Islands, the Marshall Islands, Nauru, Palau and Tuvalu. With less than eight million inhabitants, the PNA countries live on the lowest per capita income in the world. But thanks to its numerous small islands that all have their individual 200 miles zone, the PNA has a common EEZ stretching to 4.5 million square kilometres: this colossal surface area is almost as big as half of the European continent. Towards the end of the last century, the poverty-stricken island countries realised that in their seas, they held a vast capital in the shape of tuna, especially skipjack. So it was important to cherish these fish stocks as the most important marketable commodity they possessed. A policy of sustainable fisheries was pure self-preservation in order to safeguard income sources for future generations. The PNA marked the birth of the biggest global production power in tuna, which, as a production cartel, was to become an important supply-driven force to develop sustainable tuna fisheries. What OPEC is to oil, the PNA likes to be in tuna [8, 53]. Extraordinarily the PNA, as 'owner' of the biggest singular global tuna stock, acts creatively in its politics towards making tuna fisheries more sustainable, with introducing management systems like the Vessel Day Scheme and MSC certification [2, 132]. In doing so the PNA has to cope with vested economic interests in the tuna global value chain that are not particularly happy with a production cartel challenging their own commercial interests and power in the market. The PNA does not have an industrial tuna fleet of any significance itself and the issues at stake are primarily the interests of the distant water fleets, the powerful trading parties, the important tuna brands and global consumer markets. You could call it a Pacific tuna war: big financial interests battling over sustainability and geopolitical influence (see Flipper Wars). This war plays out primarily in the area managed by the WCPFC, the Western and Central Pacific Fisheries Commission, the regional management commission tasked with managing the tuna stocks in this part of the Pacific. The WCPFC's policy measures

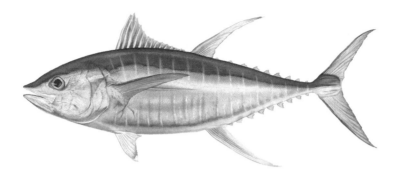

Fig. 29.2 Yellowfin Tuna (*Thunnus albacares*, Esp. Aleta Amarilla, Fr. Thon Albacore, Chin. 黄鳍金枪鱼, Jap. キハダマグロ, Ind. Gantarangang) (Flick Ford)

are eclipsed to an important degree by that of the sub-regional group PNA, which tends not to want to wait for the less decisive resolutions from the WCPFC, but rather drive policy changes. The PNA ranks as a fairly unique agreement amongst developing countries who have sufficient clout to be able to enforce their own policy, premised largely on their waters delivering 70% of the WCPFC skipjack.

Yellowfin tuna (Fig. 29.2) probably is the best-known cousin of bluefin tuna [26, 81]. It's also called 'ahi' in the Pacific, the original Hawaiian name for the fish. Its Latin name occasionally causes confusion, because *albacares* strongly resembles 'albacore'. The albacore tuna is a different tuna species: the *Thunnus alalunga*, described below. The yellowfin is easy to tell apart from albacore tuna: it does not have long pectoral fins. Juveniles have short stiff pectoral fins. Mature fish grow a remarkably long, yellow dorsal fins and a similar anal, hence its name, yellowfin. This species is primarily fished in the Indian Ocean and the Pacific, and is an important species in the market, with making up around a quarter of all global tuna catches. Although smaller than the bluefin, it can reach a considerable size, weighing around 180 kg and measuring some 2 m in length. Adult specimens are usually caught with purse seines and longlines.

Juvenile yellowfin is a major bycatch in purse seine and poll and line skipjack fisheries whenever floating objects, the fish aggregating devices (FADs), are used. Although purse seine catch is typically processed for canning, well handled purse seine and longline caught yellowfin is also suitable for the premium sushi-market and steak trade. In behaviour too it resembles the bluefin, with two major differences. Firstly, up to 20 kg yellowfin and young bigeye tuna, form mixed schools [103]. This phenomenon is one of the important causes for overfishing the yellowfin and big eye stock wherever FADs are used. Schools of skipjack mixed with juvenile yellowfin (distinguishable from the skipjack because of its vertical stripes) congregate underneath these floating rafts. When the encircling nets of the skipjack fishermen haul up such school, juvenile yellowfin and bigeye is inevitably included. This is also the case with anchored FADs used by pole and line fisheries.

As we know from experience with the bluefin, preserving young, non-sexually mature tuna works out very well in management plans for recovering the entire stock. Yellowfin research by the French FAD specialist François Dagorn concludes it is not clear if catching juveniles as such has a stronger negative effect on the stocks than catching adults [72]. But the fact of the matter is that the management to control the use of FADs as a highly efficient fishing tool until now has been proven difficult. As a result the juveniles catch with FADs is complicated to manage and can cause important damage to the stocks through overfishing, just like overfishing adults does. Stopping FAD fishery though would be especially to the disadvantage of local, artisanal fishermen in countries like Indonesia and the Philippines. On the other hand these local fishermen see the population of the yellowfin tuna plummeting as soon as the large industrial skipjack purse seine and longline fisheries mobilise in their open seas, targeting mature yellowfin.

A second important characteristic of yellowfin tuna has far-reaching consequences for sustainability as well. For reasons marine biologists are not yet able to explain, dolphins swim above large adult yellowfin tunas [71]. It is thought they associate to jointly hunt prey. Remarkably enough this pattern occurs only in the eastern part of the Pacific, approximately off the coasts of Mexico and further south along Central and South America. At the end of the 1970s this behaviour led to a fishing technique whereby purse seiners actively pursued dolphins, knowing that this would take them to yellowfin tuna, which they would then be able to catch. This resulted in an annual slaughter of hundreds of thousands of dolphins in the 1980s [45]. Public indignation about this was wide-spread, particularly in the United States, where different generations of viewers had grown up with the congenial TV dolphin Flipper. The general consensus was that Flipper had to be saved. In 1988 a Dolphin coalition of environmental groups launched a boycott against the three big brands of canned tuna. Thus the Dolphin Safe logo was born in the US for yellowfin tuna that had been caught without injury or death to dolphins. Dolphin Safe became a huge success: within a few years, mortality of the dolphin in the eastern Pacific had been virtually reduced to zero and new fishing techniques had been developed which left dolphins undisturbed.

That was great, of course. Less great was that as soon as it was introduced Dolphin Safe appeared to be a useful business tool in the hands of the same three big American tuna brands, allowing them to protect their market against competition from the Mexican industry. The Dolphin Safe label thus became a handy device for the tuna industry, which on the one hand was able to keep up appearances of a responsible catch and yet on the other conspired to safeguard its market interests in the value chain for canned tuna from overseas competition (see Flipper Wars). I should add that sustainably caught yellowfin tuna is now available with the MSC label. Yellowfin is managed by the various regional fisheries management organisations (RFMOs) of the associated oceans and is an issue primarily in the eastern Pacific, where significant fishery closures now occur with calls for more for stock recovery and the Indian Ocean, where currently there are calls for 15% cutback in mortality.

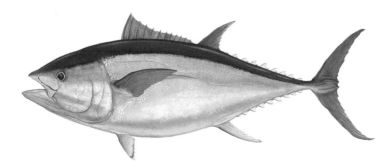

Fig. 29.3 Bigeye Tuna (*Thunnus obesus*, Esp. Patudo, Fr. Thon Obése, Chin. 大眼, Jap. ビッグア
イ, Ind. Mata Besar) (Flick Ford)

This tubby tuna is identifiable by its somewhat oval eyes, which give it its name. The **bigeye tuna** (Fig. 29.3) has a large appearance and the shape is akin to the bluefin and the yellowfin. Again confusion with the name. Just like the yellowfin, bigeye is also called 'ahi' by Hawaiians, which does not really help when making distinctions in the tuna family. The bigeye is nonetheless easily recognisable for its shorter dorsal and anal fins and the longer feather like pectoral fins. The larger head relative to the body and (indeed) the big eye completes the picture [24]. With a maximum weight of some 130 kg and an average length of 1.8 m it is also a touch smaller. It is caught using purse seine and longlines, but also with pole and line. The market demands its high-quality meat for sushi and sashimi when well-handled at sea. You see it in great quantities of longline caught bigeye on the Tokyo market, as frozen torpedoes or air-freighted fresh fish. After the bluefin tuna, the bigeye ranks as the most threatened tuna in the world seas. It is the biggest concern in the management policy of the Western and Central Pacific Fisheries Commission (WCPFC). This is partly to do with fishing techniques: as with yellowfin tuna the juvenile bigeye has a tough time of it as skipjack bycatch in the purse seine fisheries using FADs. In global capture, bigeye represents around 10% in volume, quite a substantial proportion considering the far from favourable status of the populations.

The **albacore tuna** (Fig. 29.4) [25] is undoubtedly the tuna that leads to most confusion. Albacore not only resembles the Latin nomination for the yellowfin tuna (*albacares*), but its Spanish name *bonito* (*del norte*) is used in other languages to designate skipjack. And if this wasn't confusing enough, there is also the Atlantic bonito (*Sarda sarda*), a smaller tuna species from the bastard branch of the Scombridae also fished commercially. So you have been warned. But don't worry: the albacore is quite easily recognisable on account of its extremely long pectoral fins. In its movements, when flapping his long fins, it looks a bit like a penguin, especially when it is line-caught out of the water, as happens in the north of Spain for instance. Once canned or as steak, the albacore is also easily identifiable because of its white, relatively dry meat, which is more like chicken. This is why the fish is often commercialised under the name of white meat tuna or chicken of the sea. The steak tends to be less valued, because many consumers expect tuna meat to be red and

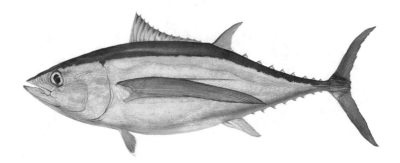

Fig. 29.4 Albacore Tuna (*Thunnus alalunga*, Esp. Bonito del Norte, Atún Blanco, Fr. Germon, Thon Blanc, Chin. 长鳍金枪鱼, Jap. アルバーコア, Ind. Albakora) (Flick Ford)

more moist. Mistakenly, as the white meat is of high quality and tastes good. The eater will realise at once that albacore is the real thing. The major difference with its cousins (bluefin, yellowfin and bigeye), apart from its clearly recognisable long pectoral fin, is its smaller size. At approximately 1 m in length and with a weight of 16 kg it is the junior amongst the tuna family's main core. Albacore is caught using pole and bait, but also by longliners and trolling, whereby the fish are lured with bait thrown overboard and caught with multiple lines trailing behind a boat. It can be found in all the major oceans and is a true migratory fish travelling in shoals. The albacore population, running to around 6% of global tuna capture, is under pressure in certain places because of overfishing, but is available sustainably with the MSC label.

This is it: **Pacific bluefin tuna** (Fig. 29.5) the Japanese *maguro*, the original engine behind the sushi and sashimi boom. Its popularity in Japanese cuisine—and with it its conquest of the global markets—is, as has been said, to some extent a coincidence. But beware of so-called old Japanese traditions: the fish was only discovered for raw consumption in Japan around 1840. No wonder: previously, bluefin tuna had been known primarily for being highly perishable, making eating it

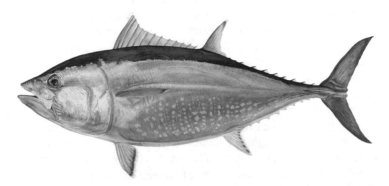

Fig. 29.5 Pacific Bluefin (*Thunnus orientalis*, Esp. Atún Rojo, Fr. Thon Rouge, Chin. 蓝鳍金枪鱼, Jap. クロマグロ, Ind. Tuna Sirip Biru) (Flick Ford)

raw somewhat perilous. It was only with new freezing techniques at ultra-low temperatures, introduced from the 1970s onwards, that the danger of food poisoning was properly kept in check and consumption took off. So if Japanese people start advancing their long traditions with bluefin tuna as an argument for carrying on fishing without restraint, keep in mind that as far as sushi and sashimi is concerned, this tradition isn't all that old.

Classing the Pacific bluefin as a separate species from the Atlantic bluefin tuna is the subject of endless debates amongst tuna scientists. It's probably less of an issue to non-biologists. But even from a biological point of view, the differences are somewhat negligible. In terms of characteristics it is indistinguishable from its Atlantic twin. The reason for listing *Thunnus orientalis* separately is more to do with the fact that is a separate stock form its Atlantic family. Also the economic power and the sustainability issues playing out around this giant tuna are specific. Pacific bluefin can live up to 26 years, although the average lifespan is about 15 years. It reach maturity at approximately 5 years, adults are approximately 1.5 m and weigh about 150 kg.

Unlike its Atlantic brother, and its remarkable recovery after 2010, alarm bells continue to ring for the *Thunnus orientalis* [193]. The guilty party is not too difficult to trace: despite the population's dramatic decline, the Japanese market is still heavily addicted to the *maguro* from its own waters. It's difficult to induce the Japanese fishermen to keep to the tight restrictions on catching the fish. And the Japanese government for its part finds it difficult to control its fisheries, which means that the rules that are supposed to guarantee sustainable capture of Pacific tuna are not sufficiently enforced. This might change in the near future, but it remains a 'wicked problem'.

Like its Atlantic counterpart, the Pacific bluefin is used to covering enormous distances (Fig. 29.6). It spawns to the southwest of Japan, in the Sea of Japan between Okinawa and Taiwan and near the Philippines. The females secrete millions of eggs, in essence enabling the fish to restore stocks if the right natural circumstances allow it. As a juvenile, before it's a year old, the tuna migrates from its spawning ground to the North American west coast and the east coast of Australia, where it will stay for a few years to mature. If it hasn't been hauled up by the Mexicans in Baja California or by the Americans off their west coast, it returns to Japan to spawn. Many of the sexually mature tuna are caught there on their return. If we are to believe the official catch figures, in 2013 a total of just 11,000 tonnes was landed, with Japan (5362 tonnes) and Mexico (3154 tonnes) responsible for the largest proportion. The US captured 10 tonnes. During the 1960s and 1970s this was still 15,000 tonnes per year, which says something about the lamentable state of the Pacific bluefin today. The Pew Charitable Trusts estimated in 2016 that the population was just 2.6% of the original unfished stocks once roaming the Pacific [194]. Almost all fish that was caught was small and unable to spawn. Pew believed fishing activities were three times the level that would be responsible under sustainable management. Saving *Thunnus orientalis* from extinction was probably the biggest challenge in the second decade of the twenty-first century. Things were not getting easier because Pacific bluefin is jointly managed by the

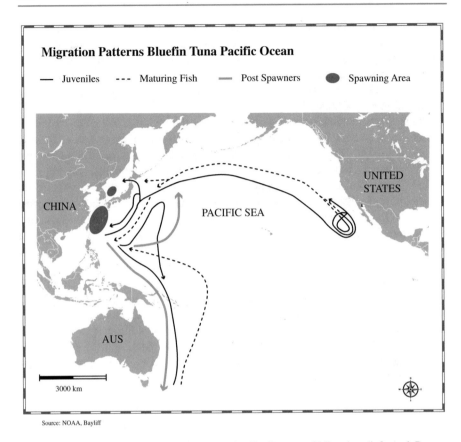

Source: NOAA, Bayliff

Fig. 29.6 Bluefin tuna also likes to travel in the Pacific. Patterns of Migration. (infograph Ramses Reijerman)

Inter-American Tropical Tuna Commission (IATTC) and the Western and Central Pacific Fisheries Commission (WCPFC) and has been allocated to a separate WCPFC subcommittee, the so-called Northern Committee, in which Japan manages to dominate the consensus. The Northern Committee has its own scientific advisory committee, whose recommendations for a more sustainable catch tend to produce little effect. All this does not benefit the implementation of necessary measures. It was all the more remarkable that the committee came to an agreement in September 2017 to rebuild the stock to 20% of the original level in 2034. That was roughly a sevenfold increase of the estimated 1.6 million Pacific bluefin still left in the ocean [177]. The NGOs like PEW and Monterey Bay Aquarium were excited. "The really big, exciting thing is they have all agreed to a 20% target for recovery. It's the level at which you can say this population really has a chance," Amanda Nickson, director of Global Tuna Conservation at Pew Charitable Trusts was quoted. If the fish are allowed to survive and reproduce, she expected the population to bounce back. And still some fishing activity could be allowed. We have to wait to 2034 to see if this was to good to be true. But it might well be that the tipping point for Pacific bluefin finally was reached.

Even though it is not immediately obvious from its appearance, the **southern bluefin tuna** ((Thunnus maccoyii, Fr. Thon Rouge, Chin. 蓝鳍金枪鱼, Jap. クロマ グロ, Ind. Tuna Sirip Biru) is considered by many a separate species of bluefin tuna [56]. Its habitat as well as its genetic profile make this clear. It might meet its northerly cousin from the Pacific off the east coast of Australia, but this bluefin tuna's habitat extends considerably further south and thus covers not only the Pacific but also the Indian and Atlantic Oceans. A notable difference is its second dorsal fin, which is longer than the first and has a reddish-brown colour. It closely resembles the Atlantic bluefin, but is a little smaller (225 cm and 200 kg max). It can live up to the age of 40. But the likelihood of meeting such an old tuna is fairly slim because the average weight of an individual caught has dropped to seven kilos. Its spawning ground is relatively small and located off the north-western coast of Australia and the Indonesian island of Java. The females can release some 14–15 million eggs per season. Juveniles ride on the Leeuwin Current towards the west coast of Australia, where they remain for around 5 years, by which time they have grown large enough to swim into the southern oceans and even as far as the Atlantic.

It was only in the 1930s that southern bluefin was caught for canning, but fishing with poles did not take off until the 1950s, when the Japanese discovered unknown stocks of large tuna in the waters south of Indonesia [45]. During the 1960s a top annual catch of some 80,000 tonnes was recorded. From then on things nose-dived. The population of the southern bluefin was squeezed between Australian purse seine and pole and line fishers that caught massively the sexually immature juveniles and the Japanese longline fleet that hunted the mature fish.

An additional problem is that the southern bluefin is considerably slower in sexually maturing than its Atlantic relation (estimates run between 8 and 10 years). This might explain why, despite a recovery plan, the growth of a sexually mature population is slow to get going. It is estimated that following intensive fishing since the 1950s, just 7–15% of the original population is left, good reason for the International Union for the Conservation of Nature (IUCN) to put southern bluefin on the red list as a 'critically endangered' species in 2011 [147]. That sounds very tough but has little practical value. The IUCN red list is different from the so-called Appendix 1 list of the Convention on International Trade in Endangered Species (CITES), which was once sought in vain by Monaco for the Atlantic bluefin tuna. The CITES list prohibits international commercial trade. The IUCN red list is an inventory of all species at risk of extinction, judged from vulnerable to critically endangered. Inclusion on the list aims to warn the public and policy makers of the precarious state and the urgency of the species needing to be saved from extinction. It does not necessarily result in any regulations or sanctions. In the 2012 red list the Pacific bluefin was nominated as 'vulnerable', whereas the Atlantic bluefin had been labelled 'endangered'. This meant the improvement of the population had not been considered for the latter, while the deterioration of the former had yet to be factored in.

The southern bluefin's wellbeing is watched over by the Commission for the Conservation of the Southern Bluefin Tuna (CCSBT), the regional commission monitoring purse seine, longline and pole and line fishing. The CCSBT is the only regional management organisation which is not geographically organised, but set up for the management of a specific tuna species. Japan, Australia and New Zealand created the CCBST in 1993 and were later joined by South Korea, Taiwan, South Africa, Indonesia and the EU. The Philippines has meanwhile been designated a 'cooperating non-member'. A management plan to restore the population to 20% of the original volume by means of catch quotas therefore appears to be producing few results as yet. There is a management plan, but because the catches are deliberately kept (too) high, the recovery is sketchy and the population of southern bluefin tuna is still only just ticking over. The problem of the CCBST is described as '... lack of political will, disparate national agendas, divergent economic priorities, different time horizons and scientific uncertainty', according a OECD study on fisheries policy on rebuilding stocks [179]. That was a handy definition of the 'wicked problem' in managing the global tuna resources. Or if you like, a management nightmare in a nutshell.

In a sense, the management of southern blue fin and its failures learn some important lessons. In 2006 a mayor scandal broke when a review of Japanese market statistics revealed that the Japanese longliner fleet had systematically exceeded the official reporting of their southern bluefin catches. The fraud had been massive: during the 1990–2005 period Japanese longliners had fished at least double the amount of officially reported catches [197, 211]. This meant that the unreported catch also greatly affected the reporting on other tuna species like big eye tuna, which had been reported in the logbooks to hide the illegal southern bluefin catches. The data had been a mess over a 15 years period. Conclusions: never trust the captain logbooks as a tool for data if they are not verified one way or the other by independent observers. Something to keep in mind also when it comes to all kind of self-certification of sustainable fished tuna. Other conclusion: longliners are difficult to control. And third: verification of the catches is pivotal for any trustworthy data that support a credible sustainability policy.

The Rest of the Family

The family of tuna and tuna-like species is even more extensive. A host of distant cousins and in-laws roam the seas and crop up once in a while. We have already mentioned the Atlantic bonito (*Sarda sarda*), which features in the warmer waters of the Atlantic and in the Mediterranean. There is the longtail tuna (*Thunnus tonggol*) which travels the Western Pacific and the Indian Ocean and which is also called northern bluefin, to make matters even more confusing. We have the Australians to thank for this, in distinguishing their southern bluefin. The longtail tuna has nothing to do with the *Thunnus thynnus*, and with its length of 1 m and weighing 25 kg it can decidedly be called small. Its meat has little flavour and this means it's only suitable for canned tuna and not for the sushi market. The western Atlantic off the Brazilian coast is home to the small blackfin tuna (*Thunnus atlanticus*), in terms of shape a tuna strongly resembling bluefin which only grows to 1 m in length and is not of great immediate value commercially. It's identifiable by its black back. For the sake of completeness we should mention the kawakawa or mackerel tuna (*Euthynnus affinis*) from the mackerel family of Scombridae. It has the typical characteristics of a real tuna, but with its length of just a metre is also called 'the little tuna'. It can be found in a wide area of distribution in the Indian Ocean and the Pacific.

Fishing Gears

30

So now we know the tuna family that emerged on out plates over the course of the twentieth century, we also should have a look at the different fishing methods to understand the sustainability challenge for tuna in the third millennium. The time when industrial fishing was limited to the classic *almadraba* nets, anglers and line fishermen is long past. Modern tuna fishing used different techniques. If you look at a can of tuna or read the label on a fresh or frozen piece of tuna in the supermarket you might see the following.

Purse Seine The encircling nets (purse seines) are perhaps the biggest innovation in industrialised, large-scale fisheries in the new tuna era. The technique developed in international tuna fishing during the 1950s and expanded in the 1980s to become the main method used on the high seas [45]. Skipjack in particular, but also yellowfin and bigeye, are the dominant catches. Without purse seine boats, our cans of tuna would remain empty and bluefin tuna farming would never have been such a success. On the minus side of the large purse seiners is overcapacity, which would never have been able to proliferate to the extent that it did. From the end of the 1960s the method became controversial in the Eastern Pacific Ocean yellowfin catch, as dolphins were actively targeted as well, with an estimated killing of hundreds of thousands dolphins a year. Soon after 2010, parts of the purse seine industry were once again contentious because of their use of FADs and the associated bycatch.

Purse Seine Fisheries Free School Set This application of an encircling net is often seen as one of the most sustainable fishing methods. This is primarily because there is little bycatch. Fishing free schools has more of hunting than harvesting. Free schools are often tracked by spotting birds who fly over, hoping to join in the feast when a school of fish is brought to the surface. The fishing boats target and encircle the mostly mature tuna they want to capture, close the nets and haul in the catch. When performed in a correct way, a 'set' like this contains few unintended fish swimming among the target catch. This method is used primarily for skipjack and

© Springer Nature Switzerland AG 2019

S. Adolf, *Tuna Wars*, https://doi.org/10.1007/978-3-030-20641-3_30

yellowfin tuna. The MSC label for this fishing method offers the consumer a reliable stamp of sustainable consumption.

Purse Seine Fisheries FAD Set Fish Aggregating Devices (FADs) are floating man-made objects used by fishermen because tuna gather underneath them. The unmanaged use of FADs is considered a problem because of the bycatch of juvenile bigeye and yellowfin and other fish, which is said to undermine vulnerable populations. There is additional bycatch, and FADs are creating a growing mountain of waste drifting about in the sea. Since FADs are widely used, good FAD-management is considered one of the priorities in sustainable tuna fisheries. With exeption of the socalled PNA-countries, FAD-management still is not common in the word seas.

Purse Seine Fisheries Dolphin Set This method is applied in yellowfin tuna fisheries in the eastern Pacific, especially by Mexican fisheries, but sometimes also by small scale handline fishermen that catch tuna one by one in Indonesia. The fishermen set up their nets around dolphins because they know that the bigger yellowfin tuna swim along underneath these groups. The technique incited fierce protests during the 1980s and 1990s because of the carnage caused among the dolphin population. Effective action by—among others—the regional management organisation IATTC has pushed back dolphin mortality almost completely. In 2017, this method was awarded the MSC sustainability certificate; not without contro-versy, however (see Flipper Wars).

Longline Fisheries A longline is, as the name suggests, a long horizontal line with floats stretching several kilometres, towed by a ship. Suspended from the line at intervals are hooks which can reach hundreds of metres in depth. It is used above all for bigeye and yellowfin tuna, but also for albacore, bluefin and swordfish. In wide open ocean areas, a longline can drag a few thousand hooks, luring everything that bites: tuna, swordfish, sawfish, shark, halibut and cod. But also albatrosses, cormorants, seagulls, seals, dolphins and sea turtles. The use of circle hooks can prevent much undesired bycatch, such as that of sea turtles. (The bad news is that circle hooks catch more white tip sharks, so the progress is relative). From the 1950s onwards, Japan conquered the high seas using longliners after its own waters had been fished dry [177]. In the 1960s the fleets of Korea and Taiwan, and later China, followed suit. Together with purse seine fisheries it is the prime tuna capture technique, even if it appears that the latter is still gaining ground vis-à-vis longliners.

Plenty of fishermen operate with integrity, but longliners as a whole do not have a particularly good reputation. As we have seen, the southern bluefin tuna stocks in the Pacific were systematically overfished by the Japanese longline fleet, who faked their officially reported catches. The ships are seen as uncontrollable lone wolves amongst the fishing boats. They can be on the open seas for years on end without calling into port. Many fly flags of convenience and transfer their catch onto reefers illegally on the high seas, without any checks. This can become a problem, espe-cially for vulnerable populations such as bigeye. Likewise social abuses, whereby crews are kept on board, almost as moderns slaves, for extremely long periods seem to occur above all on longliners. Longliners frequently crop up in scandals around IUU fishing. Killings and disappearances of crew members and inspectors add to a bad reputation.

Pole and Line Fisheries Pole and line, pole angling on a boat, became a worldwide industrial fishing technique in the course of the twentieth century. The Japanese developed pole and line in the 1930s for their tuna fishing activities in what was then Japanese territories in the Islands Pacific [31]. The Californian based US tuna fisheries developed their own pole and line fleet likewise. In Europe after the WWII pole and line was originally introduced by Spanish and French fleets, first in Basque country and later as part of their expansion towards the West African coast in the 1960s [45]. Fish caught using this method include skipjack and albacore, but also yellowfin and bigeye. The technique continues to produce results, but on a much more modest scale than purse seine or longline. Because of its gentle-looking, small-scale image, pole and line is regarded as a form of sustainable catch and quickly caught on. The UK canned tuna market especially threw itself *en masse* on this sustainably caught tuna, actively supported by Greenpeace campaigns. Often a blind eye was turned to pole and line's less sustainable features such as the use of bait to lure tuna, the poor working conditions of the fishermen and the substantial bycatch of juvenile yellowfin as a result of FAD use. The limited production of pole and line fishing and their limited range makes it unsuitable as a large-scale alternative to large distance sustainable free-school purse seine fisheries.

Drift Nets Not sustainable and preferably prohibited. Traditionally, driftnets—or drifting gillnets—were used for small-scale, passive catches, such as herring in the North Sea. From the 1970s onwards, these nets were made with new synthetic fibres by the Japanese and South Koreans, which meant that they could be towed in lengths of up to 50 km. These 'walls of death' did full justice to their name. Whales, dolphins, sharks, sea turtles, everything was dragged along indiscriminately only to be tipped overboard as drowned bycatch. In 1992 the United Nations banned drift nets from the international waters and 6 years later the European Union followed suit. In other national waters, such as those around the United States, driftnets are still allowed. It continues to be popular illegal fishing gear, amongst the Italians in the Mediterranean, for example [178]. Even worse, gillnets are still being used on a certain scale in the Indian Ocean, with a 650.000 ton of officially recorded tuna catch in 2010. This is only what the official data says. We do not have any data on the amount of dolphins, turtles, whales, sharks and other bycatch that get entangled in the nets and drown [102].

Trolling and Handline The smallest, often artisanal method for catching tuna. Trolling is the fishing method whereby lines with bait are dragged along. Handline speaks for itself. A boat, a line and a hook. Explicitly small-scale and sympathetic, very suitable for feeding your family and the village, but less practical for feeding the world. That does not mean this small scale fisheries represent an insignificant economic and social impact. Many small handline fishers have an important contribution in large coastal states like Indonesia and the Philippines. The first Fair Trade tuna in the world (even fair trade fish) was from a small Indonesian handline fishery from Ambon.

All these fishing methods have their own consequences for sustainability. Some were also controversial or even the cause of all-out tuna wars within the fishing industry, and between industry and NGOs. Everything revolves around the question: which factors are most important to the customer in terms of sustainability? The degree of sustainability could relate to ecological footprint and to the consequences for the tuna population and the wider environment, but above all for bycatch. In tuna fishing there was a harmful bycatch in the shape of dolphins, sharks, turtles and birds, but also of immature tuna, the so-called juveniles, which are not the direct target of the fisheries (although it might suit them), but were nevertheless scooped up in the nets. This bycatch had everything to do with FADs, a fishing method used alongside nets to track and catch tuna more easily.

FADs

The average tuna eater is probably not aware of this, but FADs are among the most important challenges in sustainable tuna fisheries management in the twenty-first century. FAD is an acronym for Fish Aggregating Device. The principle behind FADs is simple: throw a floating raft into the ocean and for some reason fish will soon gather around and underneath it. Is this due to its protective shade, is it the fish's curiosity, does it resemble a floating carcass? Biologists have not found the definitive answer yet, but fish are attracted to it, including tuna. Paul van Zwieten, Dutch scholar and specialist in tropical fisheries and aquaculture told me that a FAD act like a pointer to food in the area as floating things tend to aggregated where ocean currents meet. These are very productive areas due to mixing and upwelling and there is usually a lot of food for fish and thus for tuna that hunt on these fish.

Fishermen in the Pacific had been using this trick for a long time by throwing driftwood or bamboo rafts into the sea. These were anchored down and thus became gathering points for fish. The advent of large, industrial fleets from the mid 1990s, saw FADs further developed for intensive use. Bamboo became pontoons, the rafts became bigger. They were deployed, both anchored near shore as free-floating, in the oceans as gathering places where fish would concentrate. After the turn of the century, FADs were rapidly improved and became ever more sophisticated; the Spanish fleet acted as trailblazers, taking up FADs on a massive scale in conjunction with their purse seiner fleet. Originally FADs location were tracked by attaching VHF radio buoys to them with a limited transmission radius. Once powered with solar cells, FADs were fitted with satellite and later sonar equipment so that they could be located and the fish which had gathered underneath could be

(continued)

analysed with regard to depth, species and quantity. The FADs made fishing a great deal more efficient: the ships were now able to chart a course before they hauled in the tuna. FADs thus fundamentally altered the nature of the tuna industry: instead of chasing fish, it moved towards harvesting gathered fish in a 'cherry picking' way. The fisherman changed from being a hunter into a kind of farmer. The next step might well be to equip FADs with engines as a kind of drone-FAD, so that they can sail, complete with the tuna underneath, to the boats instead of the other way round.

With these rapid developments it emerged that FADs contended with a number of extensive sustainability issues. A study by Martin Hall and Marlon Roman from the IATTC [125] noticed that bycatch in the eastern Pacific included vulnerable species of shark, ray and sea turtle, but most of all bycatch of tuna species that are discarded because they are too small. In the 1993–2009 period average bycatch was 5% of the total capture, and 86% of that bycatch came from sets on floating objects, against only 10% from free school sets. The authors recommended a licence fee to control the expansion of the fast growing amount of FADs. Research into FADs, such as that by French researcher Laurent Dagorn and International Seafood Sustainability Foundation (ISSF) scientist Victor Restrepo, show a nuanced picture [72]. In their view, the FADs are not an 'ecological trap' but rather a coming and going of fish in a way that has not yet been properly charted. But it is indisputable that there is a problem with bycatch.

The FAD moreover appeared to be a favourite hangout for small, not yet sexually mature tuna. Substantial quantities of undersized juveniles were cast overboard again as bycatch, dead and gone. Or ended up mixed with catch in the cans. A blow for the future generations of tuna. Different types of tuna juveniles seemingly enjoy hanging out with the skipjack gathered around the FAD. The large-scale purse seine skipjack fisheries which use FADs the most, could, as a result of excessive juvenile yellowfin bycatch, inflict damage to their populations and this could even lead to a substantial shrinkage in the biomass. It might well be one of the causes why the yellowfin tuna population in the Indian Ocean, where the big Spanish distant water fleets deploy FADs in a massive way for skipjack fishing on a large scale [159], had ended up in the red after 2010. There was further inconvenience: after two decades of use many hundreds of thousands of FADs had been dumped in the oceans, sometimes with nets, plastic and lines reaching 50 m deep still attached. This is especially controversial because of juvenile yellowfin and bigeye tuna. An immense floating waste mountain of FADs was developing in which, in the worst case, an unknown number of fish, birds and animals were ensnared and drowned. Many FADs end up impacting reefs and foreshores in our oceans.

We are not talking about a minor issue here: around half of all global tropical tuna are caught using FADs. According to various estimates the

(continued)

number of FADs worldwide that end up in the sea during purse seine tuna fishing might well be around 100,000 [192, 217] in the second half of the second decade, chiefly in the Indian and Pacific Oceans and the Atlantic. This alarming amount accumulated year on year. The FAD became a rewarding topic for protest. Greenpeace was campaigning against FADs. Pew and WWF, too, pushed back against their negative effects and pressed for a policy on FADs. The fishing industry, via its organisation ISSF, was doing its best meanwhile to show that it was taking the problem seriously by declaring FADs a leading issue in their sustainability research. ISSF, for instance, developed biodegradable FADs with attachments that sharks and turtles would no longer get entangled in. Within a 10 year period ISSF spent 12 million dollars on research into making FADs more sustainable. Despite all good intentions, ISSF had little success with voluntary introduction of a tighter FAD policy amongst its members [27]. For the big fishing corporations, continuing to fish efficiently using FADs would weigh more heavily, until they were forced to comply with a different policy. ISSF tried to resolve this debacle by appointing RFMOs as the obvious organisations responsible for curbing FADs. Little came of this, either, not least because fisheries lobbies within these RFMOs tend not to be much in favour of reigning in FADs.

With approximately 70% of the tuna that was supplied to the main Bangkok tuna market was caught using FADs, fishing without FADs developed into a marketing instrument for sustainable tuna. Unfortunately, it only added to the general confusion around sustainability and did not make the conscious consumer's choice any easier. All kinds of 'FAD-free' labels appeared on cans of tuna. Most tuna buyers had not a clue what they referred to. At any rate, the labelling was not particularly reliable due to the absence of a serious, independently checked guarantee that no FADs had indeed been used in the capture of the tuna. The National Fisheries Authority (NFA) of Papua New Guinea, the biggest country in the PNA tuna cartel, felt compelled to label the FAD-free claims as nothing short of 'illegal': 'merely self-certification, with no validation' [13]. This said, many considered FADs indispensable tool in future tuna fisheries. Like any gear, it needed to be managed, so much was clear. Some FAD-fisheries, that offered enough guarantees of a well-managed fisheries, started to receive approval through MSC certification. But a serious FAD policy from the RFMOs was still absent in the second half of the decade. The most far-reaching restrictions were a maximum number of 550 FADs per boat in the Indian Ocean and a ban on FAD use for particular periods of the year in the Western and Central Pacific, but thorough monitoring of these policies was still out of the question. FAD management became one of the biggest sustainability challenges facing the tuna sector.

Monster Boats

Unlike FADs, which mean little, if anything, to most tuna consumers, gigantic tuna boats invoke spontaneous revulsion. The images of 'monster boats' hauling their enormous encircling nets from the water and disgorging their floundering cargo create the impression fairly effectively that we are witnessing something terrible. In a cold and superefficient way, the tuna stocks are plundered, a population throttled and the artisan fishermen robbed of their livelihood. No wonder that the giant super ships were a favourite target for Greenpeace. In 2014, this NGO published a review of 20 of the largest 'monster boats' under a European flag or owned by a distant water fleet which was subsidised by the EU [118]. These were colossal factory longliners and super seiners with a gross minimum tonnage of 4000 tonnes, able to transport at least 3700 m^3 of fish.

The industry rejected the accusations that their monster boats were not sustainable. The size of the boats as such was not a problem, the counter-argument went. Javier Garat, the Spanish president of Europêche, the industrial umbrella lobby group within the EU, contended, 'Whether the vessel is large or small really does not matter since it is all about how it conducts its activity. All EU fleets, both large and small, are under strict management systems. We could not feed the seven billion world population on small-scale [fishing] alone. Big vessels are able to exploit stocks not accessible to small vessels. Many areas throughout the chain of production depend greatly on the large-scale fleet. The larger vessels are also strictly controlled through satellite systems monitoring the location and landings, which allows for the collection of more accurate data and helps in the fight against illegal, unreported, unregulated (IUU) fishing' [87].

This was true. It is easier to control catches by big vessels, guaranteeing enforcement of a sustainability policy more easily. They possess better and more efficient freezing compartments and storage. They definitely offer better living and working conditions for their crews. Not long ago, corporate social responsibility had been given its own spot within the definition of sustainable fisheries after *The Guardian* in 2014 [96] had described the scandalous working conditions in the shrimp fisheries of Thailand, including the existence of exploitation verging on slavery. The Spanish tuna fleet, famous for its super seiners, was quick to set down its own working conditions as a component of its corporate social responsibility policy. The Marine Stewardship Council decided to include social conditions in the sustainability criteria that form part of assessments to qualify for the MSC label.

And yet the super seiner continued to be a kind of large-scale icon of evil, in contrast to the 'small is beautiful' image of methods such as pole and line, with their small vessels and quaint traditional fishermen. But in terms of ecological footprint, especially carbon dioxide emissions caused by use of fuel, tuna capture with the big purse seiners is by far one of the most ecological friendly ways to produce animal protein. Tuna caught like this consumes significantly less energy than farmed salmon, let alone protein from cows or pigs. As long as it is not produced in a can, it has the ecological footprint in terms of greenhouse gasses of a chicken egg [126].

An argument that cut more ice was that the unrestricted, heavily subsidised construction of monster boats obviously contributed more quickly to the growing worldwide overcapacity of distant water fishing, mostly in favor of the industry in developed countries. As early as 2006 a revealing study led by Rashid Sumaila and Daniel Pauly calculated that 26 billion dollars was spent annually on fishing subsidies, excluding fuel subsidies [231]. The governments of Europe and Southeast Asia in particular supported their fisheries. At least 15 billion dollars went into global capacity expansion, such as the construction of new ships and fishing fees. The true amount paid in subsidies must be considerably higher, as they aren't always granted transparently.

The tuna industry knows full well there is something wrong with this unbridled capacity expansion. 'If we're not able to agree on some kind of moratorium on new ships we'll be heading for a difficult time. Then we'll only be able to lose, all of us together,' veteran Italian tuna industrialist Adolfo Valsecchi proclaimed, when I met him at the 2016 Infofish World Tuna Trade Conference in Bangkok. Valsecchi was not exactly the prototype of a radical preacher for the sustainability revolution. In his made-to-measure suit and Italian shoes he cut a distinguished figure from the classic industrial tuna aristocracy. For decades, he had held sway as CEO over the big Italian canned tuna brand Mareblu. But in the central-Asian tuna capital of Bangkok, where in 2016 he chaired this key global tuna industry conference, tough words came from his mouth. No one seems able to push back the enormous overcapacity of the large-scale tuna fleet, Valsecchi told his audience. Not even the ISSF, the powerful lobby representing 75% of the global tuna trade, of which Valsecchi is notably one the founders. Led by the ISSF, the industry had designed a regulation under which its members were 'urged' to only build new ships if they derigged a comparable capacity in existing ships and take it 'completely out of service' [143]. While ISSF urged its members to stop expanding their fleets, large-scale fishermen quickly ordered new vessels just in case the regulation might get more teeth. In this way, some 270 new purse seiners and longliners were under construction at the time, including 17 super seiners with a length in excess of 78 m. These would not shrink the existing catch capacity in 2025 but expand it by some 1 million tonnes to 6.3 million tonnes. Added to this would be 600,000 tonnes from the so-called ghost fleets, Valsecchi warned. These were boats which ended up not being scrapped, but which more or less disappeared off the radar after they had been sold for a song and carried on fishing for a non-reporting flag state.

Some believed this was a conservative estimate, so that the total capacity within a decade or so might far exceed seven million tonnes. Perhaps 50% more than the total tuna capture taking place at that time according to the records. Many countries do not give reliable data. From 2014 it appeared that China handed out big subsidies for its distant water vessels rolling like hot cakes off the docks in the tuna landing port of Xiamen. As a new emerging tuna power in the region putting a sizable fleet on the map in a short space of time, it mattered a great deal to China. A persistent rumour circulated that China often commissioned three identical new tuna seiners at once. This was meant to sow potential confusion, especially around the new tracking and monitoring through satellite systems.

Be that as it may, subsidies increase overcapacity, and overcapacity put pressure on populations through illegal fisheries, upset competitive relationships and thwart a sustainable fisheries policy. Everyone essentially agreed about this. But in the absence of effective agreements which could be implemented on an international level, a solution was a long way off.

The State of Sustainable Tuna

31

Now that we have charted what kind of tuna we're eating and how they can be caught, we can move on to the next question: what is the state of the various tuna populations, captured using different techniques? This is indispensable for the management of endangered tuna stocks, to bring them back up to maintenance level in future. As we have seen, this is the work of each one of the five regional tuna commissions (the RFMOs) which were looking after the global tuna populations and whose objective it was to pursue a scientifically based public sustainability policy, rather like some kind of United Nations for tuna.

In the past, choosing which tuna you wanted to eat was a lot easier. Charlie the Tuna, the talking cartoon figure generations of Americans grew up with, was introduced by the American canning brand StarKist for its advertising and marketing campaigns in 1961. Charlie was a hipster avant la lettre. Perched on his head was a Greek fishermen's cap, extremely cool in the 1960s, and big, Onassis-style glasses; in his fin he clutched a can of StarKist. This gimmick could easily feature in a successful advertising campaign by Mad Men's Don Draper. Charlie believed that, hip and cool as he was, he had good taste, so should be canned by StarKist over anyone else. But StarKist would forever reject him because they were not looking for a tuna with good taste, but a tuna that tasted good. So Charlie was turned away with a laconic 'Sorry Charlie'. The punchline became a common expression. 'Sorry Charlie' performed for StarKist for more than 20 years, disappearing off stage only to return around the turn of the century, this time to advocate StarKist's canned tuna as healthy food. It appeared American consumers wanted tuna that not only tasted good but was also good for your heart, with plenty of beneficial protein and Omega-3 oils. Healthy tuna is still a sales argument but sustainable tuna appeared on the American mainstream market in 2018 as well. Chicken of the Sea, and Bumble Bee, two of the big three US tuna companies, eventually signed a purchase contract for sustainable, MSC certified PNA tuna. But unlike other brands globally the companies did not switch a single product line to MSC, just introduced it under a new high priced brand carrying the MSC logo. Before, the big three tried hard to prevent the introduction of

© Springer Nature Switzerland AG 2019
S. Adolf, *Tuna Wars*, https://doi.org/10.1007/978-3-030-20641-3_31

MSC certified tuna. Eventually Charlie too, will be called upon to do something with sustainability. He will probably start loving sustainable tuna and wants to be canned. But the only tuna StarKist likes is sustainable tuna. Something like that. 'Sorry Charlie'.

Delicious, healthy and sustainable into the bargain: it's not getting any easier for the average tuna consumer. Which species of tuna is fished unsustainable in volumes too large to endure in the long term? Which population is in a worrisome state? What about the bycatch? Which tuna is caught and processed by companies that respect the rights of their workers? Is the tuna legally ok? No issues of coloring or health-endangering treatments? What about the carbon footprint? Straightforward questions if you're standing in a supermarket isle with a can of tuna in your hand or see a tuna steak on a restaurant menu. But at the beginning of the third millennium finding simple answers to these questions was not always that easy. Instead of a solution, sustainability had become part of the 'wicked problem' around tuna capture. It had become a tool in the commercial battle within the industry, between companies and governments, and between NGOs and other stakeholders in the international tuna community. Sustainability was the new key word. Everyone had their own ideas about it. But what it meant was far from obvious.

The best place to start for someone wanting to understand something about sustainable tuna is Rome. It is home to an authoritative source of information: the Food and Agriculture Organisation (FAO) of the United Nations. The FAO regularly publishes statistics in its report, 'The State of the World Fisheries and Aquaculture' [91–95], considered the most important global data source with respect to fishing. It's also a place to find information about the numbers of tuna being fished out of and still present in the sea.

Counting tuna: since the Dukes of Medina Sidonia began their archives and Martín Sarmiento carried out his monastic toils, we have known this is no easy task. We do not see what is happening under water, only what fish is being hauled out of it. A dependable long-term recording method would provide an initial information basis to work with. Since the 1950s, the FAO has been trying to compile statistical material. The first thing that stands out is that tuna has a high ranking in the top ten of global wild fish catches [91–95]. At number three, right after Alaskan pollock and Peruvian anchovies, we find skipjack with more than 3 million tonnes in 2014. Unlike numbers one and two on the table it is clearly still on the up. If the trends continue then it is a matter of time before skipjack becomes the most widely caught wild fish in the world. Number seven in the top ten features yellowfin, with just under 1.5 million tonnes in 2014. Tuna, in short, is a top fish for the global fisheries. From 1950 to 2015 there was steady growth, with a peak of around 5 million tons in 2015. Within the tuna family, skipjack is responsible for the biggest catches by far (roughly half), followed by yellowfin, bigeye and albacore. Bluefin now brings up the rear, with an ever thinning line in the graph (Fig. 31.1).

This FAO data, the global basic administration of tuna, also reveals that purse seine fishing has become the most significant catch method by some distance. This is important to bear in mind: if you really want to make tuna consumption more sustainable you should concentrate on this catch method, which is responsible for

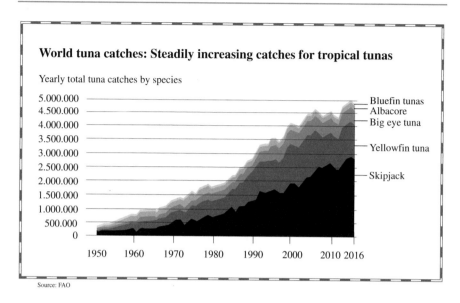

Fig. 31.1 The ever rising estimated world wide catches of tuna based on FAO figures. Is the sky the limit? (Infograph Ramses Reijerman)

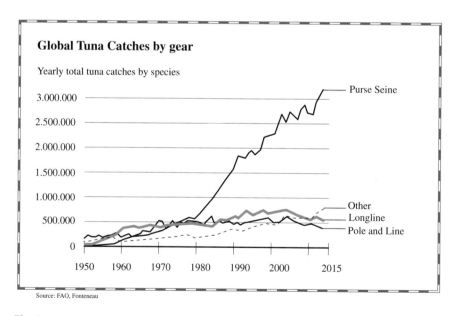

Fig. 31.2 To put things in perspective: purse seine is the main tuna catching gear. The net that should feed the world in a sustainable way (Infograph Ramses Reijerman)

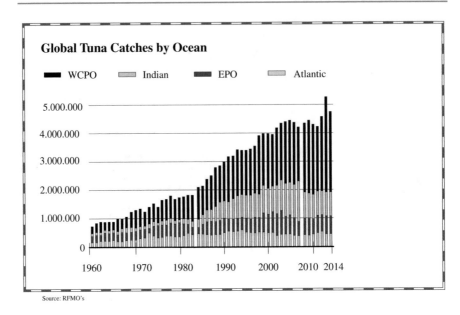

Fig. 31.3 The Western and Central Pacific became the main area where our tuna should be caught in a sustainable way, followed by the Indian Ocean, the Eastern Pacific and the Atlantic (Infograph Ramses Reijerman)

the bulk of tuna catches (see Fig. 31.2). Then there has been a major shift in the location where the highest volume of tuna is caught: the focus has clearly become the Western and Central Pacific Ocean. This is where the bulk volumes of tuna are caught, while catches continue to grow. The eastern Pacific, the Indian Ocean and the Atlantic contribute less as a catch area with growing concerns over sustainability and decreasing biomass (Fig. 31.3).

What does this amount to in terms of money, when are we talking about the global tuna catch? In 2016 the Pew Charitable Trusts Global Tuna Conservation Department made an ambitious attempt to put a price on the global tuna market [242]. It looked at seven tuna species in the tropical and subtropical oceans, representing the biggest commercial value: skipjack, albacore, bigeye, yellowfin, Atlantic bluefin, Pacific bluefin and southern bluefin tuna. These are the most important tuna stocks for the large consumer markets of Europe, Asia and North America. According to the calculations a consumer market value of 42 billion dollars was paid on the end markets for the approximately 5 million tonnes of tuna. The price at which the tuna caught was landed in the ports was around 10 billion dollars. In the value chain towards the consumer market its value had roughly quadrupled. That tells its own story: the greatest increase in value happens after the tuna has been fished out of the water. Countries that manage the tuna's waters would do well to penetrate the value chain more deeply with tuna processing and trade if they want to benefit from an increased share in this growth in value. For that matter, this assessment had not taken into account recreational fishing and the importance of tuna to the various ecosystems. It was a conservative estimate, Pew stated, which nonetheless showed that tuna fishing represented a substantial share in the global economy. By comparison, in 2016 the FAO estimated the world trade in seafood to be worth 142 billion dollars. Tuna therefore represented roughly a third of this economic value. According to Pew estimates, Indonesia was the biggest tuna fishing nation in the world when it came to how much was caught: more than 600,000 tonnes. When you look at the end product, the market for canned tuna with 30 billion dollars represented the greatest value by far. Of all tuna landed, almost 80% disappears into cans. Skipjack is the most important ingredient. At the other end of the tuna value spectrum was bluefin tuna. Even though it did not constitute even one per cent of the total catch in weight, bluefin represented 2.5 billion dollars on the global market. Bluefin was worth its weight in gold.

The above figures give an idea of how much was caught and of tuna's significance, as well as how and where it is caught, but they do not say much about how the various populations each develop in the oceans and to what extent we can talk of a sustainable way of fishing. To find an answer to these questions an entirely new player had appeared on the world stage: the scientist. The help of fisheries biologists was called in to work out the size of the fish populations, how they developed over time and which tuna species was overfished. And because sustainability was becoming increasingly indispensable in the marketing of tuna, science turned into a major tool, and sometimes even a weapon, in the battle for the global tuna markets.

Making Invisible Tuna Visible

32

There is something magical about it: making something visible that is largely invisible. A kind of metaphysics. Because one thing has not fundamentally changed since the ancient Greeks or Brother Sarmiento: throughout its entire lifetime tuna lives under water. What exactly it does there we barely know. This is not exclusively a problem with tuna but applies to all wild fish. It may be possible to make a more or less reliable estimate of the size of populations in a closed system such as that of freshwater fish in a lake. We can do something similar for groups of fish in the sea when we know that they live their fishlives more or less entirely in fixed places around coral or in rocks.

Yet 'stock assessments', whereby an estimate is made of the size of a population, are a common phenomenon amongst fisheries biologists and oceanographers. A few calculation models have been developed which look like an advanced form of mathematics and are used throughout the world. They are based on catch data. The results of the models are used to design sustainable fisheries strategies. These ingenious schemes and impressive calculations can not prevent tuna scholars from arriving at completely different conclusions about the results. If you follow the quarrels between Daniel Pauly and Ray Hilborn, both eminent figures in the field of fisheries science, you know that things can be pretty dismal amongst tuna scientists. That is the problem with estimates: they are not about certainty. There is therefore a great deal of margin for error which tends to be quirkily, and not always scrupulously, interpreted. As a result it does not take very much for a discussion to degenerate into rather noisome rows, replete with personal attacks on bias and conflict of interest.

S. Adolf, *Tuna Wars*, https://doi.org/10.1007/978-3-030-20641-3_32

In order to understand something about these quarrels we need to make a detour into fisheries science. To those readers who prefer to run away and skip this part of the tuna discovery journey: go ahead. I don't blame you. For those who stay: it might de useful to impress people as a starter for a table conversation during a dinner with tuna served. I will try to keep it simple.

If you want to create a sustainable policy for tuna fishing you need to be able to ascertain two things with some certainty: the size of a sexually mature population (spawning biomass) and catch volumes (or fishing mortality) of immature and mature fish. A healthy spawning mass can cope with an increasing catch. On the other hand: a shrinking sexually mature biomass plus an increasing catch should set alarm bells ringing. Sustainable policy is about monitoring and coordinating biomass and fishing mortality.

First biomass. An important tool to estimate how much fish is swimming in the sea is the so-called Virtual Population Analysis (VPA) model, born in the UK in the sixties [221]. This model presupposes that the population of fish from the same year class, i.e. fish that are born in the same season, gradually decrease in numbers because they are caught or die a natural death. You start with 100% of the population of a year class, a proportion of which will disappear every year until this year class has vanished in its entirety. If you have kept decent records of all catches and are able to assess the ratio of fish mortality for each year class due to natural causes, called natural mortality, you can get an estimate of the total population present now by adding up the numbers of all year classes still alive. 'You always need catches in order to find out something about biomass,' Dutch fisheries scientist Paul van Zwieten told me as he tried to explain the method to me. This method can be a bit tricky: it depends on the comprehensiveness and accuracy of the catch data. If there is large-scale illegal fishing or insufficient or no catch data, you've got a problem. It gets even seriously worse with the yearly rates of natural mortality at age of each cohort. Scientists assume that they know it, but in many cases this is not at all the case. As a result the errors in natural mortality rates produce cascading errors in estimated size of stocks.

Paul van Zwieten: 'The enormous attention there is on juvenile catches stems from this problem—if juveniles are caught and thrown dead overboard, then we do not know the actual fishing mortality on these small fish (new cohorts)—so then we underestimate the fishing mortality on this part of the stock, and then overestimate the total stock ... as we think that fish are dying through natural causes. This has been a very serious problem in earlier assessments, and a big area of contention between scientists who wanted to know about discards (because of the statistical numbers, not because of sustainability issues) and fishers who did not see the point of collecting these data. And this is also why Indonesia and the Philippines tuna catches are such a problem.'

The second downside is that VPA works like a rear-view mirror in a car: it is only possible to tell with some certainty what the original volume of a particular year class was when that year class or cohort of tuna has been fished clean to its last member.

Biomass can only be ascertained with some certainty in retrospect. This means a serious obstacle because a sustainable fisheries policy is for now and the future, but not for the fish that was. It is like driving when you only look in the rear vision mirror. Not a recommendation for a safe ride.

Then there is mortality. The scientific way to measure a sustainable level of fishing, or a catch that can be called responsible, got off the ground during the 1950s. The so-called Gordon-Schaefer Model [41, 215] is a 'bio-economic' model created by Scott Gordon and Anthony Scott, two Canadian economists. Please note: economists, not biologists. They designed their model as a biological application of the micro-economic principle of diminishing returns. Milner Schaefer further developed the model, actually published in a 1954 paper for the Inter-American Tropical Tuna Commission (IATTC, one of the tuna RFMOs), so we can say it was tuna that gave birth to the most used model for fisheries biology. The model sounds more complicated than it is: it defines the theoretical maximum quantity of fish that can be caught without jeopardising reproduction and the size of the population declining permanently. This maximum yield is known as the Maximum Sustainable Yield or MSY. It is the biggest volume that can be removed indefinitely. The concept of MSY as a reference point in the model was immediately taken up by the American Fisheries Bureau who saw it's potential and very shortly after the concept was framed into American fisheries law.

In the model economics comes into play and assumes that all other conditions or variables, such as environmental factors, stay equal. Economists love this kind of principle, they even have a name for it: [ceteris paribus]. Holding other things constant. The idea is that when the catch of a population stops growing at a particular point in time, and even begins to decline when more intensively fished, the point of maximum yield has been reached. MSY is a reference point against which fish can be caught sustainably until the end of time, without jeopardising the reproductive capacity needed to maintain the population. If you catch too few fish then there is more room for fishing and the maximum sustainable catch capacity has been underutilised. The stock is underfished. If you remove more than the sustainable optimum then the amount of fish caught drops automatically, because the population's reproductive capacity cannot follow suit and the sexually mature population declines. This is overfishing. If fishing is continued without restrictions then the population could eventually be seriously affected, leading to collapse. The practical outcome of the Gordon-Schaefer Model is in fact staggeringly simple: the size of the MSY population, that is the population swimming about in the water when the optimum amount of fishing takes place, is approximately half of the theoretical population in its unfished state. This is an easy rule of thumb that even a child can understand. So can we.

There is a second reference point for a sustainable catch that is used in the Gordon-Schaefer model: the so-called Maximum Economic Yield (MEY) [41]. This is the total volume of fish that can be caught with a maximum theoretical financial yield for the fishermen. This is achieved in a lower total catch than the MSY, which makes sense because the more closely the maximum yield is approached, the more effort it takes to catch more fish (diminishing returns) while the cost per unit caught is at least the same and will probably rise. The catch effort per fish increases in order to catch the last few fish as you approach the maximum yield. Many NGOs prefer the MEY as reference point in a fisheries policy, as it means less fishing and therefore a more cautious approach from an ecological point of view. And, the attendant argument goes, MEY creates a higher financial yield for the industry and resource rents for coastal states. This is true, in theory at least, but the industry itself takes usually an entirely different view. Fishermen prefer the total allowable catch through MSY, because it allows them to fish beyond the total allowable catch under MEY. When we suppose that the fishermen by their desire to maximise individual profits, it is easy to understand why: the industry as a whole can make more profit under MEY but that is of little interest to the individual fisherman. He would rather continue to catch as much as possible and is not that interested in what his colleagues are doing, certainly under all the uncertainties that are part of the model. The mortgage has to be paid, that is much more certain, just like the loan for his new boat and his car. Catching as much as possible increases his individual likelihood of achieving higher profit. That this is at the expense of the common good is of less concern to the individual fisherman in short term. It is a variant of the tragedy of the commons: capture as much as you can, that's what it's about for every single fisherman. It explains why fishermen tend to react in an irritated way when NGOs propose to use the MEY instead of the MSY as the maximum allowable fish quota. Just leave the yields to us, the fisherman thinks, and mind your own business. Operators in the WCPFC however encouraged limiting catch and upping resource rents, as this reduces supply, increases market prices. It results in profits for efficient operations and force out inefficient capacity. This has seen the demise of some fleets and the modernisation of others.

Another method for measuring sustainable catch that is directly related to the Gordon-Schaefer model, is the simple catch per unit effort (CPUE) [96]. Roughly speaking, this is about dividing the total catch or harvest by the efforts (fishers, or boats or costs). From an economist point of view it is the average product of the fishing effort. If a stock of fish is steadily caught by the fleet, the effort will steadily rise in proportion to the total biomass and the average catch per unit effort will steadily become smaller. It implies a linear relationship between the catch per unit effort and the total biomass, so to speak. The lesser the population or total biomass, the higher the efforts or costs to get it harvested. An implicit highly comforting effect of this model is that there never can be an extinction of the fish population. It's the economy stupid! Just think of it: when the population or biomass goes to zero, the average cost of harvesting will increase to infinity. That makes no sense to any fisherman. But it also seriously indicates the problem of this model: it is relatively easy to apply, but if certain assumptions or underlying data are not correct, the outcomes can be dangerously misleading. This is not a theoretical threat: one of the reasons of the infamous implosion of the Northern stock of cod at the North-American east coast in the 1980s was that the Canadian authorities had been in the comfortable belief that the stock was not under threat because of a stable CPUE. This was wrong: after the harvesting went into sharp decline new stock assessment methods learned that the biomass had been grossly overestimated during years. As a result the CPUE had been consistently miscalculated. After this shocking revelation the scientists recommended a drastic limitation of the quota, but that was not accepted by the industry. The parties arrived to a compromise of higher quota than were recommended. It was too little, too late: the Northern stock of cod collapsed [41], causing one of the most traumatic experiences in the history of sustainable fisheries management.

Despite this kind of setbacks, the Gordon-Schaefer Model and its reference point MSY, together with the VPA based stock calculation, have acquired over the years the status of an inviolable foundation on which most fisheries management systems are built. The entire idea of allocating quotas in fisheries, the so-called Individual Transferable Quotas (ITQs), is based on a distribution of the amount of fish available that can be caught sustainably, dividing this up over the individual fishing fleets. It is beyond doubt that the VPA type of models to estimate fish stocks might be the best that we have. But it also creates problems, because fishery scientists, like economists, have to work in a real world where the outcome of their fine theoretical models are being used by other forces—politics, industry, NGOs—to defend their own objectives. Every stakeholder underlines the importance of a science-based fishery management. And then starts looking for outcomes that serve their particular purpose. But the outcomes of the models are seldom hard facts. They produce the different options of possible pasts and futures in a sea of uncertainty. Reality tends to have unknown risk distributions and countless preconditions and circumstances which are not immediately recognised, let alone being [ceteris paribus]. Any change in assumption can change the outcome. In a world where nobody likes uncertainties, all the beautiful mathematical cathedrals of the models can turn into giants with feet of clay, tumbling under sloppy assumptions, insufficient data or simply circumstances no one had seen coming.

So no doubt, both biomass estimation through stock assessment and catch estimates or fishing mortality are of major importance for deciding on a sustainable fisheries policy. But as said, this science comes with many assumptions and uncertainties. That is also one of the explanations for the colourful quarrels between scientists Pauly and Hilborn (see The Battle of Science) [135]. Both men found quite a lot of important hidden data, but both from different perspectives and assumptions. Pauly concentrated on the catches as data source and discovered that much more fish tended to be caught than would end up in the official FAO statistics. His conclusion: because of the systematic lower reporting than the real catches, the mortality of a particular type of fish was much greater than had been assumed and in general a much more rapid exhaustion of the fish population loomed [189]. Hilborn rejected the catch data as the most trustworthy source of information about the state of the stocks ('A low catch compared with previous records does not necessarily mean fewer fish'). He found, taking into account stock assessments, that the fish's biomass tended to be systematically underestimated and was therefore much greater than always had been thought [257]. This was especially the case in the high seas and in particular for tuna. He concluded that there was much more tuna swimming around than was usually assumed and that overfishing wasn't such a big deal.

It was a fierce debate between scientists that both knew the uncertainties and assumptions in their field. At least Pauly and Hilborn agreed on one thing: more data were to be collected to improve the outcomes of all models. But in the sound and fury of their sea battles these words did not get much attention. The policymakers concentrated on the final conclusions and made their pick of the most convenient one.

Now if you really want to be able to look like you understand the world of sustainable tuna you need to have a look at one last model more: the Kobe Plot. It sounds like the title of a hard-boiled detective novel ('Bernie Gunther and the Kobe Plot') and it looks like an abstract composition of the painter Piet Mondrian (1872–1944). But the Kobe Plot is a practical diagram that summarize a sequence of several years of the results of the Gordon-Schaefer and VPA models, in which the spawning biomass is offset against the fishing mortality. In this model it is quite easy to see which tuna populations are overfished and which population is in a sustainable state or on the road to recovering.

The diagram owes its name to the first joint meeting of the world's RFMOs, which took place in 2007 in the Japanese city of Kobe. The idea behind the convention was to align the activities and approaches of the RFMOs globally. This would not be complete without a common sustainability diagram for tuna. The result was a tuna abstraction of the purest kind, a work of art as it were. Take the Kobe Plot of the Pacific bluefin tuna in de Western and Central Pacific (Fig. 32.1): the horizontal axis indicates the amount of sexually mature tuna by a particular biomass, also called Spawning Stock Biomass ('SSB'). The vertical axis shows the degree to which tuna is fished (fishing mortality) in relation to fishing mortality that produces the maximum sustainable yield ('F at MSY'). The four fields thus give a picture of the state of tuna in an indexed form. The colours function like a traffic light. The tuna population in the green field in the bottom right-hand corner is doing fine; the biomass of sexually mature fish is sufficient for sustainable reproduction and fishing is below MSY, so there is no overfishing. The yellow (sometimes orange) field in the top right-hand corner is doing less okay: sustainable reproduction of the population

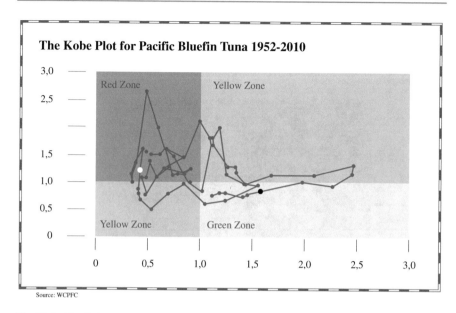

Source: WCPFC

Fig. 32.1 The Kobe-Plot for Pacific bluefin tuna over the period 1952–2010 under the explicit but unconfirmed assumption that the population was still healthy in 1952. The black dot denotes the start in 1952, the white dot is the situation in 2010 (Infograph Ramses Reijerman)

is not in danger, but tuna is removed over and above MSY and is therefore overfished. The yellow field at the bottom left is also doing not so great but more encouraging: reproduction is below what is necessary for the fish to survive sustainably but there is no overfishing and in principle the stock is on its way to recovery if the fishing pressure remains low. The red field top left is the graveyard for tuna: there is insufficient sexually mature fish for sustainable reproduction of the population and overfishing continues unabated. Tuna that does not manage to escape the red field is most likely doomed.

The dots in the diagram indicate how tuna, in this case the Pacific bluefin tuna, develops year on year (Fig. 32.1). In the case of this example the situation isn't all that rosy. Back in 1952, it was situated safely in the green corner (black dot). Then it ended up in the red danger zone top left, via the yellow top right-hand corner (overfishing by the Japanese). After a brief return via yellow bottom left (insufficient recruitment, but low fishing pressure) to the green bottom left, the fish found itself long-term in the red and was still there in 2010 (white dot). The Kobe Plot is intended as a tool for the regional tuna commission; as soon as a particular tuna arrives in the yellow/orange zone, action should be taken to protect the number of sexually mature specimens and to prevent overfishing of the population. It's a beautiful diagram, based on a sea of assumptions and hypotheses, all inherent from the models discussed. And the outcomes are subject to the power game around tuna. The aim is to return the derailed tuna stocks to the green field through effective management of the fisheries, as in Fig. 32.2.

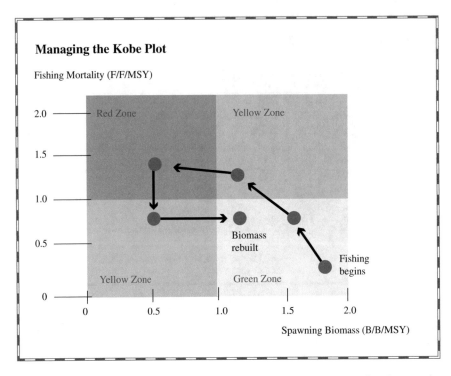

Fig. 32.2 This is how we like to see a good Kobe Plot developing. Fishing begins, the spawning biomass declines, mortality increases and before you even realise it, the stock is out of green into the yellow zone and further into the red zone. The tuna stock alarm goes off. As a result, we hope, stock management policies kick in. Mortality slows down, the spawning biomass increases and through yellow we come back into the green and can fish tuna happily until eternity. (Infograph Ramses Reijerman)

The Kobe Plot as a model is a joy to behold. It seems almost childishly simple to save a tuna population sustainably from extinction. Unfortunately, reality is more complicated. Because reliability of the data gathered plays tricks on the model here as well. Let's take another look at Pacific bluefin tuna during the period 1952–2010. But now let us assume that stocks from the start in 1952 had been overfished, but that we did not know this because catch data were not well recorded? This produces an entirely different picture of a tuna population that has never really left the red field during the past few decades. Assumption make quite a difference and can raise new questions. In this case: can we overfish an overfished stock for more than 60 years? We saw that in these years he RFMO was powerless to pursue an effective policy on Pacific bluefin tuna. But the stock was not in total collapse. Overfishing an overfished stock for decades: that seems hardly possible. Unless the stock is for some reason much more resilient than most scientists always assumed. What the scientists essentially admitted in their recommendation based on this Kobe Plot: too few tuna were born, catches had to be restricted further, perhaps especially those of young fish, in order to reverse the downward trend. Otherwise a complete collapse of the population loomed. Maybe for ever.

Unknown Past, Unknown Future

'They are estimates at best, made on the basis of extremely limited catch data and countless uncertainties'. Those wanting to know more about restrictions and the way in which science is used in sustainable tuna policy should consult Alain Fonteneau. The veteran French tuna expert does not beat around the bush. 'In my view, many of the projections of tuna populations are totally unrealistic,' Fonteneau believes. As so often, a smile played on his lips. Talking to this tuna expert is never boring. But this is a tuna expert we should take seriously: he has spent most of his working life at the French Institut de Recherche pour le Développement (IRD) researching tuna. We meet Fonteneau for another time in Brussels at the 2017 Atuna biennial European Tuna Conference. The scientist gives a presentation in which he makes mincemeat of the population statistics. 'Now it turns out that many of the estimates done over the past few years fall wide of the mark. But despite that they are still used to forecasts. How can we produce a forecast about the future of tuna if we don't even understand the past properly? As Kierkegaard said, "Life can only be understood backwards, but it must be lived forwards,"' Fonteneau explained. Whether the Danish philosopher had tuna in mind when he said this, is highly unlikely. But according to Fonteneau the dilemma applies effortlessly to our fish.

For years, Fonteneau has been exposing the sensitive taboo between fisheries management and science. What do all those splendid calculation models and Kobe Plots have to do with the real world? And is science able to guarantee its independence in a sector in which big money exerts such huge pressure? Fonteneau: 'The first problem is the data we use. With tuna in particular it is extremely difficult, if not impossible, to design models to read the exact status of the stocks with what we fish out of the water. For some major tuna stocks, the stock assessment analysis done by scientists are based on limited data, and the serious basic problems are not visible in the stock assessment results'.

'Take the Indian Ocean: we know nothing about the catches in the waters of Iran or Sri Lanka. Indonesia is a black hole. We're talking about large parts of the ocean and important fisheries. Or take skipjack: you never know where exactly the fish is in its enormous habitat of millions of square kilometres. Its migration routes change constantly. It has a flexible eating pattern, pops up and dives down again where it wants and breeds incredibly quickly. As a matter of fact, its true biomass continues to be a mystery,' says Fonteneau. Another example mentioned by Fonteneau is the state of bigeye tuna in the Western and Central Pacific. 'How come we have been seeing a growing catch of this tuna for years together with overfishing, as the WCPFC suggests? We

(continued)

saw that for 25 years two to three times the MSY was caught! How is that possible? This means that the MSY has been set ridiculously low. The Kobe Plot depictions are extremely questionable and probably simply wrong. I have little hope that the results are a realistic representation of the populations in the past and present.' (In this case the WCPFC scientific committee made a new assessment following Fonteneau's historical views. The results were totally different, with a much higher and more realistic MSY for bigeye.)

The second problem is the use of tuna statistics. Little remains of scientific nuance or the complexity addressed by scientific committees once the directors of the regional management committees get their hands on it. Politics plays a key role here, after all: depending on the power relations an estimate is selected that is most suitable for designing the most politically convenient management measures. 'Are we scientists or dancers?' Fonteneau wondered publicly when he visited the WCPFC as the head of the EU scientific delegation. Because the scientific results are often tacitly slotted into the picture that suits the fisheries best. A good example, according to Fonteneau, are the estimates of the skipjack population published in 2014 by the Indian Ocean Tuna Commission (IOTC). They looked excellent, no overfishing. But in subsequent years the catches actually appeared to be plummeting. While the FADs of the Spanish fleet became ever more technically advanced, with geolocation satellites and underwater sensors, there was a drop in the volume of skipjack caught per FAD instead of the rise you would have expected with all those efficient electronics. This negative trend was a widespread problem. Smaller skipjack catches and smaller fish: the typical signs of overfishing of a population. The IOTC's official conclusion that there was no overfishing remained highly questionable, although nobody really could say with much certainty what the real situation was.

Not that Fonteneau is opposed to evidence-based sustainable management, on the contrary. But it has to be well-founded science, based on sufficient data and done by researchers able to work independently. In 2006 Fonteneau rang the alarm bell about the Atlantic bluefin tuna's situation in an open letter to the bluefin group and scientific committee of the ICCAT. The fish was at high risk of being actively wiped out in the Mediterranean, he warned. But this was impossible to tell from the ICCAT scientists' reports. 'The lack of a firm and clear scientific message' combined with 'very poor fisheries statistics' is how he described the reports. Many member states did not fulfil their obligations to supply statistics including the three major EU countries active in the Med purse seine fisheries (Spain, France and Italy). So data were simply wrong or missing.

The mess at the tuna farms with their sloppy record-keeping, which was far from bona fide, only made matters worse. Enormous overcapacity of the purse seine fleet and a complete absence of efficient policies by ICCAT seemed to

(continued)

make the situation virtually hopeless, Fonteneau wrote. Now, well over 10 years later, bluefin tuna had survived, partly due to Fonteneau's pressing letter. He regards the strict quotas that ICCAT finally adopted as the most important cause of the strong come back of bluefin tuna, along with the introduction of a minimum catch weight of 30 kg, which gave the tuna a chance to recover quickly. It worked, although the estimations of the stock had been on the wrong side.

Management by the big regional tuna management organisations could therefore be effective for bluefin and other tuna, in the shape of catch quotas, a ban on fishing non-sexually mature tuna, close surveillance of compliance with the measures and combatting IUU fisheries. 'We need management of the tuna stocks more than ever, in particular since it looks like we are losing our grip on the use of FADs, the floating rafts on the sea that are used to make the fisheries more efficient,' Fonteneau warned. Management is above all plain common sense, the tuna scholar argued: a cautious approach in the management of stocks, especially as the figures are less reliable. The reality is that such an approach often bites the dust under pressure of the fisheries' interests. But Fonteneau is optimistic. He is convinced that the tuna populations are much more robust than had previously been assumed. The situation of the bluefin tuna in the Pacific underlines the resilience of a tuna population. 'Up to now we cannot speak of a disaster in tuna stocks,' this veteran tuna scientist states. 'None of the 21 most important tuna stocks was completely destroyed. Tuna has proved to be a stronger fish than we ever thought. We were just lucky. And the tuna was lucky.'

Eco-claims for Sustainable Tuna 33

At the end of the twentieth century private tools emerged and were applied in the market alongside the public one to achieve more sustainable tuna fishing. It fitted the *Zeitgeist* of an era of retracting governments and growing importance of the market [175]. The idea was that the market was able to design a self-regulating principle which would ultimately succeed in serving the sustainable consumer through certification, propriety labels and other eco-claims. Judging by the number of certificates and eco-labels, eco-claims can be called an unequivocal success. When you look at a can of tuna in the supermarket, study the packaging of frozen tuna or even read the wrapper around fresh (or thawed) tuna, you encounter various labels for the sustainability certification programmes for tuna. Some with credibility, even backed by a extensive chain of custody, others purely fictitious. According to the socalled 'Theory of Change', [209] the labels serve consumer power driven by demand, providing them with a tool to steer the global value chain towards sustainable fishing. You can choose what you eat. Meanwhile the system is in danger of becoming a victim of its own success. There are so many different claims that consumers have lost track of what they stand for. A brief inventory.

MSC label

Certified Sustainable Seafood from the Marine Stewardship Council ranks as the gold standard for sustainably produced fish. This obviously includes tuna. The certificate was introduced in 1996 on the initiative of food giant Unilever and the WWF but soon acquired the status of an NGO. It could be considered a partnership between industry and environmentalists in the implementation of a sustainability policy. The structure of the label guarantees a high degree of independence. As an organisation, the MSC sets the standard for a sound and sustainable policy against which a fishery is judged in order to qualify for the certificate. The assessment itself is carried out by an independent team of experts. Stakeholders (NGOs, companies and other interested parties) also have an opportunity to be involved in defining the standards and the assessment through a comment and objection procedure. The standard itself comprises three core principles: the sustainable level of a particular fish stock overall and in the unit of certification, minimising environmental impact and effective management committed to sustainability according to the rules and regulations in force. A fourth principle looks set to join the MSC standard in the shape of socially responsible fishing. A system of grades is used to judge a total of 31 criteria. For a fishery to be able to adorn its fish with the coveted blue logo it must have passed a detailed test with a minimum score. Less known is that the certification applies to both the fishery itself and to the so-called 'chain of custody', the transparency of the production chain which must guarantee that the sustainably caught fish is the same that ends up on the consumer's plate. In the case of tuna, which has one of the most complicated production chains, with countless processing stages, movements and transfers, tracing sustainably caught tuna within the chain is a far from insignificant issue. In view of the growing attention to illegal fishing and food safety tampering, traceability and transparency of the supply chain has become a major, if not vital issue. The MSC is expected to pay more attention to this as well, especially because enforcement of traceability and transparency is not always all that tightly regulated.

The MSC is a dynamic standard, which adapts to developing insights with regard to sustainability. The certificate is examined every so often and can be revoked prematurely if the particular fishery no longer meets the criteria. This can be the case when the tuna population has declined and is at risk according to the regional management organisation. The MSC, which has its headquarters in London, is a robust organisation with a strongly international character. It is financed through sponsorships and a licence payment for use of the logo. Certification is not cheap: the procedure can easily cost a fishery more than 100,000 euros and takes several years to complete. Certification has nonetheless experienced impressive growth: out of the six million tonnes of tuna caught globally in 2016, 16% was MSC-certified.

The first tuna fishery to get the MSC certification was the relatively small North Eastern Pacific Albacore trolling fisheries of the US-fleet, that was certified in 2007. Following a long series of challenges and adjudication, the MSC stepped into the really large-scale tuna market in December 2011 with the certification of the Pacifical-branded freeschool skipjack (see Second Flipper War). After completing the chain of custody of the Pacifical-branded PNA MSC freeschool skipjack it finally entered the market later in 2013. Later Pacifical freeschool yellowfin was certified in February 2016, followed by the big tuna market player Tri Marine achieving MSC certification in mid 2016, all from the waters of the PNA [2]. This provided the label with the potential to bring a substantial volume of sustainable tuna onto the market. This was followed by a long line of additional tuna fisheries, including Japanese pole and line skipjack and albacore, pole and line skipjack from the Maldives, pole and line skipjack from the north-eastern Pacific, and Fijian longline-fished albacore, as well as yellowfin caught in the eastern Pacific. In the European and Australian markets MSC-certified tuna has thus become a familiar feature, with about 40% market share under John West brand.

Although the MSC ranks as one of the most thorough private and voluntary sustainability standards, it is subject to criticism. From the outset, Greenpeace has repudiated the label as it believes the sustainability criteria are not stringent enough, especially in trawl fisheries. In its certification of some specific fisheries—such as krill from Antarctica—the MSC is regularly lambasted by other NGOs. Scientists such as Daniel Pauly also ask more fundamental questions with respect to the independence of the certification, which is ultimately paid for by the companies themselves. A bit like a butcher inspecting his own meat. The MSC counterargues it is no more than an organisation that lays down the rules for its certificate's sustainability, it only sets the standards. The actual admission procedure is outsourced to independent, commercial assessment firms which are contracted in to assist with certification. These firms are paid by the companies applying for certification. 'After reading over a hundred assessments and related documents, we could not help feeling these assessors were biased in bending the rules in favor of their clients,' a scientist wrote in 2012, having investigated several eco-labels [104]. Conflict of interest can never be ruled out entirely in private enterprise. But the MSC does try to limit the potential for this as much as possible by including extensive involvement from stakeholders, transparent appeals procedures and a Peer Review College of independent experts which guide the procedures through the definition of the MSC standards and even independent adjudications. Amongst the eco-labels for tuna, the MSC therefore offers the most guarantees for credibility and independence.

Since certification on sustainability has become an important marketing tool, the criticism on MSC can also get strongly commercially motivated, with attacks from competing tuna fisheries that want to promote their own sustainable tuna. This was to a significant extent the case in 2017 with a challenge to MSC recertification of the PNA freeschool skipjack tuna lead by the chairman of the International Pole and Line Foundation (IPNLF, see below), a leading UK tuna importer himself. Under the flag of the UK 'On the Hook' organisation the PNA recertification was attacked with the help of several NGOs and even scientists. Although the objections were thrown out by an independent adjudicator in New York early 2018 and PNA recertified for 5 years subject to annual audits, the professionally run campaign against PNA and MSC continued in the UK, including sponsored questions in Parliament and a parliamentary hearing, debating the MSC and PNA certification. Although all but three PNA countries are UK commonwealth, the PNA was never invited to participate in the parlementary debate.

The Fisheries Improvement Projects (FIPs) generally do not have a label and the consumer will only find them mentioned in the small print on the packaging of some tuna steaks or cans, if at all. It is more directed to retailers who want their fish to be sustainable or at least moving towards sustainability. FIPs are a new kind of quality mark with which producers and fisheries try to establish a sustainable reputation. Unfortunately enough, many see it as "MSC light". What is involved? The idea behind a FIP programme is to fund a particular tuna fishery for the expensive improvements and procedures associated with becoming more sustainable and ultimately MSC certified. This might be because they are too small-scale. FIPs exist in order to help these fisheries some way towards becoming more sustainable. It usually involves a partnership between NGOs, scientists, companies and funders who assist the fisheries. The Sustainable Fisheries Partnership is one of the organisations that helps coordinate FIPs. Although not all FIPS are small artisanal fisheries, it usually concerns fisheries in developing countries which are not able to access sustainable consumer markets in the West. There are hundreds of FIPs, including quite a few for tuna. A good thing for sustainable tuna, you'd think. But there is a snag. A FIP tends to derive its credibility from the notion that it acts as a stepping stone to full certification, as issued by the MSC. But in practice this very often is not the case. FIPs are difficult to control, often showing little transparency and inadequate monitoring, which makes it hard to assess whether the sustainability objectives are being met. The published route to MSC certification often takes place without clear checks or a regular time-frame so that no one actually knows what the status is. Reports suggest that only one FIP in tuna ever reached MSC certification. That was the reason why in 2015 a group of scientists published an article in *Science* [213] warning that FIPs risked becoming a 'race to the bottom', a relatively cheap way to compete with sustainability labels like that of the MSC, but which in the end would not lead to the kind of improved sustainability that can stand up to scrutiny. This could ultimately lead to a form of 'greenwashing' and undermining sustainability. Only clear and strict conditions can prevent this. Although well-monitored tuna FIPs exist, it remains somewhat difficult for the average tuna consumer or retailer to distinguish between a good FIP and a bad one. Consumers have to rely on the critical judgement of the retailer who opts to stock FIP tuna.

Dolphin Safe labels

Across the globe there is an array of Dolphin Safe and Dolphin Friendly labels of varying quality and significance. They have one thing in common: dolphin safe does not say anything about tuna, but instead they purport to indicate that no dolphins were killed or injured during tuna capture. This unilateral sustainability standard for one species, outside a wider consideration of the ecosystem, makes the label highly outdated. And even more important: the labels' credibility varies greatly.

The best known and most widely used is the black label with the word 'Safe' below an image of a jumping dolphin from the NGO Earth Island Institute (EII), which administers the logo under the US Dolphin Protection Consumer Information Act. EII was at the vanguard of a successful approach to the massive dolphin mortality in the north-eastern Pacific. This label is borne 'voluntarily' by seven hundred fisheries and canning companies, but is controversial for several reasons. Firstly, serious question marks could be placed around the voluntariness of the label and the fees involved, about which more later. Certification is not monitored in a standard way by independent on-board observers of the vessels. Verification that no dolphins are being encircled, injured or killed depends to a large extent on statements by the ships' captains or head office. This makes the label distinctly weak. Another unhelpful factor is that the label is awarded in areas where dolphin fishing is not systematically practised. The label wrongly creates the impression that the fishery has taken steps to protect the dolphin. This could be considered a form of consumer deception. The World Trade Organization for its part regards the label above all as a tool for the American tuna industry to keep Mexican competition out [247]. The label does not have any independent, scientific support for its procedures, nor does it have serious stakeholder engagement. The data is based on information supplied by the companies themselves. In essence, this label guarantees no more than the companies self-declaring that they are not guilty of killing or injuring dolphins [169], leaving room for all kind of unsustainable practices.

Considerably more credible is the Agreement on the International Dolphin Conservation Program (AIDCP) of the Inter-American Tropical Tuna Commission (IATTC), the regional commission which looks after the north-eastern Pacific where dolphins are caught alongside yellowfin tuna. The label enjoys 100% verification by independent on-board observers who ensure that no dolphins are killed or injured and that they are released after encirclement. Infringements are punishable by sanctions. This makes the label appreciably more robust than that of EII. EII on the other hand campaigns against the AIDCP label because it permits deliberate chasing and encirclement of dolphins. The EII label claims it does not permit this but omits in many cases independent verification of the catch process, making this guarantee rather inconsequential.

The Dolphin Safe label is tinkered with to quite an extent. A widespread logo in Spanish supermarkets resembles a jumping dolphin. This logo connects to a webpage with information about sustainable tuna fishing which appears to be linked to a page with some general information about sustainable tuna capture from the International Seafood Sustainability Foundation (ISSF). This umbrella lobbying partnership of the big tuna canning industry, together with the WWF, pretends to develop science-based sustainability measures. ISSF is in the process of developing a sustainability policies for its members but cannot issue certificates and does not issue an eco-label for the safety of dolphins.

Friend of the Sea label

Started in 2006 in Italy, Friend of the Sea (FOS) is the European arm of Earth Island Institute (EII) Dolphin Safe Project. With a small staff (six people including its founder and director), FOS works in a similar way to EII, i.e. no cover by independent on-board observers, relatively few scientific criteria, minimal stakeholder participation, quick and superficial procedures and opaque financial accountability. An important distinction between EII and its dolphin label is that FOS not only certifies all wild fish but also all farmed fish and even aquariums and shipping lines, which could be called quite a miracle considering their limited staff resources. The sustainability criteria for assessment are not all that exhaustive, to put it mildly. For tuna you need to be an EII "member". Species that are on the IUCN red list or are exhausted, overfished or recovering according to the FAO are off the FOS list, and only a restricted level of bycatch (8%) is allowed, though not monitored. Moreover, the fish cannot be caught illegally or by vessels sailing under a flag of convenience. A lack of independent observers means FOS-certified fisheries are not given a particularly hard time in practice. Within 2-4 weeks after submission of some papers to the Italian certifiers the eco-label is issued for a small sum, plus an annual license fee for use of the logo. Fundamentally, the 'eco-label' says little more than that the company self-certifies that it fishes sustainably and legally. It says: we declare that we are no crooks. This isn't much help to the sustainable consumer. (As a business model the label can be considered successful though, according to the FOS's annual report; its turnover for the year 2016 was 1.1 million euros, 25% more than for the previous year. 'A healthy growth,' the report stated.) Its director likes to report that the FOS is growing five times as fast as the MSC and is therefore the only true major certifier of sustainable wild fish. The label has even been expanded with a 'sustainable shipping standard' for transhipments of fish, cruise lines and whale watching boats. It is also doing pretty well in the tuna fishoil business. A healthy growth indeed.

FAD-free labels

Capitalising on the idea that capture without the use of Fish Aggregating Devices (FADs) can be more sustainable, so-called FAD-free claims appeared on the market from about 2012 onwards, notably in Europe. For most consumers is it probably a non-existing label: most people buying tuna do not have the faintest idea what FAD stands for, or mistake it for low-fat products from the diet sector. It is not really aimed at the consumer, and more at business to business level in the tuna industry itself between trade and the retailers as a self declared guarantee that suggests sustainability. Unlike MSC labelling, it FAD-free almost comes free of cost. The 'certification' mostly ends up with own-brand cans of skipjack tuna for a number of German supermarkets. The fish is said to come from fisheries from Equador and the Philippines. In 2015 the European representative body for fishermen Europêche stated that this was pure self-certification on the part of the companies and therefore did not offer any guarantee that no FADs had been used. Europêche stated that this could be 'unregulated, fraudulent and highly misleading to the consumer' [86]. This criticism from inside the industry itself was not surprising; the big Spanish fisheries that Europêche represents fish for the most part with FADs and are therefore not happy with the label. But it did stand up to scrutiny. This was also evident from a response by the fishing authorities of Papua New Guinea where some of the so-called FAD-free skipjack appeared to have come from. Without the MSC label the FAD-free label is of no value whatsoever, the authorities in PNG declared. The claims by EII and FOS that they cooperated with Papua New Guinea to allow independent on-board observers to monitor FAD-free fishing was described as 'absolutely not correct' [13]. Philippines fleets within industry circles validate their freeschool claims with that they have fished freeschool for decades. A claim that further has no validation.

FAD-free has been designated by some 'the new Dolphin Safe' because of its rather non-mandatory character and the lack of independent supervision. Others call it part of a 'race to the bottom' [213] as far as tuna eco-labels are concerned and simply a form of 'greenwashing' or cheap deception of the market.

Pole and line labels

Not a tuna eco-label in the strict sense of the word, but an indication on the can that the fish is claimed to be caught using a pole and line gear. This claim is often accompanied by MSC certification, like for instance with tuna coming from the Maldives. An important distinction between this and FAD-free is that much pole and line fish is actually caught using this method, in particular when the fishermen are members of the eponymous International Pole and Line Foundation (IPNLF) [144]. Pole and line as a sustainable catch method is promoted heavily by Greenpeace (this organisation, on the other hand, is not a supporter of the MSC) and is especially popular on the British market. There are drawbacks of this gear however: there is far too little volume of pole and line caught tuna to meet the potential global demand for sustainable fish. Coupled with the lack of independent observers in the chain of custody this is an important weakness in the sustainability guarantees the claim suggests. The other weakness is the fact that pole and line, despite its eco-friendly image, uses FADs, causing a significant bycatch of juvenile tuna, in particular yellowfin. There is an issue with the amount of bait the boats use to attract tuna: live bait fishery has shown significant adverse impacts on reef ecosystems. Last but not least: working conditions on the small-scale operation pole and line boats in many cases are questionable.

Industry promoted schemes

Though not considered by any means certification organisations, several fisheries industry partnerships are presenting a kind of sustainability claims through their sheer membership that are sometimes difficult to distinguish from sustainable fishing certificates. The idea is that members are in some way complying self-imposed standards that pretend to make its fisheries sustainable. The International Seafood Sustainability Foundation (ISSF), a partnership that claims its members cover 75% of the global processing capacity in the industrial canning industry, on many occasions suggests that membership in itself is a guarantee for promoting sustainable fisheries. In the words of ISSF-president Susan Jackson: 'When it comes to the global tuna fishing, industry must play a leading role in assuring the long-term sustainability of global tuna stocks' [143]. Its members like to echo this idea: buy from ISSF members and it is sustainable. Reality, however, is much more complex. It is without doubt that ISSF with its Proactive Vessel Register (PVR) contributes to transparency in the market in the fight against IUU. The promotion of non-entangling FADs helps to improve sustainability. But this cannot be taken serious as a form of real certification with an independent assessment. The fact alone that compliance to the ISSF 'standards' is seldom enforced, is telling. Certification is a term that ISSF itself carefully avoids. The logo is never found as a label on cans. With reason. It is not up to the industry to check the industry. It can even be harmful for consumers when an imposed standard is used as an excuse to close markets for outsiders.

Less restrained in this regard is the responsible tuna fishing standard (APR), introduced in 2017 by the Spanish Organization of Associated Producers of Large Tuna Freezers (OPAGAC). The APR was announced to be displayed with a label on the cans. During a presentation at a European Parliament act in Brussels OPAGACs energetic managing director Julio Morón explained that together with a FIP that OPAGAC is developing with WWF the APR is 'the best ways to achieve a full sustainable concept: socio-economic and environmental'. A closer look though [181] learns that APR is foremost a guarantee that the boats of tuna fleets of the seven big Spanish fleet-owners comply a standard for labour and social rights on board. The standard is assessed by the Spanish Association for Standarization Aenor, and is the first in the global fisheries that guarantees on-board compliance of the labour conditions that are required by the International Labour Organization. This is without doubt an important step forward in social sustainability. But concerning the environmental aspect, the standard is remarkably less demanding. It requires the use of non-entangling FADs, just like ISSF. There is a training of vessel skippers and captains on the status of fishing resources and the impact of the fishing activity on the marine ecosystem. All this is verified with 100% observer coverage on all the vessels and the use of electronic monitors, states OPAGAC.

The APR label is more of a message to the Spanish distributors, Morón explained. During its presentation in Brussels the OPAGAC managing director also told his audience that as far as he was concerned APR should serve as a referent standard for EU fish import. That sounded familiar. Considering the ongoing efforts of the OPAGAC lobby in Brussels to close the EU market for third country competitors (because of 'lack of level playing field' = lack of any standards and thus dishonest competition by these flag state countries) the APR seemed to serve this purpose well.

EU catch certification scheme and carding system

This is not a (private) certificate in the sense of a label on a product, but an important public initiative in the global fight against IUU fishing [148]. Regulation of fish imports by the European Union, the biggest import market for tuna, plays a trail-blazing role in making fishing more sustainable. The EU employs a catch certification scheme for its fishing fleet, including distant water fleets, which should guarantee that the fish has been caught in accordance with all legal and specific management rules. Nine countries outside the EU have a structure in place that supports the required catch certificates. These certificates show where and how the fish was caught. A weak link in the certification is that it involves a great deal of easily forged paperwork. If you run a paper certificate through the photocopier a few times little remains to verify. A second problem is the number of certificates, 40,000–60,000 per year for all fish imports in the big EU countries, which makes effective control only possible through random checks. Digitisation of certificates is under way and the implementation of new techniques such as blockchain may well lead to compelling improvements.

But a first significant hurdle in the long and complex value chain to transparency and traceability of tuna has been taken, in particular in the fight against IUU fisheries. Carding consists of a warning system using yellow and red cards to combat illegal fishing in countries that export to the EU or have a fishing fleet under their flag which supplies fish for the EU market. It involves the European Commission checking out to what extent the country concerned meets the international fishing and conservation requirements as defined by the EU, and whether it is a full member of the required tuna RFMOs and adequately controls its fisheries. Developing countries can receive technical and administrative support from the EU for capacity building. If the cooperation does not run smoothly or if the requirements are not adequately met, the country receives a yellow card. It is a warning that improvements must be introduced in the near future. If this does not happen a red card is issued, which involves tough sanctions such as blocking of access to the EU market and a ban on European vessels fishing in the waters of the country concerned, pending a solution to the problems. 'A red card is the end of the relation with the EU and trade stops. For many countries that is a major issue. And even if there is no big impact, the stigma alone can be damaging,' former director of DG Mare Stefaan Depypere explains.

Repeated infringements lead to hefty fines. A country that can prove that it is meeting the requirements is given a green card and renewed, unrestricted access to the European market. The carding system and the catch certificates constitute a creative partnership between states, whereby the EU uses its monopoly position on the demand side of the value chain as a lever to stimulate global sustainable tuna capture. An analysis conducted in 2016 by a coalition of NGOs (WWF, Oceana, Pew and the Environmental Justice Foundation (EJF)) [80] concluded that the system was handled not very consistently in each of the main importing countries (Germany,

France, the United Kingdom, the Netherlands, Spain and Italy), especially when it came to checking catch certificates from high-risk countries with a yellow card. But the system was also defined as 'one of the greatest achievements (. . .) to further on-the-ground improvements in fisheries management standards in third countries'.

With regard to tuna, quite a few firm yellow cards were issued [148]. Tuna superpower Thailand, home of the main canning and trade, decided to comprehensively clean up its act following a yellow card it received in 2015 for illegal fishing coupled with persistent reports about slavery in its shrimp industry. Taiwan, one of the largest suppliers of tuna to Bangkok, submitted willingly and without hesitation to EU inspections following a yellow card in 2015. A red card could cost almost 250 million dollars in annual export damages, the country had worked out. That same year, the Philippines succeeded in ridding themselves of their yellow card, which had threatened their most important export market, after making substantial investments to tackle its illegal fisheries. South Korea was given a yellow card for illegal fishing off the West African coast by its distant-water fleet. Once their fishing vessels had been equipped with a monitoring system, South Korean fishing boats returned to green. Ghana saw its EU exports running into danger following a yellow card for illegal fishing in its EEZ in 2013. Having introduced better controls at sea and traceability of its products, the country achieved green again. Vietnam pulled out all the stops in 2018 with port inspections after its 'blue boats', notorious for their illegal catch, had been given a yellow card in 2017. It is worth noting that virtually all countries that had received a yellow card showed full willingness to cooperate to get rid of it as soon as possible.

Who labels the Labels?
The fights about the labels had exposed a structural weakness in the new way in which sustainability of fisheries was organised. In a world in which certifications were beginning to play an ever-increasing role as a guarantee of sustainable production quality for the consumer, there was no organisation to take regulatory action in potential conflicts. In the end, the only driving force behind these private labels was the market. The market could be capricious and unreliable when it came to quality standards. The United Nations FAO offered a directive with which eco-labels had to comply if they wanted to be taken seriously. The MSC seemed to come closest to meeting this, but there was no formal supervision of compliance with FAO requirements. The way in which the standards were laid down, the credibility with which the certification procedures were implemented and enforced, the transparency of the organisations hiding behind the certificates, the way in which various stakeholders were involved in the certification process, the robustness of the chain of custody, possible objection procedures or sanctions for the certificate holders: it was all left to the free market and public debate as to how this was dealt with.

(continued)

The consumer, who soon seemed lost in all the technical details of procedures and standards, stood empty-handed. National governments never had much of a grip on sustainability certification and had no mechanism to scrutinise and correct programmes, unless they were dealing with outright fraud, deception, illegal tuna or a danger to public health. Outside government, there were initiatives in the making. There was the global ISEAL Alliance for eco-standards and certificates, whose members committed themselves to a code of good practice [145]. The MSC was the only wild fish label to be a member, which was all well and good, but it did not get the consumer much further in the forest of certificates. Another initiative was the Global Seafood Sustainability Initiative (GSSI) which had been set up by companies, NGOs and governments and was supported by the FAO to develop a benchmarking tool [121]. This was intended as an independent, global touchstone for certification programmes, using the FAO Code of Conduct for Responsible Fisheries as its starting point. But testing proved to be rather problematic. At the beginning of 2017, only three certificates had stood the test of criticism. By far the most important one was once again the MSC. Because of the sensitivity of the topic, the GSSI, alas, did not publicise which labels had failed, so this meant continued guesswork and lack of transparency for the buyers who wanted to know more about which labels they could trust and which were better avoided. Sustainability certification remained the Wild West in tuna fishing to some extent, without a sheriff to put things right.

Last but not least there are some useful tools in the shape of various apps to help out sustainable consumers who would like to know what's what when shopping in their supermarket. The best known are those introduced by the American Monterey Bay Aquarium Seafood Watch, the VISwijzer seafood guide by the Dutch Good Fish Foundation and the Good Fish Guide of the UK Marine Conservation Society. All can be downloaded from the App Store or Google Play. The apps feature different tuna species and regions, as well as other species of fish, which denote the degree of sustainability of the particular fisheries using a traffic light system (red, amber, green). Seafood Watch/ VISwijzer are not eco-labels, but they are a practical and reliable guide for helping people make a more sustainable choice. They have their limitations though, when it comes to the supply chain and knowing that the sustainable fished tuna is really the one on your plate (the so called transparency of the chain of custody).

Sustainability had become a new mantra in the global tuna market at the beginning of the third millennium. This sounded rather ambitious for perhaps the most important wild fishing industry in volume worldwide, with a turnover of tens of billions of euros and operating largely on the wide open oceans with no oversight. Millions of people were dependent on it for their food, work and income. This was a global industry, netting different kinds of tuna using different gears. In this complex environment, sustainability had grown into an indispensable marketing argument and had become the stake in an often heated battle between business, governments and NGOs. Regional management organisations and private certification labels tried to steer tuna into more sustainable waters with their policies. Tuna scientists challenged each other with arguments that might be diametrically opposed. The EU, the biggest market in the world, tried to tackle illegal catches and social ills via a clever system of import controls. It had become the stage for entirely new wars around tuna, hybrid sea battles around sustainability and business interests, full of 'wicked problems' the solutions for which tended to harbour the beginning of new troubles. Some of the sustainability policies started to bear fruit, it seemed. So management of tuna stocks was not a lost cause after all. That was a positive message. The more uncomfortable question was if enough action was being taken to cope with the devastating threats that a growing number of 'tuna wars' were causing on a global scale.

In the next part of the book we will give an overview of some of the most important hybrid tuna wars that were fought at the start of the third millennium and that were always linked with the issue of sustainability.

Tuna Poachers' Wars

34

It's party time in General Santos City, better known as GenSan. The port at the tip of the southern Philippine island of Mindanao is celebrating its annual Tuna Festival, a kind of carnival, but all focused around tuna. It includes a colourful parade with floats displaying spectacular giant plaster and papier-mâché tuna. People sing and dance in the big festival grounds in the centre of the town, which are kitted out with fairground booths and food stalls. Everything revolves around tuna here: tuna flags, colourful, tuna-shaped hats, T-shirts depicting tuna. In the port, men sprint off the boats along terrifyingly narrow gangplanks carrying hefty yellowfin tuna on their shoulders in an attempt to break the Guinness World Record with the largest haul of tuna ever landed at a fish auction (Fig. 34.1). It all culminates in a procession through the city with a prize going to the float with the most striking giant tuna.

GenSan has been the 'Tuna Capital of the Philippines' since the 1970s. It was yellowfin tuna above all to which the city owed its rapid advance as the country's major tuna port. The fish swam in big shoals in the coastal waters nearby. GenSan became a boom town, thanks to the plantations introduced by the American fruit giant Dole and thanks to tuna. Within a few decades, the sleepy fishing village on the bay expanded into an industrial city of considerable stature with half a million inhabitants. The port became the centre for fresh and frozen production of sashimi-quality yellowfin tuna and home to seven large canneries and processing plants with big names such as Gentuna canneries of the Century Pacific Group, RD Fishing Industry, San Andres Fishing Industries Inc. (SAFII), PhilBest, Ocean Canning, Seatrade Canning and Alliance Select Foods, brands that would become well-known far outside the country's borders. Ronnel Rivera, son of RD founder Rodrigo Rivera, was the city's mayor. The families that control tuna fishing are in charge of the economy and the administration of GenSan.

During the festival a major tuna conference is organised by the tuna companies in a well-secured hotel. Violence is always a genuine threat in the Philippines. At the entrance to GenSan's big shopping mall there are signs politely asking visitors to leave their weapons with security. The island of Mindanao is plagued by the radical

Fig. 34.1 Hard work to land Yellowfin Tuna in General Santos Harbour. Afterwards there is the Tuna Festival (Photo: Steven Adolf)

Islamist separatist movement Abu Sayyaf, a small but aggressive gang of criminals notorious for its abductions for ransom money. But among the delegates of the tuna conference today, held in the hotel opposite the mall, there is a mood of cheerful anticipation. The reason is clear, trade exports to the European Union could pick up considerably if the special zero tariff is applied to the Philippines within the Generalised System of Preferences Plus (GSP+) agreed with the EU. This would mean 20.5% less in import duty. The Philippines is the only country in the ASEAN region that has been granted this beneficial export settlement and as a result has seen 35% year on year growth in EU fish export for Rules of Origin fish.

There is a further reason to celebrate: after years in which dwindling tuna landings forced the canning industry to buy half of its tuna in Manila and overseas in order to keep the factories open, catches entering the port have risen sharply. The bounty of tuna has not been coming from the once so rich tuna waters of GenSan but from much further afield, from the waters of Indonesia and more still from Papua New Guinea where the Philippine tuna fleets have been fishing for years. During the conference, NGOs such as the WWF and Greenpeace tell me GenSan no longer deserves the moniker 'Tuna Capital of the Philippines', because most tuna landed is from elsewhere. Comments like these are not particularly appreciated by the local large-scale fishing businesses. Mayor Ronnel Rivera: 'From the moment it is landed here in GenSan it is Philippine tuna. We don't look at where it comes from' [82, 83]. Insiders, and that includes everyone here, understand the rub. Increasingly, there are issues around tuna being caught by Philippine fleets in overseas EEZs. Stories about illegal catches and licence tinkering crop up frequently. No one enjoys being reminded of this at the Tuna Festival, least of all Mayor Rivera, whose father was the pioneer with the foresight to move his fleet to the waters of Papua New Guinea. He even built a factory in the PNG port of Madang for processing skipjack and yellowfin tuna in cans or as frozen loins for production factories overseas. No one knows exactly which part of the GenSan catches are from the Philippine fleets that have been diverted to Papua New Guinea. But one thing is clear: the factories are desperate for fish to process and much of the Philippine waters have been all but fished dry. All shipments of tuna are most welcome. Most people couldn't care less about its precise origins, and critical questions about this are not appreciated at the annual tuna fest.

It is not the only dissonant at the festival. Nearby, next to the entrance to the shopping mall, men are protesting with red banners. Leaflets are handed out from the Philfresh tuna factory workers' action committee, a cannery owned by the Citra Mina conglomerate of tuna businessman Jake Lu. Lu is a powerful man in GenSan. Citra Mina is a big tuna player, the Philippines' second largest exporter of chilled, smoked or frozen fresh tuna and one of the leading companies in the Asia Pacific tuna industry. So now this protest, at this inopportune time. 'Philfresh is "Philtrash",' says one of the demonstrating factory workers standing next to the banner. 'Trash, yes, that's how they've treated us for years.' In 2013, when they demanded better working conditions in the factory they received personal threats, he recounts. More than 240 people, the majority of union members, were sacked. The matter would eventually even reach the Philippine parliament.

Owner Jack Lu, a prominent presence at the conference, is visibly affected by the protest. It's put a dampener on the tuna festival, which should have been a showcase for local industry. As if the case of the Philfresh workers was not enough, Citra Mina is up to its neck in another dispute, around the 43 crew members of the tuna boat Love Merben II [37]. They are unable to attend the tuna festivities in GenSan after they were arrested by the Indonesian coast guard at the end of August for illegal fishing in Indonesian waters. The tuna boat is alleged to have cast its nets even though its fishing licence had expired. The men were known as the *abondonados*, Spanish for 'abandoned'. They ended up spending 6 months in an Indonesian prison before being allowed to return to GenSan. According to the fishermen's union which endeavoured to have them released, Citra Mina supposedly sent the ship purposefully on an illegal trip to Indonesian waters. The company denied this. According to spokesmen, they had warned the fishermen not to fish outside of Philippine waters. What's more, the boat had nothing to do with Citra Mina, the argument went. They had been fishing under contract, that was all there was to it. The crew saw it quite differently. Their ship had been supplying tuna to Citra Mina for years and was even financed by the company [38]. Citra Mina had simply despatched them to the grounds, without the required papers. Now they had lost everything. Their boat had been detained, chained up in an Indonesian port; they were left in debt.

One of the biggest wars over sustainable tuna fought in the twenty-first century was undoubtedly around IUU catches. The battle involved all the world's seas, but primarily the Pacific where the FAO believed 60% of all tuna was caught. These conflicts had plenty of causes: the picture of overlapping claims on EEZs in sea areas underlined the fact that there were large sections where several states at once laid claim to the tuna. Quite a job to try and work out what is in fact illegal. It was in the tuna's nature to disregard the EEZ's limitations, not exactly simplifying controls. But even if the area claims had been more or less formally agreed, it still mattered that the fishermen kept to these agreements, and that was not the case. In the Western Pacific, in particular, poaching was more the rule than the exception.

This was partly because the fishermen had never been used to recognising borders at sea. There was constant pressure to catch tuna to supply the industry with enough raw material. Crews were usually paid on commission: no catch, no income. The fleet owners tended not to see it as a problem that the tuna was caught illegally. Tuna was a migratory fish, after all. The claims these countries believed they had on certain populations was arbitrary anyway, was the implied thought. What did not help either was a total lack of any supervision over the vast waters of the many archipelagos found in East Asia. Researchers believed IUU fishing had developed into a 'major factor' threatening the sustainability of the fisheries [232]. Poaching within the EEZ was the most prevalent factor, but illegal fishing on the high seas, which no one was able to lay claim to, was practised profusely too. Fishing without licence or with forged papers, fishing in prohibited areas, fishing of stocks where the quotas were already exhausted or for which a seasonal closure was in place, fishing with banned gear, fishing of undersized fish or banned species or with banned bycatch. This included sharks whose fins were cut off, while the rest of the otherwise worthless creature was thrown back into the sea, still alive, facing a gruesome death.

Often illegal fishing morphed seamlessly into illegal trade, with landings at banned ports or with forged catch papers. Consignments that were swapped in the big cold stores without any control. It was a smorgasbord of violations, which carried on throughout the value chain. The driving forces behind poaching were smuggling, corruption, slavery on ships, quick profits and indistinct borders. Above all, there was an enormous overcapacity in vessels which would be lossmaking if they stayed in port. 'serial depletion', the tuna variant of serial murder, entered the scene. As soon as the population in a country's own coastal waters became overfished, the fleets moved to their immediate neighbours to pursue the catch there illegally. It appeared that yellowfin, bigeye and skipjack in particular were subject to poaching. At the end of the first decade poaching in East Asia was estimated to extend to around 1 billion dollars and between 4 and 8 million tonnes of fish. Illegal fishing might amount to 16% of the total Pacific catch. Tuna forms a substantial part of this [4, 163]. These were highly approximate estimates: illegal fishing is by definition hard to quantify. There are no reliable figures. Deficient supervision or enforcement only makes matters worse. But the picture was clear. The 2016 figures in the Western and Central Pacific, the richest tuna grounds in the world for poaching and other tinkering with catches, were estimated to be 277,000–338,000 tonne annually, or a sum worth between 700 million and 1.5 billion dollars [225]. Not entirely coincidentally, fraud involving bigeye tuna, possibly the most endangered but also the most expensive tuna after bluefin, proved to bring in the biggest share of illegal tuna money. More than one in three bigeye tuna were fished illegally, estimates showed. In particular, it was impossible to control longliners, which tend to stay at sea for more than a year without calling into port, and were mostly involved in tinkering with illegally caught bigeye and yellowfin. The majority of violations on purse seiners seemed to be around fishing with FADs and illegal transhipments at sea. Catch reporting irregularities in purse seining became less significant with 100% observer coverage, VMS-tracking, in port transhipping and validation of declarations. Longliners remains a significant concern of IUU in the WCPFC.

It was during this particular incarnation of the tuna festival, therefore, that attention was focused on the arrest of the crew of the Love Merben II with their illegal tuna catch. They are no exception, however: thousands of Philippine fishermen had been apprehended since the Indonesian Minister of Fisheries Susi Pudjiastuti had declared war on illegal fisheries in Indonesian waters during the course of the second decade of the twenty-first century. Tuna poaching is a problem in the Philippines itself too. Small-scale fishing businesses in many Philippine coastal waters, for whom tuna tends to be a basic source of food and income, complain that their catches dwindle due to the invasive big industrial fleets. This is alleged to haul out juvenile yellowfin tuna illegally, as a result of which the fish's stocks in coastal waters have declined sharply. The tensions between the big commercial companies and the small coastal operations thus increased after 2010.

Tackling poaching was complex and often a case of laborious management of tuna capture. The immediate coastal waters up to 15 km offshore fell within the jurisdiction of the coastal municipalities, while the Ministry of Marine Affairs and Fisheries, formally part of the Ministry of Agriculture, manages the EEZ between 15 km and 200 miles. The municipalities tasked with issuing coastal fisheries licences often simply lacked the manpower. 'This means that 80% of the small-scale fisheries fish without a licence,' says Jonah Van Beijnen, an independent fisheries advisor from the Netherlands who supervised several sustainable fish projects in the Philippines. Eighty per cent of the smaller fishing businesses: that adds up to a lot of tuna in the extensive Philippine archipelago. The bigger commercial boats, for their part, tend to have a ministry-issued licence. But they do not tend to keep to the 15 km boundary of nutrient-rich coastal waters, home to both adult tuna and juveniles. There is no control and, under pressure from commerce, local politicians in the municipality issue special licences to the bigger fleets. So the bigger boats encroach on the coastal waters, including the spawning grounds. Beijnen frequently saw with his own eyes how the encircling nets scooped up masses of juvenile tuna far below the minimum legal size of 250 g.

The problem is that everyone overfishes in each other's waters to some extent. The Philippines complain about the Chinese fleets from Hong Kong with good reason. These began to fish in ever greater circles outside their own waters in search of sea cucumber, abalone and groupers. But the Philippine fleets did the same with tuna in the waters belonging to the Indonesian island of Sulawesi. In 2014 the countries reached an agreement about boundary allocation but the Philippine fishermen took little notice. Together with Vietnam and Malaysia, they became a major target for Indonesia in its campaign against tuna poaching. In Papua New Guinea the Philippine fleets were accused of shipping their tuna catches to their production facilities in the Philippine home ports instead of landing them at the local production factories in Papua New Guinea for canning and processing, contrary to the agreement. Further EU allegations were made of packing in Philippines and laundering though PNG plants to get duty free access.

Wherever it takes place on such a large scale, combatting IUU fishing is a protracted affair.

What's more, IUU fishing is often closely linked to other illegal activities such as arms smuggling, human trafficking, terrorism and other forms illicit trade and organised crime by the local warlords. That's why some people argue for making IUU fishing a serious criminal offence [124], as a form of cross-border organised crime. This might particularly help in areas where illegal migration, human trafficking and terrorism are a potential issue. But whether it would be effective in removing the causes of the illegal tuna trade remains to be seen. In the case of tuna, making its value chain more transparent might be a better remedy. Just as in food quality, it was becoming increasingly important to know where tuna came from; the importance of traceability expanded appreciably. Traceability helps to blockade illegal goods. It provides consumers with a guarantee of the tuna's provenance so that they can make up their own minds as to whether they want to buy it or not. This works preventively; nobody in the chain, least of all the big retailers, can afford to offer IUU tuna on a long-term basis. That is the driving force of demand within the chain which turns on the heat when it comes to making fishing more sustainable. Traceability thus became another priority in the global tuna battle against IUU fishing. It was a big challenge, especially for this particular fish, with its long and complex value chains.

The Pacific Tuna War

35

A long white beach, coconut palms and crystal clear water. We are snorkelling in the Ironbottom Sound, off the north coast of Guadalcanal Island, 30 minutes' drive away from the Solomon Islands' capital Honiara. On our way to the beach, navigating the potholes in the road, we were greeted by dancing dark-skinned children with sun-bleached afro hair. The peaceful scene is a contrast with a violent past. A big war memorial in Honiara keeps the war memory alive here. But aside from that, it is difficult to imagine that right here a massacre took place during the invasion of August 1942 when the American ground troops and marines attacked the Japanese occupying forces. Photos from that period show the beach where we are now swimming littered with the bodies of dead soldiers. Now nothing recalls these war atrocities, apart from a small war monument in the shape of the wreckage of an American landing craft in the crystalline water. Multi-coloured fish have found a home here and swim around remnants of rusty metal.

The new tuna war is less visible and thankfully less bloody. In Honiara, atop a hill with views of the port, we find the headquarters of the Pacific Island Forum Fisheries Agency (FFA). The agency, which was founded at the end of the 1970s, helps 17 countries in the region in their joint efforts to make tuna fishing more sustainable and thus safeguard this important fish for the future. These waters are home to by far the largest stocks of tuna worldwide, so this is an ambitious attempt to save fish stocks from extinction at a global level through a common policy.

The Solomon Islands, a forgotten archipelago in the vastness of the ocean, are once again a place of strategic importance. During WWII, the islands were a crucial hub for securing the Allied supply lines between the United States, Australia and New Zealand. The Japanese fleet used the islands as an operational base. With the occupation of southern Guadalcanal, the capture of Papua New Guinea and the attacks on the southern Bismarck Archipelago, the Americans isolated the Japanese fleet base. The Solomon Island campaign was to be decisive in the strategy by which Japanese sea power in the Western and Central Pacific was paralysed.

The nowadays strategic battle was all about access to the vast surrounding ocean area for the distant water fleets and the related attempts to make the tuna catch more

© Springer Nature Switzerland AG 2019
S. Adolf, *Tuna Wars*, https://doi.org/10.1007/978-3-030-20641-3_35

sustainable. It involves billions of dollars, and just like the military marine battle more than half a century ago, the outcome is of strategic importance for the future of the area.

The images of waving coconut palms, white beaches and a crystal clear sea persist in our collective memory of the Pacific Islands. The population dressed in colourful clothes, flower garlands in their hair. An idyllic environment of colour, plenty and pure harmony. The painter Paul Gauguin promulgated these images when he lived in Tahiti. The Norwegian adventurer and explorer Thor Heyerdahl chronicled his adventures following a stay on the island of Fatu Hiva in the Marquesas Archipelago. Singer Jacques Brel spent the last years of his life here. This romantic image was largely defined by Polynesia. The hard reality of the majority of the other island groups in the Western and Central Pacific looks considerably less captivating. Here, poverty prevails, with the deprivation of former colonies which were too small and situated too far away to become the focus of attention. It takes days to reach your destination, clusters of islands in the big blue void. Most island states have to contend with the lowest GDP per capita in the world. There are very few commodities to trade, exporting industry and tourism is barely developed, if at all. Too far away, too poor, too complex, with just one true resource in abundance: vast amounts of tuna swimming in a huge ocean.

The Solomon Islands form part of the poorest Pacific island states. Approximately 650,000 inhabitants have to manage on an average of scarcely 3200 dollars per year. Most people living here, primarily Melanesians, but also Polynesians and Micronesians, with 70 languages in addition to pidgin English, walk the streets barefoot. Seventy-five per cent of the population catch fish, grow their own food and live outside the monetary economy. The infrastructure on the islands is old and neglected, as we experienced on our pot-holed track from Honiara to our snorkelling beach.

An unfortunate misunderstanding caused the Solomon Islands, consisting of 6 large and around 900 smaller islands, to be named after Solomon, the legendary, immensely rich and extraordinarily wise Jewish king of Israel. Solomon was alleged to have obtained his wealth in part from a treaty with the Phoenician King Hiram of Tyre, who helped him develop a merchant fleet in the Mediterranean. It is believed that the Israelites joined the Phoenicians on their voyages, right up to Cádiz, and so must have become acquainted with bluefin tuna fishing. But it would have been too apt if tuna had been the reason for naming these islands in the Pacific after the rich and wise Solomon. The first Europeans to set foot on the islands where Spanish explorers. They might have landed on the very beaches where I snorkeled amid the American wrecks. The Spaniard Álvaro de Mendaña de Neira (1542–1595), whose statue stands at the entrance to the national museum in Honiara, thought he had discovered the gold treasures of El Dorado and would become as rich as Solomon. The explorer had sailed in a westerly direction from Peru, looking for the gold which, according to Inca tales, was located far away in unknown lands, said to be 'Terra Australis', the mythical Australia the Spaniards hoped to discover.

It was all a big mistake. Sailing westwards, he first came across a Polynesian island group which were christened the Marquesas after the expedition's sponsor, the Marquis of Cañete. Then, in 1568, the two Spanish ships reached the southern

Solomon Islands. It should have dawned pretty quickly on Álvaro de Mendaña that King Solomon's wealth was nowhere to be seen. There was barely enough to eat and drink. The population fished, cultivated the odd crop and raised pigs. It must have reminded Mendaña of the poverty in his native village Congosto in the north of Spain, only a bit worse. Relations with the natives, who specialised in headhunting, could not exactly be called cordial. It often came to quarrels and bloodbaths with the pigs as a stake. Legend has it that the (indigenous?) mood hit rock bottom when the Spaniards deeply insulted their hosts by demonstratively refusing to eat a piece of a roast boy.

A second expedition with almost four hundred colonists who were due to settle on the islands turned into an even greater fiasco. Again no riches were discovered, nor King Solomon's wisdom, not least because the Spaniards began to fight amongst themselves. Hate, greed and jealousy poisoned the atmosphere in the Spanish camp. When malaria struck and the unfortunate leader Álvaro de Mendaña himself died, the decision was soon taken. Led by his wife, the remaining quarter of the original travellers set sail for the Philippines where they landed not far from today's General Santos and continued their journey on to Manila. The story of the Spanish colonists was so calamitous that the writer Robert Graves (who, ironically enough also wrote the story of tuna emperor Claudius in '*I, Claudius*') turned it into a novel with the potent title *The Islands of Unwisdom* [116]. The Spaniards had neither brought wisdom, nor found riches on the islands.

Tuna, the Solomon Islands' true wealth, remained hidden. The annals of the Spanish colonial occupiers mention fishing with spears and lines from the canoes used by the indigenous people, but exactly what type of fish was caught escaped their attention. All energy must have been invested in the quest for 'the golden mountains'. One wonders what might have happened if the Spanish King Philip II had sent the duke of Medina Sidonia as expedition leader to the Solomon Islands instead of appointing him commander-in-chief of the Armada? The tuna duke would almost certainly have spotted the shoals of different tuna species in the rich fishing grounds. He might even have attempted to set up an *almadraba* and discovered the true riches of the Solomon Islands. There was gold swimming in the water here, after all. But things turned out differently. The unfortunate Spanish conquest did not immediately incite colonial imitation, that much was clear. The Solomon Islands acquired the reputation of a poor colony where little could be extracted. It would take until 1893 before the British declared the archipelago their protectorate.

It wasn't until the second half of the twentieth century that the perception of a vast sea area with few resources and an underdeveloped economy began to shift. This had everything to do with tuna and the geostrategic redistribution of power in the Pacific. After the Treaty of Versailles, which sealed the end of the WWI, Japan was given control over many of the colonial dominions in the Pacific, including Palau, the Federated States of Micronesia, the Marshall Islands and the Northern Mariana Islands [31, 112]. Japan set about developing the areas as part of its strategic food policy. It began to invest in tuna fleets, chiefly pole and line. At the beginning of 1930, 116 Japanese pole and line vessels were active in the Pacific. The first canning factories appeared, but the majority of production comprised skipjack processing for the traditional grated tuna flakes of *katsuobushi*. Japan also began to operate longline boats in the Pacific waters.

After WWII, once many of the Japanese colonial dominions north of the Equator had come under American governance, large-scale tuna fishing took off properly, building upon Japanese katsuobushi fishing operations in the islands post WW1. The recovering Japanese tuna fleet now sold its tuna to the emerging American canning industry. The American tuna fleet had also discovered the Western and Central Pacific since tuna fishermen and their boats had been deployed by the American navy in WWII. StarKist and later Chicken of the Sea established themselves in the region in the 1960s with factories in Pago Pago and Palau. The Japanese reinforced their pole and line imperium in Papua New Guinea, the Solomon Islands and Fiji in the 1970s. The Hawaiin canning industry pole and line fleets fished well into Micronesia. During the 1980s, the Americans increasingly moved their purse seine tuna catches from the eastern Pacific to the rich fishing grounds of the Western and Central Pacific. The strong effects of the shifts in sea currents, known as El Niño, also meant that more tuna could be caught in this area during some years.

The tuna expansion was fertile ground for innovations in fishing. Under pressure from burgeoning Japanese competition, tuna fisheries in California had switched from pole and line to the less labour-intensive purse seine in the 1950s. Both the US and Japan invested a great deal of money in research and improving fishing techniques in tropical waters. The introduction of powerful hydraulic power blocks and the application of synthetic fibres in the nets allowed for upscaling and paved the way for further enlargement of the purse seiner fleet. The Japanese gratefully copied this method during the 1960s and 1970s. A major discovery was the fact that tuna likes to gather under large blocks of wood which could make capture much more efficient. The nascent Philippine tuna fleet expanded this discovery into floating or anchored bamboo rafts or *payaos* and thus the Fish Aggregating Device (FAD) was born in the Pacific. The new Japanese longliner fleet produced a significant shift in the market from the 1970s onwards; Japanese fishermen now focused on their own tuna market for sushi and sashimi and *katsuobushi*. Following the discovery of super freezing technology which chills tuna to $-70\ °C$, Japanese longliners were able to remain at sea for months at a time and had their frozen cargo transported back to their home market.

Industrial tuna fishing thus shifted much of its activity towards the Western and Central Pacific as the most important source of raw material for the global tuna chain. Skipjack, yellowfin, bigeye and albacore were hauled out of the sea in bulk quantities for the canning market, as well as fresh and frozen products. Around the new millennium, more than half of the registered global tuna production came from the Western and Central Pacific and this proportion only seemed to continue to grow. Following in the footsteps of America and Japan, it became ever more crowded with purse seine and longline fleets from South Korea, Taiwan, China, New Zealand, Indonesia and the Philippines. The Spanish distant water purse seine fleet, the largest in Europe, and their fleet based in Latin America likewise became active in the region with monster boats and mass deployed FADs. Long after their ancestors had searched in vain for King Solomon's gold, they finally discovered the riches in the waters of the Pacific Island countries.

There was one significant difference with the past. Like elsewhere, after centuries of oppression, the colonial powers had lost much of their influence; the majority after WWII, others sooner. The island nations in the Western and Central Pacific now stood on their own two feet. The realisation began to dawn that the enormous tuna stocks in their waters were among the few resources at their disposal. The new Law of the Sea Convention and the Exclusive Economic Zone (EEZ) 200 miles limit, in particular, were eye-openers. Tuna was not a fish to take any notice of borders or limits, but as long as it swam in the vast area formed by their common waters they were able to exploit the big foreign fleets, thanks to the fishing rights they possessed in their EEZ. This realisation proved fundamental for the establishment of the PNA: the largest partnership between states for the creation of a production cartel for tuna which strove for greater sustainability in the catch (Fig. 35.1) [89].

Sourcew: Secretatiat of the Pacific Community

Fig. 35.1 The EEZs of the Island states in the western and Central Pacific form a huge patchwork in the Ocean that defines the power over the world's biggest single tuna stock (Infograph Ramses Reijerman). This is basis of the creation of the Parties to the Nauru Agreement (PNA), the partnership of eight small island states that represent a global tuna production cartel: the Federated States of Micronesia, Kiribati, the Marshall Islands, Nauru, Palau, Papua New Guinea, the Solomon Islands and Tuvalu. The autonomous island territory of Tokelau that forms part of New Zealand has been a 'cooperative member' since 2012

The Tuna Production Cartel

36

Outside the realm of tuna, the PNA is virtually unknown. But within the tuna world everyone knows what the initials stand for: the Parties to the Nauru Agreement. As we mentioned before: what OPEC is to oil, the PNA is to tuna: a production cartel of eight island states with the largest stocks of this resource. The PNA was set up in 1982 by the Federated States of Micronesia, Kiribati, the Marshall Islands, Nauru, Palau, Papua New Guinea, the Solomon Islands and Tuvalu (Fig. 35.1). The autonomous island territory of Tokelau that forms part of New Zealand has been a 'cooperative member' since 2012. The PNA is one of the most inspiring agreements when it comes to increasing tuna stock sustainability through fishing policies. Just think about it: eight island states, with less than 9 million inhabitants (of which 8 million in Papua New Guinea), dispersed over thousands of islands and atolls, with a combined EEZ comparable to the size of the European Union before Brexit, joined forces. With an estimated half of global skipjack catches in their seas and a quarter of all total global tuna catches, the PNA was the most important source of raw material for the global canned tuna market. Of the 590 big purse seiners from the distant water fleets which were fishing the world seas from 2010 on, at least 250 could be found in PNA waters.

The PNA agreement was launched practically simultaneously with the US fleets discovering the tuna-rich waters in the Western and Central Pacific. That was hardly a coincidence. The UNCLOS maritime laws with the 200 mile zone opened new opportunities for countries to make legitimate claims on the tuna in their waters. However, whilst these nations declared their EEZs, the US fleet was transferred from the Mexican coast to the Western Pacific with total disregard for the new 200 miles zone of the Island nations. The PNA nations prime ministers came together and in 1982 established the PNA partnership, premised upon the principles of concern over large industrial fleets fishing near their beaches with total disregard for their rights over tuna and over the sustainability of their stocks. Individually the small island states could not take on the super powers. As a partnership they could take a

© Springer Nature Switzerland AG 2019
S. Adolf, *Tuna Wars*, https://doi.org/10.1007/978-3-030-20641-3_36

leadership role in conservation and management of their stocks in order to guarantee the future in a way that was unrivalled.

The question was how. During the 1970s the last island states had swapped their colonial status for independence. The big tuna fleets in their region, on the other hand, belonged to developed states which for decades had habitually regarded the Pacific as a rightful extension of their own waters. The Japanese and American fleets initially refused to pay an access fee in order to fish in the Pacific 200 mile zones. What certainly helped the young independent island states to take action was the coastguard incident in the waters of Papua New Guinea (PNG) in 1982. The PNG coastguard vessel came into action against the American fishing boat Danica which had been caught fishing for tuna within the 200 mile zone [199]. The boat was initially brought in, but was soon able to head out to sea again after the American tuna industry coughed up compensation. Then the Solomon Islands coast guard's only patrol boat (on loan) brought in the tuna vessel Jeanette Diana from San Diego in their waters. This time a firmer approach was decided upon. The crew was allowed to leave, but the boat was confiscated. The Solomon Islands put the vessel, including its helicopter, up for sale for a few million dollars. The Americans were outraged. Without delay, under the Magnusson Act, an American trade embargo was imposed on the Solomon Islands and Papua New Guinea. The affair escalated when the Solomons and Kiribati hit back with a tried-and-tested Cold War tool: the Soviet Union was invited to sign a fishing agreement. This proved to be extremely effective: if there was one thing the Americans hated, it was the Russians in 'their' Pacific Rim, the ocean they regarded as their backyard. This diplomatic storm began to resemble a return of a kind of Pacific War. Australia had to mediate between the parties. In the end, the US capitulated. An agreement was signed regarding tuna capture by the US fleet in the EEZs of the Pacific island states. This US Treaty confirmed the tuna fishing rights and would play an important role as a policy instrument within the PNA over the coming decades. It was the first step towards managing ownership by producer countries of the tuna fisheries in the Pacific. The island states were still loosely, almost informally organised; during the initial years PNA meetings usually took place in a bar. Nonetheless they had won the first battle over tuna. Now it was a matter of settling the entire war around tuna to their advantage. That proved to be a hard battle.

These were the years in which the former colonial territories regarded OPEC with admiration, as an example of how a production cartel for a commodity was able to take a stand against the big powers in the developed industrial world. Here, the colonial model was turned upside down: instead of the resource's wealth largely disappearing overseas, a production cartel was created at the beginning of the value chain which was able to set the prices and claim a larger share of the pie. Juxtaposing the oligopoly of a limited number of big fishing companies and their distant water fleets, a partnership of states now formed a countervailing power able to impose supply conditions itself. What had worked for oil could work for tuna as well, was the idea. But when it came to tuna, unlike oil, rendering production more sustainable was as important for the PNA countries as achieving optimum yield. If it was managed well, the thinking went, tuna could be an everlasting, inexhaustible production supply which generated income. Coordinating a tuna cartel which pursued both objectives was no easy task for a group of small island states with limited means and administrations too inexperienced to be able to take a stand against the powerful fishing companies and countries in which their fleets were based. So the cartel started with a humble but significant step: the introduction of a cap on the number of vessels operating in the combined EEZs of the PNA countries. The idea was to keep the number of purse seine vessels on 205 through a licensing system that was formalised in the Palau Agreement in 1995.

The existence of licences or rights for large-scale tuna fishing was far from new. There had been fishing rights with respect to tuna since antiquity. The Carthaginian family of Hannibal Barca had acquired—better say conquered—its tuna fishing rights, first in Sicily and later on the southern Spanish coast. The difference was that they did not pay for the right to fish. Mediaeval Spain knew a more formalised privilege of the *almadrabas*, which were granted to Guzmán el Bueno for services rendered to the crown and which became the basis for the dukes of Medina Sidonia's tuna empire. You could call it the first primitive model of an agreement in which the state transferred the property rights to a private party under licence, in this case a noble family. This licence would remain theirs for centuries. The fishing activities of the tuna dukes were integrated vertically, just as they had been in ancient times, involving tuna conservation, trade and distribution in addition to the fishing itself. With the advent of the new tuna industry during the nineteenth century, no longer organised in families but in companies, the global value chain for tuna became segmented. There were fishermen, the tuna was transferred in boats and other forms of transport, and processed in factories; traders and middlemen emerged; shops and later a complete industry of retailers sprung up, with large stores selling their products to the consumer. Over the course of the twentieth century this industrial segmentation was partly reversed; vertical integration was instigated by big fishing companies which established their own factories, brand manufacturers of canned tuna with their own fishing fleet and large trade conglomerates which succeeded in binding fishing companies and tuna farms through contracts. Government once again played a part as an actor in the value chain with new marine laws and the economic zones where they were able to assert fishing rights.

With the rise of sustainability as an increasingly significant issue, the group of different stakeholder in the value chain of tuna grew. The non-governmental organisation or NGO appeared, ready for battle for the protection of the fish, the marine environment and social conditions in the sector. Parallel to this development, the traditional role of the state in cross-border production chains for food was being rapidly and increasingly undermined. This was partly an ideological shift; in the liberalisation of the markets, with deregulation, privatisation and commodification, the influence of the state was seen as inadequate and undesirable [175]. Globalisation of trade had made it more and more difficult for nation states to get a grip on transnational production: core ingredients were extracted on the other side of the globe, processed and traded in different places and transported to their final destination, where they ended up on someone' splate. New models for managing sustainability and food security were sought with partnerships between new stakeholders and with new power relations in the industry.

Managing tuna policies in a global environment had become much more complex. Returning to the question of how to make distant water fleets pay for fishing licences: a system of checking catch quotas seems an almost impossible task to monitor. The ocean is a big and difficult space to supervise and the island states hardly had an administrative organisation to enforce such a policy. So other methods were sought. Fishing quotas are 'output-based' controls, involving looking at the amount of fish that can be caught and divided in quota amongst the countries and fishing operations. The PNA decided to go for an 'effort-based' control, monitoring the 'input' from fisheries in their waters in the shape of the efforts they exert in order to fish. That was more easy to handle. The first step in this direction was taken with the maximum of 205 purse seine skipjack vessels into the PNA region in return for payment. The idea was that this served as a kind of limit on the total capacity of fishing in the zone which could be considered sustainable [131]. By gradually reducing the total capacity of the number of boats, the price of fishing licences would hopefully rise of its own accord whilst the tuna stocks would increase. Exactly like OPEC oil: increased income, but with curbed production which should vouchsafe the survival of this industry in the future.

But it wasn't that simple to fight the powerful foreign fleets that still had to get used to the idea that they did no longer have a self-evident right of access to the resources. The system was far too noncommittal for the affiliated countries. The income from fishing licences was only rising in dribs and drabs, while the catch capacity of the ever larger purse seiners was increasing dramatically. The effects were not in the least sustainable. More the reverse: the amount of vessels fishing in the PNA EEZs was still growing and new, bigger purse seiners entered the area. The PNA countries got frustrated with the limited economic returns. Alternative options had to be explored. In the inaugural meeting in December 2004, PNG tabled a proposal for the Vessel Day Scheme (VDS) as the next step in effort-based management.

Meanwhile, the PNA had to face the Western and Central Pacific Fisheries Commission (WCPFC), a management organisation set up in 2004 to oversee proper management of tuna populations in high seas. A regional management organisation is by definition a body in which contradictory interests have to work together: the countries in whose waters the tuna was caught and the distant water fishing countries that are home to the tuna fleets. Although they had the same objective in name—sustainable management of the tuna populations—the PNA of partnering nations was much more unified in its interests and aims and therefore able to act more decisively than the regional management organisation of which the PNA countries were just one of the members. This became most evident in 2007 when the Vessel Day Scheme (VDS) took off as new and innovative licensing system intended to lead to more sustainable tuna capture and more income for the countries [132]. The VDS was built on the idea of an effort-based system; fishing licences were not paid based on volumes of fish caught (quotas) but in the shape of a number of days a boat had the opportunity to catch fish (fishing days). Based on the target of catch volume in weight of a particular fishery and catch per unit effort, a 'total allowable effort' in fishing days was estimated. These were divided over the PNA countries which, in turn, could be sold for the year to vessels wanting to fish in their EEZ. From 2012 on, a minimum benchmark price was set annually for the fishing days sold to the big fleets to avoid small island nations being played off.

The VDS system became one of the biggest and most complex fishing management systems ever deployed, a gargantuan network of fishing days and licences, governing the biggest single tuna fishery in the world. More than 1.6 million tonnes of tuna became involved at a production value of around 3 billion dollars. A total effort of approximately 45,000 tradeable fishing days are distributed over eight countries and sold on to numerous fleets comprising hundreds of purse seine ships. A special satellite-controlled monitoring system keeps an online eye on these ships' movements to automatically counting days at sea. Independent observers monitoring the catches have to be on board all vessels, that is a 100% coverage. The PNA scheme also includes closure of two 'high sea pockets': sizeable areas of international waters enclosed by the PNA countries' EEZs as part of a raft of conservation measures including FAD closures and catch retention to help rebuild bigeye tuna stocks identified as being overfished. No less important was the fact that the fishing days were linked to a scheme to curb use of FADs through a periodical ban on FAD fishing, termed FAD closures, during particular months of the year. The PNA hoped that this would limit the negative effects of FADs, especially where bycatch of juvenile yellowfin and bigeye tuna were concerned, due to their lingering under the FADS together with shoals of skipjack; the bycatch of these species was linked to population decline. The measure seemed to work: in 2017 the bigeye tuna population showed the first signs of recovery with all four PNA tuna stocks in the green.

But the VDS system worked first and foremost as a financial cash cow which played a decisive role in the success of the PNA. Competition to get hold on the limited amount of fishing days and an increased number of vessels under foreign flags interested, put a strong upward pressure to the vessel day values. The price per vessel day rose spectacularly from 2010 onwards. What had once started as low as 500 dollars per fishing day rapidly multiplied. Within a matter of only 8 years 12,000 dollars for one fishing day was not exceptional. The high prices resulted from increased use of tenders to sell the vessel days to competing parties. Some fleets like the one from the US, which initially had suggested that 400 dollar a day was the absolute maximum the industry could afford, had to change in line with the others. Over 8 years, between 2009 and 2017, the income for the eight island states included in the VDS system had multiplied from 60 million to 500 million dollars. The VDS had evolved from a pittance for decades under catch based schemes into a substantial source of income for one of the poorest regions in the world. This success led to the PNA wanting to introduce a VDS for longline fisheries in 2018, alongside the existing system for purse seiner fisheries. However, longline fishery is more high seas based rather than EEZ based, so it does not have the leverage of purse seine. More importantly a VDS scheme cements the PNA rights over the resources in zones unlike a flag state allocation.

The VDS was in particular a success for the new PNA office, that was set up by the leaders of the PNA in January 2010 with the mandate to increase the fishery value, take a bigger share and vertically integrate in the tuna value chain. Dr. Transform Aqorau, former CEO of the PNA office, commercial manager Maurice Brownjohn, and Anton Jimwereiy, VMS officer, were the original team of three that had developed the idea, implemented the improvements and were in charge of its delivery for years. 'Other countries can make choices for economic development, whereas we have very few,' Aqorau told me during our conversation at one of the big annual INFOFISH World Tuna Trade Conferences in Bangkok. 'Our only resources are fisheries. Through the VDS, we are simply establishing our private property rights over our resources, and adding value to them in ways that improve benefits to our countries. The increase in returns that we have received is really transforming our countries.'

It almost sounded axiomatic. The introduction of the system had been far from simple for the small office. The PNA fisheries cover four tuna species, multiple fleets and multiple countries. They also comprise several different types of fishing gear, the most important of which are purse seine, longline and pole and line, complemented by various types of FAD. The fishery is spread over a very large area, with many EEZs as well as high seas pockets.

The system was controversial, in particular with the fishing industry and its distant water fleets that suddenly were confronted with an increasing bill to pay for access to a resource that until recently was free. No wonder the industry ceaselessly complained about the high prices for fishing days. Some fleets, such as the Spanish fleet of super-seiners, thought the system prohibitive. Fishing in the PNA waters was no longer cost-effective compared to what they paid other developing nations under the European Fishery Partnership Agreements, they told me. There were complaints that the monitoring of fishing days was not effective. It was claimed that the on-board independent observers were in a permanent state of inebriety. It was alleged that some fleets (with the accusatory finger primarily pointed towards the Asian industry) were getting far more fishing days against big discounts, simply because they were more corrupt than the rest. The system was highly opaque; no one knew exactly how many fishing days had been issued and were being used, and what prices had been paid. Worse still, it was claimed that everyone fiddled with their own definition of a fishing day and that corruption amongst the fisheries managers and island administrators did not make matters any easier. Some claimed it would lead to an erosion in the value of the fishing rights. Why would one particular fishery adhere to a system of paid fishing rights when other fisheries so clearly disregarded it? For the European Union the suspected unreliability of the VDS itself was one of the reasons not to renew a sustainable fisheries partnership agreement with the island state of Kiribati [85, 160]. This despite the fact it was the PNA countries' most important policy instrument in the region and that Europe had solemnly declared that it wanted to support the countries in the region in their attempts to organise sustainable fisheries. For the EU the supposed unreliability of the VDS was even the reason to suspend further negotiations to continue the Fishery Partnership Agreement with the region.

Transform Aqorau was unimpressed with the criticism. 'There is no perfect system. But I don't think you will easily find another system where you are able to generate the kind of revenue we have been able to collect in a relatively short time. I am disappointed that there is so much criticism for a scheme that we developed ourselves and that delivered so much.' That was a fact: the VDS was a financial miracle for a group of countries whose administrative powers were not always that effective and, in some regions, even had the earmarks of a 'failed state'.

A production cartel is a favourite toy for power analysis by political economists. In their terminology: the PNA operates within a 'bilateral oligopoly' between mutual concentrations of power. On one side they represent a formal partnership between states which sell a limited number of fishing days, whereas on the buyer's side a limited number of leading fisheries in the tuna chain try to limit their costs per fishing days as much as possible. Within this set-up, the partnership of common states on the supply side is inherently unstable. As long as the affiliated partners preserve their unity, they will be in charge of the market, but there is always the danger that individual participants within the cartel will benefit from a commonly extorted advantage (in this case high fishing day prices) but will try to benefit more ignoring the cartel agreements (by selling more fishing days below the agreed minimum price). This works out advantageously for the country that violates the agreements (more buyers, more money), but means a loss in income for the cartel as a whole. This can ultimately undermine the cartel construction. The leading firms in the fishing industry know this and have everything to gain from playing all the participants off against each other in order to break the cartel. 'This was in fact why a benchmark price was introduced', commercial manager Maurice Brownjohn stated. 'Just to put an end to fleets playing off parties with mis information and compromises. This has worked extremely well since.' And as for the kind of cartel: one could term it like OTEC (Office of Tuna Exporting Countries) instead of OPEC, Brownjohn jokingly suggested. 'It is in fact an eight nations partnership in a "conservation cartel": If they put hard limits on effort, they can maintain stocks at very healthy levels, optimise the efficiency of the fleets and maintain prices in the market', Brownjohn explained. 'It is a win for the region. But the biggest winner is the tuna stock.'

The PNA seemed well aware of the dangers within the cartel [10]. Before the PNA came into existence, Japan, the biggest regional distant water nation, had always been very adept at playing countries off against one another and thus exacting the lowest possible licence fees. Korea and Philippines were emerging masters of this art, the PNA-office realised. The lack of compliance by individual countries within the cartel still was a weakness and point of concern. It appeared that the definition of a fishing day, was not always completely clear. Some countries did not close their fisheries when they had used up their allocated fishing days. Or fishing days were offered below the minimum price to bring in more customers. Frequently, a whiff of corruption hung around the deals, with fishing companies able to bribe the right officials or politicians to turn a blind eye. These had been familiar problems in the region in the beginning. The big difference here, however, was that the PNA countries realised that the unity in their cartel was of vital importance. This meant that it was necessary to improve the system. A tighter and purpose-built information system (FIMS) was designed to verify the distribution of fishing days and electronically monitor the use of them through satellite. The transparency of the system was improved, just like the supervision of compliance. With regards to unity within the PNA, the CEO of the PNA office Ludwig Komuro explained: 'We see parties wanting to break out alone, but once they put their head out, they instantly fight to get back in'. The original principles of setting up PNA in 1982 seemed to prevail: individually we can't achieve anything, thus the need for unity.

The tuna cartel had been extremely successful in achieving a spectacular rise in the income from tuna capture, but what was the state of affairs vis-à-vis making it more sustainable? Some claimed that sustainable fishing seemed to have taken a backseat. By opting for an 'effort-based' system, in which fishing days were traded and not the fishing rights on a particular volume that could be caught, fishermen would be incentivised to ramp up their volumes as much as possible within a fishing day. Boats that had been enlarged, more efficient techniques and more creative assessments as to what a day's fishing actually amounted to were introduced; fishing companies pulled out all the stops in order to haul in as much fish as possible within the confines of the licence. This phenomenon is known as 'effort creep'. By stretching the definitions and rules for fishing days ever more efficiently, more fish was caught within a time-limited fishing session [10]. As a consequence of this the skipjack catch per fishing day rose continuously. Not only that; the use of FADs also went up, since the floating rafts increased the yield per haul considerably. Despite this PNA maintains the stock of skipjack at 50% of the unfished biomass harvested annually, way more conservative that the limit reference point of 80% take and way above MSY. This cautious policy is unequal for any stock in any of the tuna RFMOs.

Meanwhile the limitless expansion of the number of FADs had gradually expanded into the biggest sustainability problem. It meant an exponential rise in bycatches of juvenile yellowfin and above all small bigeye tuna and other vulnerable bycatch. From 2008 on, PNA introduced FAD closures to limit the FAD impact. PNA also introduced an annual FAD deployment surveys since 2010, the tracking of FADs alongside fishing vessels, and it became involved in MSC certified free school fisheries as an economic incentive not to target FADs. The PNA office claims that by doing so the percentage of FAD sets has progressively fallen. At the end of the first decade free school sets exceed FAD sets in the PNA waters. The PNA declared attempting to bring FADs under control through new management measures one of its biggest challenges. The application of new, innovative technologies, from satellites to track FADs to underwater drones to limit bycatch, to the use of non-entangling and biodegradable FADs and FAD retrieval schemes are believed to be the next step in order to keep the PNA at the front line of sustainable development. 'From Small Island Nations to Smart Island States' became the new slogan in its fight for more sustainable tuna capture [28]. PNA-CEO Ludwig Komuro: 'Controlling the FADs is a complicated task for the PNA. We try to cooperate with the fisheries to get more information. The boats now have to give the manufacturer serial numbers of the FADs they set in the water as part of the conditions for their license to fish, so we can follow them. The onboard observers send their logbooks to land so we get an online coverage on a daily basis of the fishing activities. Collecting information in a smart way will be essential if we really want to improve our sustainable fisheries.'

One thing is certain: as a production cartel and a countervailing power against the leading firms in the tuna industry, the PNA has achieved something with tuna that OPEC had never managed with oil. Henceforth, the global tuna industry had to reckon with a partnership of producer countries demanding a rising access fee to its waters, able to curb catches, promote sustainability and generating income which would have been unimaginable previously. Where sustainability was concerned, the PNA had won another battle; but it still had to win the war.

Three Flipper Wars

37

The First Flipper War: The Conquest of the Pacific

Ric O'Barry was angry. The entire car journey from his hotel in The Hague to the Netherlands' main airport Amsterdam Schiphol was one long, unremitting complaint against Earth Island Institute (EII), the custodians of the Dolphin Safe label. Everyone knows the label on canned tuna—the image of a jumping dolphin on cans of tuna across the globe—whether they are aware of it or not. Now Ric O'Barry was furious or rather livid, with EII and its label. 'I have been fighting the tuna industry for 40 years! I don't want my salary coming from the industry! The whole thing is corrupt!' They, EII, had never told him that some of the money he was paid had come from the tuna industry, O'Barry said. The tuna industry! He was on the point of exploding. The Dutch scenery of polders and windmills passed him by largely unnoticed.

O'Barry is a kind of legend, about which more shortly. He was here for the dolphin and not for tuna, of course; that should be stated unequivocally. He had just given a hearing at the invitation of Dutch members of parliament to explain why he thought the oldest dolphin park in the small Dutch town of Harderwijk should be closed down. The dolphin parks are torture chambers for dolphins, intelligent sea mammals designed to swim freely in the sea, O'Barry had pointed out to the MPs. This should be opposed. Just as putting out nets for dolphins whilst fishing for yellowfin tuna was a form of mass murder that should be opposed. He had never regretted this battle, which made his disappointment all the greater when it transpired that his former comrades-in-arms accepted money from the very same big tuna firms he had been waging war against his entire life. EII, and especially managing director Dave Phillips, had to take the rap for it. 'I asked, "Dave are we taking money from the tuna industry?" The answer is always the same: "It's voluntary." That means, "Yes, we take it."'

It was indeed a strange business. Now Ric O'Barry was on a war footing with the EII custodian of the Dolphin Safe label, which may have been one of the most

© Springer Nature Switzerland AG 2019
S. Adolf, *Tuna Wars*, https://doi.org/10.1007/978-3-030-20641-3_37

successful instruments for saving dolphins in the Eastern Pacific and for making a major tuna fishery more sustainable. This was far more than just the irony of fate. It was a prime example of how the private sustainability management policy in tuna fishing, or what passes for it, could end in a battlefield of interests and cliques which overshot its original goal completely. This was an issue, moreover, which remained largely hidden from consumers. In good faith they thought they were making a sustainable choice by buying tuna with a Dolphin Safe label. And now O'Barry had become collateral damage in the Flipper War on tuna, one of the biggest hybrid tuna wars which started around the turn of the century and had evolved into a largely concealed scandal concerning certification labels.

This wiry, sprightly man in his seventies sitting next to me in the car on his way to the Amsterdam airport is perhaps the best known and at any rate the most radical dolphin champion. Ric O'Barry is revered by activists within the marine environment movement. Conversely, he knows he is deeply hated by Japanese dolphin fishermen, dolphin park owners and tuna fishermen. I have known of O'Barry since I was a boy, although I was not aware of it initially. Just as I had been glued to the TV screen every week to watch the documentaries by Jacques Cousteau, I would not miss an episode of *Flipper* [98]. This dolphin, popular throughout the world as the star of the 1960s television series, embarked on wild adventures with his human boyfriends Buddy and Sandy. Ric O'Barry was the Miami Sea Aquarium head trainer of the five bottlenose dolphins who by turns played the role of Flipper. Occasionally he appeared as a stunt double in the programme. The series became a roaring success in the US and later also in Europe. It was the first time the wider public got to know a dolphin at such close quarters. Everyone instantly fell in love with the clever Flipper with his touching guttural sounds and fixed smile. Flipper became an unprecedented global media success. Throughout the world, dolphin parks shot up like mushrooms. Everyone wanted their own Flipper. The Netherlands acquired the Harderwijk Dolphinarium mentioned above. We went to visit it on a school trip "to see Flipper with our own eyes". Just like elsewhere in the world, catching and training dolphins became big business. In this million-dollar industry a pioneer like Ric O'Barry was able to earn gold.

But things changed dramatically. In 1970 Kathy, the dolphin that had taken the part of Flipper most often under the care of O'Barry, dived under water never to resurface again. O'Barry was heartbroken. He had noticed that in the period leading up to her death Kathy had been profoundly unhappy in her captivity and had become deeply depressed. He was convinced that this unhappy Flipper had committed suicide. The event provoked a radical about-turn in O'Barry's life.

From then on, he would fight against dolphin captivity and the dolphinarium industry he had earlier on helped to create. With this Dolphin Project he dedicated himself to freeing dolphins in captivity and later on to the rehabilitation of dolphins released from dolphinariums. His appearance in 2009 in the film *The Cove* would put O'Barry in the spotlight anew. A year later, the film won an Oscar for Best Documentary Feature. The film tells the story of the annual dolphin drive hunt in a cove near Taiji, Japan, during which the animals are captured and killed. The massacre was filmed using hidden cameras. O'Barry plays a leading role in the film, talking about his battle to save dolphins.

At first it seemed to make perfect sense that O'Barry and EII would join forces in their fight for the dolphin's wellbeing. The institute, founded in Berkeley in 1982, had after all initiated the International Marine Mammal Project, an idea by EII staff member and later director Dave Phillips. The project dedicated itself to the protection of dolphins in the eastern Pacific. It was here that, from the end of the 1950s, a purse seine tuna fishery was launched that set its nets on dolphins in the knowledge that they migrate together with big, adult yellowfin tuna. If you targeted dolphins you were guaranteed a good catch in yellowfin tuna. Inevitably, the dolphins got caught in the purse seine nets and drowned. The resulting slaughter of the dolphin population was brutal. In the US alone, the purse seiners would on average kill at least 100,000 dolphins as bycatch per year between 1960 and 1972 [234]. The annual total mortality in the eastern Pacific during this period was estimated at 350,000 [151]. The slaughter, totalling at least 6 million dolphins during several decades, grew into a national scandal in the US after EII campaigner and biologist Sam LaBudde recorded this fishing practice on video on the Panamanian tuna boat where he had signed up as a cook. The 11-minute video was shown on all mayor American television networks in September 1988. The same generation that had grown up with O'Barry's smiling Flipper on the screen now witnessed countless Flippers being brutally massacred in the purse seine nets of the boats supplying the main ingredient for their tuna sandwiches. The outrage was huge and widespread. The video was shown at a hearing of the House of representatives Committee on Merchant Marine and Fisheries; famous Americans affirmed their support for saving the dolphins. There was even a national boycott of canned tuna. 'Sorry Charlie— StarKist kills Dolphins,' banners stated.

Now that even StarKist's cool Charlie the Tuna had become the target of protests, it was the turn of the three largest brands and powerful manufacturers of canned tuna in the US—StarKist, Chicken of the Sea and Bumble Bee at the time the three biggest in the world—to become jittery. The boycott posed a serious threat to their turnover and in the worst case even to their survival. The reputational damage was enormous; acquiring the image of Flipper killer was commercial *harakiri*. This was not allowed to happen. Of the 'big three', StarKist-Sorry-Charlie was the first to announce it would no longer use tuna that had been caught using dolphins. That same day Chicken of the Sea and then Bumble Bee followed their competitor's example. Flipper's eternal smile had unleashed a powerful force, capable of making or breaking the tuna industry, that much was clear. The big three announced that they would conform to the production of tuna which involved no capture, killing or injuring of dolphins. A Dolphin Safe label on their cans, administered by EII, would vouchsafe this from then on. Thus, the Dolphin Safe label was born from a joint action by three leading industry brands. The American government for its part introduced the Dolphin Protection Consumer Information Act. Henceforth, the label 'dolphin safe' would only be allowed on cans of tuna whereby no dolphins had been caught during the entire voyage, and this would be guaranteed by an appointed observer on board of the American ships. The legislative action, together with the big three's Dolphin Safe label conditions, led to an effective closure of the American market for tuna caught by setting nets around dolphins. EII, an NGO, became the new gatekeeper for watching over the 'Flipper Seal of Approval' [234]. Internationally, the US tightened its import regulations. Only those countries that pursued a dolphin protection policy comparable to that of the American fleet were allowed to export their yellowfin to the US.

Flipper had thus caused a lasting shift in the paradigms of power in the world of tuna. For the first time a sustainability principle, in this case bycatch which threatened a species, instigated an effective policy change. For the first time it had not been the authorities that had formulated the policy but a combination of NGOs, government and leading firms in the global value chain steering these effective measures. This happened under pressure and in conjunction with a new force in the chain. It was not the fishermen nor the traders, but the consumers and their boycott that were the determining factor. The sustainability policy change was consumer demand driven, a new paradigm later to be known as the 'Theory of Change'. In fact it was a bit more nuanced still: initially it had not been the consumers themselves, but an NGO which, by mobilising consumers, had been able to assert its power. And it was a powerful oligopoly of the three leading tuna brands in the world that had played a very active role in imposing the new rules that effectively became a necessary condition or barrier for any trader or producer who wanted to enter the US market. This created entirely new relationships with new stakeholders who henceforth would have to be involved in value chain policy regarding principles of sustainability.

Dolphin Safe became an impressive success. In a matter of years the excessive dolphin mortality had practically disappeared the map in the eastern Pacific. From an annual slaughter of more than 130,000 dolphins during the period 1986–1996 the number of dolphins killed dropped to around 2600, scientist Martin Hall concluded [125].

The real step forward, however, occurred in 2001, when the regional management organisation IATTC designed its own Dolphin Safe programme: the International Dolphin Conservation Program (AIDCP). The objective was to make fleets other than those from the US, such as the Mexican tuna fleet, likewise stop injuring and killing dolphins. It became a thorough approach, developed closely together with NGOs like WWF and Greenpeace. The IATTC built in safeguards using new techniques under strict independent supervision that would allow dolphins to escape the nets unharmed. The fishermen were trained in these new techniques and there was tight monitoring. A new label called AIDCP Dolphin Safe was to be the certificate for the new sustainable tuna. In a joint statement the stakeholders proudly presented this certification. 'For the first time, consumers will be able to purchase 'Dolphin Safe Tuna' with absolute confidence in what the label means and that it means what it says', a press release stated [5]. 'Consumers will know that the AIDCP Dolphin Safe tuna they are buying has been tracked and verified from the moment it comes out of the sea to the moment the label goes on the can.'

This would mean dolphin mortality in the IATTC zone tuna industry would be pushed back permanently. The American congress was about to recognise the new label as dolphin safe and was to allow the Mexican fleet export onto the American market. The WWF and Greenpeace were so keen on the IATTC approach that they supported the initiative.

If this had happened, EII would have gone down in history as the successful force that initiated the first reliable and sound sustainability strategy for tuna fishing.

But things went differently. EII went to court and saw to it that the proposed amendment for access for the new dolphin safe tuna from outside the US was blocked. The reason was that despite all precautions under the AIDCP policy, nets were still set on dolphins. And that was not acceptable, EII argued, even if the mortality figures had been spectacularly diminished. The American market had to stay closed to all tuna that did not meet the standards of their Dolphin Safe label. WWF and Greenpeace, that had been in favour of the AIDCP ruling, had egg on their faces. They had been inches away from being called Flipper Killers as well. And the Mexican tuna fishermen still were not able to offload their tuna onto the American market.

Who would gain the most from this? Not the dolphins, that much was obvious. The real winners were the big three: StarKist, Bumble Bee and Chicken of the Sea, who saw their American market remain closed to unwelcome competitors [246].

Meanwhile American tuna fishing had shifted increasingly to the Western Pacific. Ever more tuna originated from here. Bangkok became an important production zone for canned tuna and the tuna capital of Southeast Asia. The Philippines, a former US colony, became home to major tuna fishery and production factories and even its own regional EII office.

The shift to the Western and Central Pacific had opened up an entire new market for EII to distribute its Dolphin Safe label too. All these new Asian firms had to come up with Dolphin Safe tuna if they wanted to be welcome on the American market. With its small staff EII began to actively acquire new customers by pressuring fishers, traders, processors and retailers to subscribe to the voluntary scheme or be black listed as "dolphin killers". The Dolphin Safe label appeared on more and more cans. Producers, fishermen, traders; everyone in the value chain who wanted to sell on the American market became Dolphin Safe. Any organisation that refused, ran the risk of serious problems. As EII managed to sell its label at different levels in the chain, pressure could be exerted on both suppliers and customers to not do business with companies that were unwilling to feature the label. As major Spanish canning company Calvo experienced, when tuna suppliers would no longer work with them because they had refused to take the plunge with Dolphin Safe. Instead, Calvo made the mistake of going ahead with the AIDCP dolphin safe standards. It seemed a *bona fide* scheme to Calvo. But that was not enough for EII: encircling dolphins should not be allowed, supervision or not. There could only be one label for sustainable tuna. The AIDCP Dolphin Safe label was a 'fraudulent label', executive director Dave Phillips wrote in his annual report [77]. Only the EII label standards were acceptable as 'dolphin safe'. If you did not participate you found yourself in the dolphin killers' camp and you could face a boycott. Or in the words of EII associate director Mark Palmer, 'Our monitors also investigate any non-conformance by tuna companies to our Dolphin Safe standards and require the tuna involved to be taken off the market.' [185]

From the outset, there was something odd about the expansion of the Dolphin Safe label to the rest of the Pacific and even to the Indian Ocean. Outside the eastern Pacific, EII did not have infrastructure for on-board checking for dolphin friendliness. They simply did not have their own 'monitors' in place. This was persistently denied by EII. 'International monitors are now on board virtually all Pacific Ocean tuna purse seine vessels,' associate director Mark Palmer claimed uncontested in a blog in the *HuffPost* [184]. But the fact of the matter is that these observers are not aboard the ships on behalf of EII, but on behalf of parties like the PNA. The Papua New Guinea National Fisheries Authority (NFA) was categorical about this. It stated that their on-board observers in the Pacific were only monitoring the MSC certification and not EII's Dolphin Safe standard. NFA staff said that they had never encountered any EII observer on the purse seiners in the Western and Central Pacific [13].

In fact it was even stranger: there was no need for monitoring, because outside the Eastern Pacific tuna did not migrate alongside dolphins. Purse seine fisheries did not use dolphins for setting their nets when going after yellowfin. Skipjack was not swimming along with dolphin anyway. Apart from incidental victims amongst the dolphins, there was no fishing whereby dolphins were encircled and killed systematically and on a large scale. The Dolphin Safe label for these catches was essentially nonsense, a hollow form of certification, because there were hardly any dolphins killed or injured in the first place. It was as if you had stamped the captured tuna with a cow-friendly mark. It's true, no cow was killed during the tuna catch, but it's got nothing to do with the issue on hand. Dolphin Safe gave an empty guarantee, and began to smack of a greenwashing label; a certificate that kept up the semblance of responsible, sustainable fishing but in practice required little or no tangible effort on the part of the certificate holder to make its fishing more sustainable.

Yet there were relatively few people in the world of tuna who dared to raise an objection to this. Everybody seemed happy with the Dolphin Safe story. American consumers were happy with the label because they thought it was a guarantee that no Flipper had died during the catch. (Or even worse: a guarantee that no dolphin had ended up in the cans.) US politicians were happy too, dolphin protection remained very popular and keeping the issue on the political agenda was easy for maintaining a good image, albeit with the help of celebrities or school children who sent letters and drawings. The producers, fishermen and traders for their part were maybe happiest of all: they saw the label as a useful and relatively inexpensive marketing tool through which they could sell their tuna as a sustainable product. Other NGOs such as WWF and Greenpeace tended to look the other way when the issue of Dolphin Safe label was raised. Especially after the debacle when IATTC dolphin safe tuna had been barred from the American market, there was little zest for a public fight with the self-proclaimed Flipper protectors.

Thus the Dolphin Safe label remained a familiar and even world wide appearance on cans of tuna. EII made a transition to Europe in 2012 by creating a European arm linked to the Friend of the Sea label, which focused on the European consumer market (see Friend of the Sea). Friend of the Sea CEO Paolo Bray and EII director of the International Dolphin Safe Tuna Monitoring Program Mark Berman became an inseparable duo that jointly scoured the international tuna conferences in search of new 'clients'. Mark Berman, a rather small man with a goatee, soon acquired the nickname of 'the Bermanator' because of his coercive methods to get tuna companies to sign up to the Dolphin Safe label 'voluntarily'. If you caused any trouble, you would be excluded and subjected to opprobrium within the industry. And off he went, the Bermanator. It was indeed all *Hasta la vista* baby, 'I will make you an offer you cannot refuse'. Most in the tuna industry took the offer to avoid a contest.

No one seemed to have much of an appetite for criticising EII openly, fearful of Flipper's terrible revenge. But it did begin to cause tension. What if well-intentioned consumers sooner or later cottoned on to the fact that the Dolphin Safe label on their cans of tuna in fact had little value? And what if they started to notice that there was hardly any transparency around the procedures and financing of the label? Everything was mixed up at EII. It was at once an action group, a designer of a standard and an implementer of a form of self-certification that at best suited certain huge industrial interests in the market. And all that involving unknown sums of money.

This led to 'questions of accountability to consumers as well as the tuna industry' according to a scientific article that put the label under scrutiny [169]. 'Authority without credibility' was how the authors described the Dolphin Safe label. It was a scientific way of saying that we were dealing with a label with barely any credibility. EII challenged the Dutch university involved over the article but to no avail. Nonetheless the message was clear: EII did not care for any criticism. If you dared to call the label into question, you could be in for a tough response.

But meanwhile Ric O'Barry had lost his faith in EII. Although the three big tuna brands in the US had been at the forefront of the introduction of the Dolphin Safe label, he claimed he had never realised that the tuna industry had made funding contributions to EII. He had been too involved in his own dolphin projects to worry about this, O'Barry said. When it dawned on him and Phillips stated that there had been a 'voluntary' contribution by the tuna companies, O'Barry immediately resigned. Furious, having lost his salary and life insurance, the pioneering dolphin protector told the staff he was leaving. The leading global sustainability label on cans of tuna could kiss goodbye to his input. His wife, who also worked for EII, was sacked on the spot half an hour later. The couple were out on the street, without an income. 'The whole thing is corrupt!' O'Barry stated bitterly one last time as we arrived at the airport. He saw himself as a whistle-blower who exposed EII in public, he told me over a cup of coffee in the departure hall. And whistle-blowers do not make themselves popular. 'I was criticised by a lot of my colleagues because there is a code of silence in the animal welfare industry. It does not police itself anymore on the dolphin captivity issue. So if you speak out, you are branded as a kind of traitor.'

O'Barry's departure from EII in 2014 illustrated a much wider underlying problem around the system of private label, market-driven sustainability certification. It became increasingly a powerful commercial marketing tool involving major commercial interests in the chain. That had been the idea behind these kinds of market-driven governance instruments of sustainability in the first place. The Theory of Change: on the demand side of the market, consumers would be able to exact sustainability. Certification would help them to make a responsible choice. The consumer could decide what he would like on his plate and pay a premium. But when badly monitored, this inevitably turned certification into a tool with which big companies in the chain were able to achieve entirely different, commercially motivated goals like a nontariff barrier against competitors, which had nothing to do with sustainability as such. This danger increased when the consumer, as well as other stakeholders in the chain, stopped paying attention to what was really going on behind Flipper's smile.

The Second Flipper War: The Battle for Pacifical

Just as during the First Punic War in the Roman era the seeds were sown for a Second and Third Punic War, when, during the first Flipper war the foundations were laid for a second and third Flipper wars. Soon after 2010, approximately 20 years after the massive dolphin mortality problem around eastern Pacific yellowfin fishing had been solved, the Dolphin Safe label began to run into difficulty. At one point it had been the most important sustainability label in the tuna industry, but now serious competition hit the market for eco-labels. This applied in particular to the MSC label which used a much broader set of sustainability criteria and had not proclaimed protection of the dolphin population as an absolute and infallible dogma of faith.

The 25th of February 2010 was an important date in the history of tuna certification. We travel to Koror City on the island of Palau, some 1000 km to the east of the Philippine port of General Santos, to witness this. The area with its golden beaches, crystal clear water, stunning reefs and green rock islands, is paradisiacal. But those present at the Ngarachamayong Cultural Centre in Koror focused their attention intently on the presidents and prime ministers of the eight small island states, who had convened for the first time since the establishment of the PNA's tuna cartel in 1982. The agenda for the first presidential summit included issues you could easily regard as the essence of a production cartel: 'To develop innovative ways to maximise economic gains from sustainable management of the members' tuna fisheries.' Read: working closely together for more income and more sustainable tuna.

An expectant atmosphere hung in the air. Only one month previously the PNA had set up its new headquarters on the Marshall Islands. After 28 years of trial, and the last 5 years with one staff member employed under the wings of the Pacific Islands Fisheries Forum Agency (FFA) in Honiara on the Solomon Islands, the time had come to create an independant and separate agency office in order to manage the group's commercial interests. This step had underlined the PNA's autonomy from regional institutes such as the FFA who had different and largely donor driven agenda's. With a staff of three the PNA office, which was not dependant on donor countries or subsidies as it had to pay its own way, was a rather modest administrative basis for a powerful cartel. The new PNA headquarters could be considered an appropriate, decisive action. The clear aim in mind was to make the PNA a global tuna power that would transform the Pacific irrevocably.

One of the new PNA office's first initiatives was the summit of PNA presidents which had been convened on that day. The question was how to proceed as a cartel. One idea was to rename the secretariat OTEC (Office of Tuna Exporting Countries), profiling the PNA even more emphatically as the OPEC of tuna. On the agenda was obviously the Vessel Day Scheme (VDS). The system had to be given new impetus to succeed, with higher prices for fishing days and better implementation. There were also brainstorming sessions about other new measures which might enhance the position of the PNA in terms of sustainable tuna capture. A few international tuna industry leaders had also been invited to the conference. They had travelled to the remote island state, a journey which in some cases involved an itinerary of several days, to give their view on the future of sustainable tuna in the Pacific.

How would the PNA countries be able to guarantee sustainable tuna fisheries in the future and how could they themselves play a bigger role in the fishing industry that used the tuna coming from their waters? With the VDS an important step had been taken to earn more money issuing licences to distant water fleets. That in itself was great but it was clear that, once a licence was issued the catch was not owned by PNA. Tuna travelled a long distance in the global value chain before it ended up on the consumer's plate and during this long journey the tuna's value rose considerably without the countries of origin benefitting. Was it possible to play a more active part in this tuna chain of custody, instead of only selling fishing licences to companies which then bolted with the fish? Might the PNA be able to develop its own brand of sustainably caught tuna, so that the association of small island states would be able to market itself worldwide and share in the earnings as a benchmark for sustainable ocean and tuna management? 'Can we make the cake bigger, can we take more slices and can we put icing on top?' Maurice Brownjohn, the former industry leader and chairman of the Papua New Guinea fisheries industry and now the PNA office's commercial manager, summed up the challenge.

Earlier that year Dutch tuna trader Henk Brus received a call from the president of Palau, Honourable Johnson Toribiong's secretariat. Brus was known in the small tuna world as a man with outspoken ideas on sustainable tuna fisheries. Would he be willing to give a keynote presentation to the eight PNA presidents about his vision on marketing and sustainability? This was an offer the trader could not refuse. It seemed a good idea to focus on the strategic possibilities for the PNA countries. His suggestions would probably not be particularly well received by some within the industry or region. Brus was a bit of an outsider. He did not belong with the big, powerful market players, but in previous years he had managed to attract attention for specialising in the trading in sustainable tuna. At a time when all the industry achieved was marketing its cans as cheaply as possible, the Dutch trader had already been convinced that the future of the fishing industry would lay in sustainability. He tried to promote this vision commercially and as a result acquired some fame at international tuna conferences and meetings. His company Sustunable only traded in AIDCP certified tuna. But he liked to go for MSC certification as one of the few schemes eligible with the FAO sustainability guidelines for eco-labels. But at that point MSC in the industrial purse seine tuna fishery did not exist. 'Our goal was clear. Companies we worked with would have to be brought under MSC certification,' Brus stated.

In the rather conservative world of tuna, this vision was controversial and provocative, not least with competing labels. It immediately resulted in overt enmity from EII, which had blacklisted Brus in the 'dolphin killer' category, because of his refusal to deal with their Dolphin Safe label. Trading with Brus and Sustunable would no longer be allowed, EII ordered its affiliated companies. But Sustunable persisted. In 2009, together with fisheries in Colombia, he requested an MSC pre-assessment for the free school tuna fisheries in the Eastern Pacific. A pre-assessment is a kind of trial certification. This involves an evaluation of the extent to which the fishery complies with MSC sustainability standards so that it can apply for certification in the future. This was an extremely sensitive issue in the tuna industry, where the lead firms were not exactly chomping at the bit for a certification that was considered expensive, restrictive and cumbersome. The company performing the pre-assessment advised Brus to drop the matter. The time was not yet ripe for MSC-certification on the difficult tuna market, they said.

The assessment never materialised, but the attempt had drawn attention. On the other side of the globe, in Papua New Guinea, Maurice Brownjohn was developing similar plans for the PNA. Would tuna fishing in the PNA be able to comply with MSC sustainability requirements? This would be a pre-eminent calling card with which the tuna cartel would be able to present itself to the world. In utmost secrecy, Brownjohn had conducted an MSC pre-assessment of eight different tuna fisheries for yellowfin and skipjack. The Papua New Guinea National Fisheries Authority (NFA) provided funds to carry out this research. It was a provocative thought: what if the PNA cartel as island nations, which has the world's greatest volume of tuna in its waters, succeeded in bringing onto the market its own sustainably caught, MSC-certified tuna? It had the potential to be a giant leap for sustainable tuna. With their enormous resource of around 50% of the global harvest of skipjack, the eight Pacific Island states would able to take improving tuna's sustainability to a higher level worldwide. For both the PNA and the MSC this could be a game changer in sustainable tuna capture.

The good news came at the beginning of 2010. The free school tuna caught with PNA purse seiners stood a good chance of meeting MSC standards for sustainable tuna, the investigation concluded. It was agreed to move cautiously and start with freeschool skipjack. The tuna conference in Koror City with the PNA island presidents was an opportune moment to launch MSC certification for the PNA. The tuna fishermen and traders who had been invited to attend the summit in Palau would be the first to hear the news. It had been a good idea for Brownjohn to invite Brus to give a presentation explaining how this fitted in with a sustainable strategy for the PNA.

Brus knew nothing about the PNA MSC story. With his presentation he just wanted to give the PNA a boost on the challenge of sustainability and tuna fisheries. When he entered the hall that morning at the agreed time, the Dutch trader was shocked to see not only the eight PNA leaders but all the tuna industry representatives as well. It became an uncomfortable situation. Brus: 'I told the chairman, there's no point talking about strategic vision with the tuna industry in the audience. The hall has to be vacated. Otherwise my presentation won't happen.' The chairman understood. 'There has been a change in the agenda. Closed session. Everybody has to leave,' he told the audience. Almost 100 people got up and left the hall in rising anger. The tone had been set. If the industry ever had any future dealings with Brus, he would know about it.

When the last furious tuna industrialist had left the room, Brus began to speak to the heads of state. What was needed was a shift in mindset. It was high time the island states stopped playing the victim. They were the oil sheiks of tuna. But the PNA waters were still being emptied for a song. The big fishing companies were earning silly money on tuna. For years, there had been a true tuna boom. The fleets were expanding their catch capacity explosively across the globe with purse seiner vessels costing 25 million dollars each. The money was earned back in a few years time. The PNA had the power to ask much more for its fishing licences. The prices for fishing days had to be raised drastically.

Not only that: it would be good if a substantial part of the catch was processed in new factories on the islands themselves, instead of being transferred to Bangkok or the Philippines. The PNA could become a global household name with its own brand and its own sustainable product, certified with the MSC label. This dovetailed with the message of sustainability the PNA needed to start selling to the consumer, through its own marketing organisation. It was a daring idea, but the time was ripe for serious sustainable progress for the small island states.

Like the pieces of a jigsaw puzzle Brus' ideas fell exactly into place. The assembled presidents responded enthusiastically to the pep talk. After the presentation the president of Palau thanked his guest effusively for his contribution and invited him into another room. 'Next we will have a press conference. We as the leaders of the PNA will announce that, based on the positive outcomes of the pre-assessment, will start full certification for our free school skipjack fisheries.' Brus was flabbergasted. His hosts appeared to have pursued the exact strategy he had in mind for them. The gathering left the hall and went straight to a room where the cameras and sound equipment were ready and waiting. In front of an audience of tuna industrialists, the surprising plan of MSC-certified skipjack was announced. The mood amongst the fleet owners, traders and canners present had soured once and for all. First they had been sent out of the hall. This after the PNA presidents had told them that the low prices they paid for their Vessel Day Scheme licences were 'totally unacceptable'. The price of a vessel day would rise sharply, that much was clear. And now at this press conference they were confronted with plans for a certification which would only present more problems. MSC-certified tuna from the PNA, free school skipjack, to be caught without FADs, this could only come at the expense of the favourable *status quo* the industry currently enjoyed and disrupt the domains of traders and global brands. The PNA seemed willing to undertake a new role in the tuna supply chain as a kind of competition. That could be considered no less than a declaration of war.

The group sloped off to dinner. Tuna was on the menu.

During the first decade of the new tuna millennium, sustainability was on the rise as an increasingly important marketing tool. This effected the production chain in two major ways, one at the start and one at the end. At the beginning of the global value chain, the coastal states began to realise that sustainability was becoming a weapon to strengthen their position against the nations which came into their waters to fish with their industrial distant water fleets. The PNA small island states were a striking example of sustainability becoming a main objective. At the end of the chain, brands and supermarkets ran the risk of getting into trouble with consumers if the tuna they sold was not considered to be caught sustainably. The worst-case scenario being demonstrations on the doorstep, leading to reputational damage and ultimately declining turnover or worse. The big retailers also began to realise that they had become a major factor in the market, a powerful group comprising a limited number of big retail companies on the demand side of the value chain. If needed and wanted, they could make a strong fist with the demand for sustainable tuna.

Sustainability had emerged as a forceful marketing tool that could be used at the beginning and end of the global value chain, changing existing power relations.

MSC certification for sustainable tuna was therefore an interesting marketing tool. There was not much practical experience with the label and there were many questions. Who owned the certification for instance? In the case of MSC certification for PNA skipjack the certificate was held by the group of island states. But the reality was that the PNA did not own the tuna once it was harvested. Brownjohn: 'If you sell your fishing licence to catch MSC to a Taiwanese fishing company it will catch its MSC-certified tuna and then it becomes the product of Taiwan. How could we as PNA-countries earn anything on that sustainably certified fish or even cover our certificate maintenance costs? We would be strangled by those not fishing on free school but who fish with FADs.' Now there was a trick: by introducing a kind of lease on the MSC licence it was possible to recoup a share of the additional cost. The question was how. The PNA was a partnership of countries and not an enterprise.

A few months after the Palau meeting, with summer arriving, Henk Brus took a call from Maurice Brownjohn from the PNA office. Could they have a chat about the plans Brus had unveiled to the heads of state in February and how it might dovetail into PNA MSC? No problem. Brownjohn flew to Brus' country home near Barcelona. Over the course of 2 days of intense brainstorming a new draft strategy was drawn up for the marketing of the PNA's MSC-certified tuna. For that purpose a public-private partnership was created and christened Pacifical, a new company in which the PNA countries would collaborate with Brus's firm Sustunable. PNA officially endorsed the concept in Nauru late 2010. Pacifical was officially set up on 1 January 2011 and the logo—a Pacific girl with a flower in her long hair—globally registered. MSC certification was under assessment. If everything went well, it would only be a matter of time before the first sustainable MSC-certified PNA free school skipjack would be brought onto the market by Pacifical. 'I knew immediately that Pacifical would turn the world of tuna on its head,' Henk Brus said. 'This meant war.'

Pacifical became in essence a marketing company to plug tuna that had been caught sustainably in PNA waters. This was skipjack caught by purse seine and without FADs. The potential was huge. The world's biggest bulk volume of tuna for production and canning became available for sustainable capture and sale. The cans of sustainable PNA tuna would be identifiable by the cobrand Pacifical logo on the packaging alongside the MSC label. This had a great advantage: unlike traditional trade this would not be a brand which would have to fight for every centimetre of shelf space and require incredible investment and work force. As a cobrand, existing brands would pay Pacifical to be used alongside with their logo on the can. The slogan was: 'Pacifical, straight from Paradise'. In this way the PNA, a country group virtually unknown to the general public, would acquire its own image as a group of paradisiacal tropical island states that wanted to manage their tuna stocks sustainably. The PNA countries could penetrate the tuna chain deeper with this new joint enterprise. From being purely a producing region, the countries had set up a marketing company which could benefit financially from the added value of sustainable tuna. The leaders had directed vertical integration in the sector, this had begun with VDS as a resource rent, Pacifical in effect was a retail rent. This inevitably meant that they had potentially become competitors of the big existing global tuna companies by moving up in the chain.

The ambitious sustainability plans for Pacifical escaped the notice of the tuna-eating general public. But to many stakeholders with vested interests in the market, the leap in sustainability was more of a potential threat to their own positions. The big traders and brands had to deal with a new cobrand, Pacifical, that used the MSC certification to market its tuna. The Dolphin Safe label was suddenly faced with a competing sustainability label with an entirely different approach to sustainable tuna capture. The distant water fleets were not happy with the initiative either. Brownjohn: 'The Spaniards didn't want to be bothered by anyone 'raising the bar' on sustainability measures. In particular not if it was restricting their main fishing gear, the FADs. The companies that belong to ISSF, including traders and processors, are linked to global brands so basically not interested in sustainable tuna that might only result in competition for their own products. The big tuna companies clearly just wanted the PNA countries to go on selling them fishing days without becoming active in the market themselves.' Regional powers like Australia and New Zealand didn't like the idea of the PNA as an independent commercial player that might divide their influence in the region. PNA's Pacifical was the new kid on the block no one really wanted. A modest competitor for now, but one with the potential, if needed, to upload 420,000–600,000 tonnes of sustainably caught, FAD-free tuna onto the global market. No wonder that the slogan 'Pacifical, straight from Paradise' sounded more like 'Pacifical straight from Hell', to many in the tuna industry. Even before the first Pacifical can rolled from the production line, a mighty joint force was willing to take up arms against this sustainability project.

The first blow that could be struck was in the assessment procedure for awarding the MSC label for sustainably caught PNA skipjack tuna. MSC is a market-driven sustainability certificate with procedures open to all stakeholders in the market. This open structure means that the assessment procedure is open to input and all possible objections, including claims that might be called improbable. All interested parties— in other words the competing, leading firms in the industry—have an excellent opportunity to exert their influence on the certification assessment. The competition did not let this opportunity pass. Among them: OPAGAC, the lobby group for the Spanish tuna fleets and among the biggest FAD users in the world and foremost the International Seafood Sustainability Foundation (ISSF), the partnership of leading tuna traders, brands and factories who claim to represent 75% of the global capacity of canned tuna. All these groups took part in the assessment procedure with one clear goal: to try to prevent the certification of the PNA freeschool skipjack. Sustainability had become a weapon in the commercial fights. As a result NGOs advocating sustainability would increasingly find industry players partnering up in certain certification procedures.

Whereas the drive to participate in these procedures might have little relationship to sustainability, the questions the industry brought up were legitimate. Could the PNA countries really guarantee that their sustainably caught free school skipjack was not mixed with tuna caught using FADs? Were the guarantees for a sustainable catch robust? Reading the objections from the powerful industry stakeholders you would be left with the impression that a MSC label was simply never going to be awarded as far as they were concerned. In the unlikely event that MSC certification would take place, the intention seemed to be that not one single can of this certified sustainable tuna would ever end up on the shelves in supermarkets. In the words of ISSF President Susan Jackson, 'If approved, this fishery would ultimately have the capacity to provide hundreds of thousands of tons of MSC-certified skipjack from a region rich in this abundant resource. But the world will have to wait.' [149] So the world would have to wait, the ISSF had decided.

Things turned out differently, however. Despite intense pressure from the industry, including an adjudication hearing in London where Opagac, ISSF and their lawyers presented objections, MSC certification for the Pacifical free school, purse seine fished skipjack tuna was awarded at the end of 2011.

Now, having the MSC certification for the fisheries is one important step, but still nothing could be done without a MSC certificate that guarantees that the sustainable tuna also finally will land on the consumer's plate. The next step was the certification of the so called chain of custody (COC)—a complicated way to say that in all steps along the complicated supply chain of tuna it is guaranteed that it is indeed the sustainable product that end up on the shelves. Although started a year before the MSC fisheries certificate, the COC certificate assessment proved to be a procedure with a complexity 'without precedent', Brownjohn stated. Just take all the steps: tuna are caught on one vessel that next to free school also harvests with FADs. So the free school fish has to be physically kept stored separately from FAD fish. The catch is invariable transhipped in bulk ungraded, sometimes even twice, before landing. The checking and grading for certifiable weights is established at the coldstore door. At all points the catch and segregation had to be credibly monitored. This saw more opposition from ISSF in the assessment. In the end all objections were denied. Chain of custody was finally certified mid-2013. So the PNA had MSC certification for its tuna, the COC had its MSC certification and Pacifical had customers and was ready to run. This was the moment for the launch of a second Flipper war.

The second Flipper war was a sustainable sea battle in the tuna industry lasting several years, that mostly went unnoticed by the consumers. And once again it was the Dolphin Safe label that was used by the industry as an excuse to close the market to the new competition by the MSC-certified PNA skipjack. The battle started with EII warning the companies carrying their Dolphin Safe certification that under no circumstances would they be allowed to do business with Pacifical. Brownjohn: 'They made outrageous claims, like that Pacifical was a deal between Henk Brus and corrupt island leaders. Fishing fleets, transport ships, canneries, hauliers and importers; everyone was forbidden by EII from trading one single tuna (frozen, canned or cooked) with Pacifical on penalty of losing the Dolphin Safe label and being excluded from the trade. EII sent imperious emails demanding that the firms in the chain should alert them at once when MSC-certified tuna surfaced anywhere, so that action could be taken instantly. 'We would like to inform you that our head office is requiring reports for all MSC-produced tuna in any of your factories immediately,' Mark Berman, alias 'the Bermanator', ordered the EII company-members in an e-mail [78]. The reason? Henk Brus was alleged to have traded a batch of yellowfin tuna from the Eastern Pacific which was said to be in violation of EII's dolphin safe protocol. According to Brus, it was a batch of yellowfin tuna with the official AIDCP label from the Inter-American Tropical Tuna Commission (IATTC), the official dolphin protection programme. For EII that didn't matter: he was stigmatized as a dolphin killer that had to be destroyed.

It was not the first time that Brus had got into a conflict with EII. Brus: 'I initially joined Earth Island Institute because, without the Dolphin Safe label, no one in the industry wanted to work with me. But I soon found out that much about the label was flawed, so I asked them to remove us from the Dolphin Safe list. We only wanted certificates in accordance with the FAO directive and they did not comply with this in any way at all. There was no transparency, no control and no monitoring.'

In an unprecedented move the NGO community united to support Pacifical against EII campaign with a joint press release signed by PEW, WWF and Greenpeace. But that did not restrain the self-declared Dolphin Safe police. They now had an opportunity to strike back at its competitive MSC certification. And as with the Mexican yellowfin tuna before, the attack from the EII label was a perfect boost for the dominant tuna traders to prevent the potential PNA competition entering the market. 'This conflict between Dolphin Safe and Pacifical must be resolved before we get involved,' was the recurring argument used by the industry in those days.

Industry now claimed several reasons, all perfectly defendable if you have to run a business, why they could not participate in the Pacifical brand. There clearly was very little enthusiasm for adopting the expensive and cumbersome MSC certificate standards. The price premium for certified sustainable tuna, an important sales argument for the concept, was rejected as being barely credible in a market chiefly driven by selling cans as cheaply as possible by big brands and discounters. Moreover the fleets that were using FADs as tools in their fisheries were not happy either with an incentive promoting free school fishing as a more sustainable alternative.

So the resistance from the Dolphin Safe label served the industry well, once again. As had been the case when the American market was effectively closed to Mexican yellowfin tuna, Dolphin Safe acted as a 'weapon of sustainable mass destruction' in this tuna war. And oddly enough, the weapon of mass destruction was used without the consumer noticing anything. The story was far too complicated for a public which allowed itself to be seduced by Flipper's smile. The tuna consumers were unaware of the fact that EII had set itself up as a kind 'tuna police' (or better: dolphin police) to fight the MSC-certified tuna. Not only was the certified skipjack tuna attacked, but also the MSC itself, which meant the war was now being fought between the two competing eco-labels. This was the ugly face of a certification war: EII openly accused the MSC of promoting a standard that offered insufficient guarantees for the protection of dolphins. MSC standards director David Agnew reacted furiously. Dolphins were one of the bycatches, he wrote, just like sharks and turtles, which the MSC standard did care about. Moreover, it was ultimately about sustainable tuna capture, something EII was not fundamentally involved with at all. Contrary to the MSC, EII did not enforce independently monitoring of its own standards anyway, Agnew stated. 'The MSC has assessed (...) fisheries on their performance, not whether they sign a piece of paper,' [11].

Pacifical initially had tremendous difficulty breaking through the Flipper boycott and the unwillingness of the fleets and traders to work with their certified scheme. A Philippine fishing fleet was finally willing to fish for MSC-certified Pacifical tuna and a small quantity of sustainable certified tuna reached the West via the SPAR supermarket chain in Austria. Not without hurdles either: due to late shipping, the first cans had to be flown over from Singapore to meet the Austrian SPAR launching date. At the end of 2013, two years after it had been introduced, a mere 400 tonnes of Pacifical tuna had managed to find its way to the consumer. That was not even 0.1% of the volume deemed potentially available for MSC certification. The leading tuna companies could be satisfied. The world would have to wait for this tuna 'Straight from Paradise'.

The turnaround came at the beginning of 2015, after the skipjack prices on the Bangkok market had dropped steeply. The trade was looking at ways to increase its margins and the price premium that could be asked for sustainably caught skipjack became a more attractive option. More important still was the fact that the demand for sustainable tuna by retailers began to pick up and big global traders and canners like Tri Marine and Thai Union started to realise that they should be able to cover an increasing demand for sustainable tuna. The PNA waters, with their abundant tuna stocks, offered the only really large-scale, sustainable production area in the long term. They started to realise that their membership of EII increasingly became a problem and even a potential source for reputation damage if the fight between EII and Pacifical would heat further up. Finally Thai Union brokered a truce between both parties and EII stopped its attacks.

The first big traders, Tri Marine and the Taiwanese FCF, now switched and signed the memorandums of understanding with Pacifical. Pacifical tuna had succeeded in breaking through. In the year after the boycott was lifted, just under 70,000 tonnes was supplied to the market via outlets such as Australian canned tuna leading brand John West. The PNA had finally managed to sell its sustainable skipjack. This was a victory over the campaign of EII, but also the collective opposition by the leading firms in the value chain for canned tuna. The battle had taught them that, in the end, the decisive factor was demand for more sustainably caught tuna and that this was able to steer the market towards a sustainable product. The consumer and the big retailers had shown their might in the value chain of canned tuna. An important insight for future policy focused on sustainability.

But the changes came at a price. It required tenacity, they found, to overcome resistance and to penetrate a market with a sustainable product. The prevailing interests within the industry saw sustainability more as a threat to their established position than as a challenge of a new market for a sustainable product. Long-standing vested powers were not going to be pushed aside that easily. That meant a battle between competing business interests. But it could easily degenerate into a conflict between several competing eco-labels. Sustainability certifications, or what passed for sustainability, were weapons in the fight, a tool that could be used to conquer markets but also to defend existing interests. Industry was using the related stakeholder participation, cleverly siding with the NGOs. Eco-labels were used without the consumer understanding fully what effect this had on their demand for responsibly caught fish. It became increasingly important but also more difficult to determine which sustainability claims were justified and which were used for other, more commercial objectives. The modern hybrid tuna wars around sustainability had become complex and messy, waged between states, regional organisations, big companies and NGOs. Varying coalitions and partnerships were formed on different fronts; in the fisheries, in chains of custody and on the consumer markets. Billions of euros were at stake. It had become a complicated, treacherous and dirty sea battle around tuna.

In the end, EII did not manage to maintain the boycott of the MSC. With big traders such as Tri Marine started to move in the new market of sustainable tuna, the label soon had to back down. Dolphin Safe had always served the industry well. It was a relatively cheap way to acquire a sustainable image, whilst requiring little action. But in the end it were the big global tuna companies that payed their bills and pension funds. So when the big players started to move into the MSC-certified market, EII did better to recant. The tide was turning. EII, which had just settled another case in the court of Florida, could not afford more problems with business. The reality of a growing demand for credible sustainability of tuna kicked in.

Mark Berman, 'the Bermanator', who for years had successfully marketed the Dolphin Safe logo on the cans, died unexpectedly in May 2016 as a result of complications from an acute intestinal complaint. According to EII, Berman had managed to enlist 550 tuna companies across the world to the label. It was a good indication of how far the power of Flipper had reached [202].

The Third Flipper War: The Battle of Mexico

For the war around Mexican tuna, we make a detour to beer. So we are off to the German city of Munich, capital of the federal state of Bavaria and famous for its Oktoberfest where men dressed in *Lederhosen* down litre tankards of beer. Beer is part of an age-old brewing tradition in Munich. Consequently, Munich's beer industry has been big and powerful as far back as the Middle Ages. The population wanted good quality beer, not dross. In 1478, the then duchy of Munich brought in an important beer law which was embraced by the industry. Henceforth, only beer brewed solely with natural ingredients, namely barley, hops and water, could legally be sold. All other beers were prohibited. The *Reinheitsgebot* ('commandment of purity') became the first commodity law of any significance, a kind of eco-label avant-la-lettre, which was successfully introduced by all the states within the former Holy Roman Empire. The German beer industry clearly benefitted from the decree; competition was kept out, accused of not adhering to the strict ingredients it imposed. The market remained closed, competition was kept in check, consumers paid more and the brewers' profits increased. The *Reinheitsgebot* became a classic example in economic literature of how a market can be successfully protected by means of a certificate. Only in 1987 did the European Court of Justice find it in breach of European Union competition laws. With more than 500 years of the *Reinheitsgebot*, the combined German brewers had been able to exclude all competition and achieve handsome profits at the expense of the German consumer.

Beer is just one example. I knew from my time studying economics at the University of Amsterdam, that these kinds of sophisticated market closures happen in all types of industry, and in particular in those industrial sectors where there is a concentration of power amongst the lead firms in the value chain. After all, a limited number of companies in the market makes it much easier to put your heads together and fix things, as in the case of tuna. During the second half of the twentieth century the tuna companies StarKist, Bumble Bee and Chicken of the Sea grew into the leading brands on the US market. They would capture 70% of the American canned tuna market. Whether the big three knew of the German beer market's *Reinheitsgebot*, history does not recount. But when, as a result of the first Flipper war, they decided jointly on 13 April 1990 no longer to use tuna from fisheries which caught dolphin, an idea must have formed in someone's head.

The American tuna industry had not doing too well since the beginning of the 1980s. The canneries in southern California were no match for cheaper competition from Thailand and shut down one by one. The tuna catch volume was barely growing, and the fleet had to go ever further afield on the Pacific to be able to operate cheaply. Prices dropped. Mexico was one of the formidable competitors offering tuna at a lower cost. This pushed the US fisheries into a tight corner, to such an extent that it led to a true tuna trade war. This started when the Mexicans arrested six American tuna boats fishing without a licence in its newly introduced 200 miles EEZ [228]. It was a classical border incident that took place during the years of the introduction of the EEZs. The US simply did not recognize the international law with the argument that tuna was a migratory fish and for that reason excluded from the EEZs. After losing their boats, the US hit hard back and closed its borders to Mexican tuna imports. This was extremely convenient; without such an embargo on Mexican tuna the American tuna industry, with its closing processing plants and massive lay-offs, would receive a death blow by cheaper Mexican imports. But the big three also knew that with the winds of global free trade blowing the embargo on Mexican tuna would not last. The trade ban had to be re-negotiated. Something else had to be devised to protect the American tuna-industry.

As always Flipper was right there when you needed him. The sincere outcry around dolphin killing, up to an estimated 6 million individuals since 1950, had turned into a threat to the American tuna industry. A consumer boycott was imposed that could be fatal for the Big Three brands that owned 70% the canned tuna market: Chicken of the Sea, Bumblebee and StarKist. But a remarkable move of the industry turned around that threat into an opportunity: all three companies announced on the same day, the 13th of April 1990, that each of them would no longer use any tuna for their products that was caught in nets that also 'injure or kill dolphins' . In one day, the Big Three turned from villains into allies of the environmental groups, in particular Earth Island Institute (EII) and their Dolphin Safe standard. EII-director Dave Phillips was jubilant: he declared that this was the most important step for saving dolphins since the introduction of the US Marine Mammal Protection Act of 1972, that was intended to prohibit the import of fish products in the US that had been caught with the incidental killing or hurting marine mammels. Everybody hailed the industry decision to safe Flipper. The only not-so positive comment was from the American Tuna Boat Association of the 29 boats of the San Diego tuna fleet. It was all a 'marketing ploy to increase tuna sales', president August Fellando told the press [219].

Fellando had a point. The same day that the companies decided to go Dolphin Safe, Mr. Phillips announced that the StarKist company would join them in pressing for legislation that barred the sale of all tuna caught in association with dolphins. Environmentalists and industries joined forces in a lobby towards the state. This effectively worked out well for the industry: in close harmony with EII they pleaded for a strong legislation for the tuna imports that served as the ideal instrument for the protection of the market against unwelcome competitors. The Dolphin Safe label would become the main arm in the Mexican tuna war. It could be exploited by the American industry as a barrier that protected their market. Five months later, in August 1990, the US market was closed for the Mexican yellowfin import by the authorities, because it could not show compliance with the new, strict Dolphin Safe ruling that prohibited he setting of their nets on Dolphins [252]. The Dolphin Safe standard turned out to be the *Reinheitsgebot* for the American tuna. Although the industry was committed to other rigorous standards to avoid harm to dolphins (like the AIDCP), only the American standard for Dolphin Safe was allowed for imports. The state took the lead in enforcing quality assurance and imports had to comply, which gave the EII Dolphin Safe label a strong market position and effectively the guardian for the American tuna industry. In a way the administration outsourced an enormous power to Earth Island Institute. The big three in turn committed themselves to their Dolphin Safe rules and contributed 'voluntarily'. It was a price worth to pay.

The Mexicans were frustrated that, despite the efforts to reduce drastically the dolphin mortality in their fisheries, the were not able to sell their yellowfin tuna to buyers on the American market. Mexico then made an attempt with the newly signed free trade agreement between the North American states, NAFTA, that entered into force in 1994 once again to no avail.

In 1995 another attempt was made to break the import embargo against Mexican tuna imports through the Panama Declaration, in which the regional commission IATTC ratified the Agreement on the International Dolphin Conservation Program (AIDCP). This new arrangement legally binding for all the members of the IATTC (including the US) allowed nets to be set around dolphins, but they would have to be released unharmed, supervised by independent on-board observers [263]. Both Mexico and the US signed up to the protection programme which was welcomed with great acclaim by individuals such as President Bill Clinton and organisations including Greenpeace and WWF. This was considered an important step forwards in the protection of dolphins. Nonetheless, the US held on to the requirements of its own Dolphin Safe tuna certification. And EII decided that this new international arrangement was unacceptable: it had altered their own strict Dolphin Safe definition that even setting nets on dolphins, whether they could escape unharmed or not, was unacceptable. The EII brought the ruling to an American court after researchers of the American National Marine Fisheries Service concluded that the tuna fishery was not having any impacts on dolphin populations. With success: the judge ruled in favour of EII, maintaining the de facto embargo effective. The industry could breathe a sigh of relieve.

The Mexicans started a battle on a new front. After the US had refused to recognise its obligations towards dolphin safety under AIDCP ruling for 10 years, the united Mexican manufacturers decided to take the case to the World Trade Organisation (WTO) panel in 2009 to enforce the export of tuna [252]. It argued that the US conditions imposed on 'dolphin-safe' labelling were 'discriminatory and unnecessary', and in fact an excuse to close their market for the tuna import. It became a protracted case, in which the WTO ruled against the US time and again. Meanwhile the US amended its legislation half-heartedly but in fact upheld its

Mexican tuna embargo. During all these years the conclusions of the WTO were in favour of the Mexicans [258] and rather unfavourable for the US Dolphin Safe standards. They did not meet the sustainability objectives, according to the WTO. Dolphin Safe was more likely to mislead consumers than inform them. Moreover, there was no transparency and traceability to verify that the 'dolphin safe' tuna caught was the same that ended up on the consumer's plate, the WTO concluded. US zero tolerance of dolphin killing during tuna capture proved also to be highly selective: 95% of imports came from other areas with higher dolphin mortality, without any problems. And: the AIDCP dolphin programme which had been endorsed by Mexico had been quite effective in combatting dolphin killing in the Eastern Pacific. In a 2017 verdict the WTO ruled that Mexico could claim an annual trade sanction of $ 164 million against the United States [237]. For the WTO there could only be one conclusion: the American Dolphin Safe programme had been chiefly used as a device by the American tuna industry to keep out Mexican competitors. Mexico had reasons to be pleased by the WTO decisions. The US was now again open to Mexican imports, but many American retailers still feared the Dolphin Safe label's power and preferred not to have tuna on their shelves that where not labelled with EII's jumping Flipper.

Of course, the US used its last possible appeal against the WTO decision. Not without reason, because the tide had turned. The WTO was understaffed and hardly able to complete the complex appeal procedures. Meanwhile, with the Trump administration the outlook for free-market based decisions of the WTO turned bleaker. Trump was blocking free trade where ever he could, in order to make 'America Great Again'. There were signs that the US presidents even would pull out the WTO altogether if it came with more decisions unfavourable for the US. And the bare thought that Mexico, the country that according to his campaign promises had to pay for its own wall at the border, now could cash in $ 164 million a year was pretty enraging for the president.

It was not totally surprising that at the very end of 2018, after almost 10 years of procedures, the WTO reversed all its former verdicts in a final judgement. The WTO judged that 'setting on' dolphins with purse seine nets was likely to kill or injure them. The precautionary measures and observers that where put in place under the AIDCP ruling: all this was set aside. Even without clear evidence of such dolphin deaths, the US could keep its boarders closed for tuna without the Dolphin Safe consent [164].

So, in the end the Mexican tuna industry lost its Flipper War. For the moment. Business-wise it didn't hurt that much, since in 10 years the market situation had changed and the Mexican tuna had found clients outside the US. Meanwhile their fishing method had been positively resisted the test of sustainable criteria: in 2017 the Mexican fleet of 36 purse seiners in the eastern Pacific was awarded the MSC certification for its yellowfin fisheries [184]. United within the Pacific Alliance for Sustainable Tuna (PAST), the four leading Mexican fisheries on skipjack and yellowfin tuna were thus able to attenuate the long American boycott by putting itself through the MSC sustainability standards. PAST obtained the sustainability recognition which they had always been denied by the Americans.

EII furiously attacked MSC. 'We will be contacting the public, the press, the retailers and the companies saying that this certification is bogus, that's it's trying to put lipstick on a pig . . . this will be a big black eye to the MSC', EII director Phillips said. 'The MSC process is fraudulent.' [251] It sounded like an angry attempt to save the face of Dolphin Safe. Coming right after the second Flipper War with the PNA, this was another serious defeat.

In a way, the MSC certification of yellowfin tuna in the Eastern Pacific settled the Mexican tuna war. But still: Flipper had succeeded to keep a market closed to foreign tuna competition for over almost three decades and there was no sign that it would not continue to do so. That was a remarkable achievement for a dolphin.

Tuna Conspiracies

38

Adam Smith, founding father of the free market, warned about it: as soon as businessmen join together, there is a big chance the consumer will pay a bill for the party. 'People of the same trade seldom meet together, even for merriment and diversion, but the conversation ends in a conspiracy against the public, or in some contrivance to raise prices', Smith writes in his famous Wealth of Nations [222]. Smith's point was that businessmen can succeed in a 'conspiracy against the public' if they are given in one way or the other protection by government regulation. That was the case in the eighteenth century society, but in the new configuration of governance at the end of the twentieth century the weakening role of the state created space for other stakeholders in the market to promote this kind of protection. As we will see one of the new manifestations of this kind of protection of business were 'supranational state-like regulatory mechanism that combines science with free trade and environmental ideals' in certification schemes set up by NGOs to promote sustainable fisheries [61]. 'Tuna is the debut for a great debate between environmentalists and traders', a Venezuelan consultant for the tuna industry was quoted in the nineties. He proved to be right in many ways more than he probably ever could have imagined.

Another point Smith makes is that business conspiracies should fall apart under the pressure of mutual competition. That is: if there enough competition in the market, businesses would be too busy in looking for their own sake and with many competitors around had no ways to conspire. But Smiths idea also has a downside: in markets where the competition is not working that well—as is the case many times in real life—this free market correction mechanism can fail easily. Hitherto the likelihood of a conspiracy increases when the market is dominated by a limited number of lead firms which control competition and trade. The need to work together will be even more stronger when market demand is falling, margins are getting smaller and outsider competition is getting stronger. All these conditions for blossoming conspiracies where greatly met in the United States tuna market at the start of the twenty-first century.

© Springer Nature Switzerland AG 2019
S. Adolf, *Tuna Wars*, https://doi.org/10.1007/978-3-030-20641-3_38

Americans were eating decreasing quantities of canned tuna. Consumption peaked in the mid-eighties to a yearly 4 pounds per capita to almost half of that at the start of the second decennium of the twenty-first century. The big brands believed it was a market in which consumers only were driven to pay the lowest price for their tuna. Margins were shrinking. Competition from foreign competitors was strong and growing. No wonder the industry had welcomed the opportunity of the Dolphin Safe standard to close their market.

The joint intent to close the market borders for competitors seemed to be a stepping stone for further conspiracy. Big American retailers like Walmart (the biggest retailer of canned tuna in the United States) Kroger and Hy-Vee had noticed something was smelling fishy inside the American tuna market. The air reeked of a price fixing practice between the Big Three brands who together controlled almost 75–80% of the big American consumer market for tuna. (The Big Three had become less American than ever. Bumble Bee was in the hands of the British private equity investor Lion Capital, Chicken of the Sea owned by the Thai tuna giant Thai Union and market leader StarKist belonged to Dongwon, a tuna trader that manages the biggest South Korean fishing fleet.) The cans of tuna on their shelves were becoming noticeably smaller. The quality of the content was dropping. Tuna chunks where replaced by fish flakes floating in a sludge of hydrolysed proteins. The stuff was more popular known as 'tuna soup'. Remarkably enough the practice of the tuna soup in smaller cans resulted in consumer prices that went up rather than down. It had also struck the retailers that none of the big three had FAD-free tuna in their range of canned products. There was no choice as far as they could judge. And the top executives of the Big Three were getting far too cosy to their taste, switching jobs back and forth between the big tuna firms. The whole bunch also met each other on a regular basis in the meetings of the International Seafood Sustainability Foundation (ISSF), which they had set up together and in which they controlled 75% of the global can production, trade and marketing. All in all, to the opinion of the retailers, this was too much to be sheer coincidence. And they did not like what they saw.

Walmart, Kroger and Hy-Vee are big players at what you call the downstream end of the value chain, where the processed product finally reaches the consumer. In world markets in which the power to decide had become increasingly demand-driven, whether in relation to sustainability or other issues, retailers started to discover that they represented an important force in the market. In august 2015 the New York based food distributor Olean Wholesale Grocery Cooperative became the first retailer to take a bold step forward and instigate a civil class action lawsuit for conspiracy and price-fixing against the Big Three, their operators and their holding companies. This opened a new front in the hybrid tuna wars of the third millennium [14, 18]. One year later Walmart joined the case, followed by Kroger, Hy-Vee and a range of other retailers. They charged the Big Three on conspiracies alleged to have taken place on the American market for 10 years and concerning many hundreds of millions of dollars of canned tuna, in the form of price-fixing, working together in joined canning and the 'Tuna the Wonderfish' campaign and collusion through participation in the International Seafood Sustainability Foundation (ISSF). Walmart came up with a list of more than 50 top-level tuna executives that allegedly took part in the Big Three conspiracy [19]. It featured big part of the tuna world elite: chairman Thiraphong Chansiri of Thai Union, National Fisheries Institute president John Connelly, Susan Jackson who had swapped the StarKist vice presidency for the ISSF presidency, Bumble Bee CEO and former ISSF chair Chris Lischewski, a remarkable and well-known appearance with his long, flowing blond hair, Chicken of the Sea President Shue Wing Chan, StarKist President In-Soo Cho and Dongwon Chairman Jae-Chul Kim.

In the course of the case, it soon became clear that Thai Union (owner of Chicken of the Sea) had been the whistle blower who got of the hook by implicating Starkist and Bumbleby. In the multiple lawsuit, which included claims for damages, Bumble Bee was the first to plead guilty. It paid a fine of 25 million dollars after two of its executives admitted they had been guilty of the tuna price-fixing in the parallel criminal price-fixing probe started by the Antitrust Division of the US Department of Justice. During the period in question, Bumble Bee had had a turnover of 1.3 billion dollars, so that fine was not a big deal to the company in comparison to the amount of money involved. A former vice president of StarKist followed with an admission of wrongdoing and the payment of a fine. But this was just the beginning of what promised to become a protracted case. Settlement by the American authorities could take years. Thai Union set aside 44 million dollars to settle all civil law suits involved. The total in damages and fines would be at least 100 million dollars, it was estimated. And it was perfectly conceivable that some of the involved tuna executives would spend time behind bars. The case shocked the tuna world.

A few conclusions could already be drawn from the American tuna brand conspiracy. In a value chain in which firm power concentrated at certain levels and in a buyers-market with falling demand and margins, there was an increased likelihood of all kind of deals between the leading companies. Tuna involved just such a value chain, with a limited number of big traders and brands that acted as lead firms. Price fixing made sense in a shrinking market, but also a conspiracy around sustainability issues, like in this case the common use of FADs, appeared to be unfolding as well. A response to these kinds of practices was possible through the growing force of stakeholders on the demand side of the value chain. This countervailing power to take action shifted to the consumers and retailers. They could use their market demand as a weapon, and occasionally even fight a case in court when competition failed. This had become a new battlefield where the tuna war on sustainability would be fought in the next future.

But it proved to work also the other way around. Since sustainability had become increasingly a matter of market-driven governance instead of state-regulation, it had become a marketing instrument for the industry that could serve all kind of purposes that went beyond the idea of preserving the tuna-stocks, guarding the ocean life or guarantee the working conditions of the fishermen. It could help to conspire, close markets and rise prices. Sustainability had become a new and very effective weapon in the hands of the industry that eventually could be used to kill competition. And in the new partnerships that were formed to govern the sustainable developments, industry even could enforce this weapon with the help of partnering NGOs to cover up their conspiracy goals under an image of sustainable tuna fisheries. The American price-conspiracy case unfolded this new danger to be aware of in the twenty-first century global tuna market, and probably in other markets where sustainability and partnerships had become a common feature.

In October 2017 the ISSF, the global organisation that represented 75% of the global tuna canning industry, that had partnered with WWF, announced a remarkable new step in their cooperation. From now on, the almost 30 company members of ISSF should purchase tuna primarily from fish traders and processors that also participated in the ISSF. That meant that the companies had to comply with the conservation recommendations set out by ISSF and undergo compliance audits. 'This new measure accelerates these sustainability best practices among a greater portion of the tuna industry', according to ISSF President Susan Jackson [20]. That was one way to frame it. The other was that never before such great part of the global tuna canning industry had acted so closely together in promising to join a common purchasing strategy that left everybody else out. And that, of course, all in name of a sustainable tuna fisheries.

Who listened carefully, could hear Adam Smith chuckling in his grave.

Pirate Wars

39

Let's leave the administrative struggles and business conspiracies for what they are and return to the more physical violence of the tuna wars of the third millennium and their fall out on sustainability. Take Somalia. With one of Africa's longest coasts (without any authority to speak of and plagued by Islamist insurgents) this country had become an ideal operating base for modern pirates. The film *Captain Phillips*, starring Tom Hanks, told the true, horrifying and sad story of the hijacking of the cargo ship Maersk Alabama by Somali pirates. The distant water fleets of the big tuna fishing nations, Taiwan, Spain and France, were also frequently attacked by pirate gangs in their heavily armed speedboats.

The hijacking in October 2009 of the Spanish Alakrana, a modern, 330 ft purse seiner from the Basque tuna port of Bermeo, shocked the nation. The Spaniards anxiously followed the daily fate of the 36 crew members, including Spaniards, Indonesians and Ghanaians. An earlier pursuit by Spanish marine commandos resulted in the arrest of two Somali pirates. Their 63 accomplices aboard the Alakrana threatened to kill the crew if their colleagues were not released. These captives had by then already appeared in court in Madrid. In the port of Bermeo thousands of people took to the streets to demand the release of the tuna fishermen. The hostage drama lasted 47 days and ended, unconfirmed reports suggest, after a ransom of 3.3 million dollars was paid [153]. The crew was eventually released unharmed and the ship was able to join two Spanish warships that had been sent to the area. Three years later in Spain the two captured pirates were sentenced to a total of 439 years in prison for armed raids and kidnapping. The hijacking confirmed that the Somali 200-mile zone should be considered a dangerous area from then on. In response, the EU launched 'Operation Atalanta', in which warships escorted European fishing vessels for protection. On-board armed guards were introduced on the fishing boats. Piracy on tuna ships would ultimately disappear by force of armed protection. According to a World Bank report, the pirates were alleged to have received close to 400 million dollars in ransoms between 2005 and 2012 [243].

© Springer Nature Switzerland AG 2019
S. Adolf, *Tuna Wars*, https://doi.org/10.1007/978-3-030-20641-3_39

In the end, it was about money. But as fate would have it, these events had some fairly major consequences for the sustainability of tuna stocks in the region. Before the pirates became active the practically uncontrolled, tuna-rich waters were frequently plundered by ships that cleaned out the tuna population without any control or enforcement of tuna management rules. Around 2003 alone, Somali fisheries lost 100 million dollars poached tuna and shrimp by foreign fleets, according to estimates in a British study [70, 218]. The waters became the proverbial example of IUU tuna fisheries and a major black hole in tuna catch statistics.

Because of piracy, however, many people engaged in illegal fishing began to avoid the region. In this sense the Somali pirates proved to be more effective for the recovery of tuna stocks than the minimal management by the regional management commission for the Indian Ocean. By contrast, the successful approach to tackling piracy had a perverse, unexpected side-effect on sustainability of the tuna stocks. After the NATO and EU marine fleets chased away the pirates a fleet of hundreds of illegal tuna boats from Iran and Yemen started their illegal fishing practices, as well as the bigger ships in Chinese, Taiwanese, Korean and even European hands [227]. There were even reports that the former pirate gangs now could earn better money from tuna fishing by offering armed security squads to the illegal tuna fishing boats, chasing away any local fishermen who approached.

In the bigger plan, illegal tuna fisheries and piracy in the waters of Somalia was again part of a hybrid war in a strategic area known as the Horn of Africa. The fisheries interests were important: Spain might have fished only in 2013 16,000 ton of mainly tuna in Somalian waters according to some observers [139]. Iran and Yemen fished even more tuna, according to the estimations. The exact amounts were unknown because of the lack of proper data. But the money involved had to be significant enough to justify the efforts to secure it. Unsurprisingly it was Spain that took the lead in the marine operation Atalanta, with a fleet-headquarter at the Spanish/American marine base of Rota in Andalusia. In 2017 the Spanish government defended that the EU-marine fleet should stay in the area for a much longer period: not only the tuna interests had to be defended against pirates, the whole area became more important for maritime trade and fisheries, with increasing presence of fleets of Japan, China and India. The fleets of the US, South Korea, Russia and NATO completed the patchwork of navies active in the area. In particular China's expansion of trade and strategic influence worried the Europeans. This was reflected in tuna: the list of licenses for tuna fisheries for 2019 published by the Somalian government showed 31 boats allowed to fish in their EEZ. All were exclusively from Chinese origin.

The new successful poachers' strategy in the Horn of Africa, earning some money chasing away the smaller fishing boats using violence, was apparently copied elsewhere. In 2017 rumours intensified that the Philippine tuna fleet used force to scare away competing fishermen from the waters of Papua New Guinea (PNG). When an investigation confirmed these practices, PNG had enough. 'The Archipelagic Waters are sovereign internal waters of PNG, not waters that should be controlled by foreign companies or boats under threat,' the conclusion from the PNG authorities read. If necessary, it would take back control of its waters by force.

The threats were not the only acts of piracy by the Philippine fleet which the PNG wanted to stop. The Philippine tuna fishermen had availed themselves of a special deal with Papua New Guinea from as far back as the 1980s. The idea was that instead of disappearing over the horizon, it could be better that distant water fleets with their holds full of tuna unloaded their cargo in the ports of the PNA countries, to be turned into canned tuna or loins for export in local factories. A greater proportion of the added value of canned tuna would stay in the region, and it would create employment and investment in the industry. In 1997 Papua New Guinea, with eight million inhabitants the biggest PNA country by far and less isolated from the civilised world than other countries, was able to persuade the big Philippine fishing company RD to build the first of a few canning factories. Henceforth the catch had to be processed there. In exchange the associated fleets were given a substantial discount on fishing rights in the waters of Papua New Guinea [31] the Philippines capitalised on the deal: for a few million dollars a factory was erected and tuna came rolling off the production line for formality's sake, on a modest scale. The result was that the fishermen had virtually free access to the rich fishing grounds. But landing and processing the tuna in Papua New Guinea was a considerably less smooth affair. Ever more frequently, the Philippine boats set course for home to offload their cargo or transfer to carriers to deliver to at their own canneries elsewhere. In reality, 'Philippine' tuna increasingly came from the waters of Papua New Guinea. And now the fishing boats that dared to come near were attacked as well. The agreement with the PNG about cheap fishing days did indeed resemble a charter for tuna pirates.

But the days of free fishing seemed to be over. A quick calculation revealed that Papua New Guinea had missed out on over $100 million in discounts given for fishing days within the Vessel Day Scheme [73]. The principle of discounts in exchange for production in the local PNG factories was retained, but from then on only under a system of tight monitoring of production.

The End of the Trail, the Beginning of a New Tuna Era

40

This is where our tuna quest ends. It has been a long journey in time and space in search of the whereabouts of the fish that was caught, processed and traded on an industrial scale for millennia. I have tried to guide you through the complex and fascinating history of a fish that always has had an impact on us in many ways—ways that most of us are hardly aware of. I have also tried to explain some of the classic as well as the modern tuna wars of our time and pointed out that they are decisive for a sustainable future of this fish. At the start of the third millennium, tuna swims between hope and fear, a protagonist in hybrid wars on the seas and the markets that has become the battlegrounds for issues of its sustainable survival.

To tuna lovers who will read this in 200 years' time: the sentiment of hope did not hold the upper hand at the start of the twenty-first century. It was a turbulent and, in many ways, quite gloomy time for man as well as for tuna. Maybe not as turbulent as the years to come, but troubled enough to generate dark thoughts about the future and the way we can manage a global governance of sustainable fisheries. The world was in turmoil. In 2007 a banking crisis erupted which turned into the worst economic crisis since the Great Depression of the 1930s. Europe was weighed down by an austerity policy which for several years led to large-scale unemployment, adjustments and growing inequality of income and wealth. Economic growth in the US stagnated and was recovered again. The Asian economy, with China in the lead, was in the process of heading for world economic supremacy. The geopolitical power balance started to shift. Following a number of disastrous wars in the Middle East and civil unrest in North Africa, the world was beset with Islamist terror groups that caused much unrest. Immigration to Europe and the United States created division and xenophobia. Riding the waves of populism and following a referendum, the United Kingdom decided to leave the European Union under Brexit. In the United States, Donald Trump came to power as president with the support of dissatisfied people who distrusted everything to do with cooperation, globalisation and governance in general. International cooperation crumbled. Trade wars started, with devastating effects on global commerce. In Southeast Asia, tensions were rising

© Springer Nature Switzerland AG 2019
S. Adolf, *Tuna Wars*, https://doi.org/10.1007/978-3-030-20641-3_40

against the growing superpower that was China. In North Korea, a cruel dictator used the threat of nuclear arms to try and save his inhuman regime from collapse. In the areas of conflict, the demand for shelf-stable foods like canned tuna was on the rise once more.

All this happened against a backdrop of growing scientific confirmation that the planet was now subjected to irreversible global warming, with disastrous consequences such as melting ice caps, rising sea-levels, unprecedented hurricanes, droughts and change of sea currents the impact of which could only be approximated, not in the least the impact on migratory fish like tuna.

For those obsessed with tuna, this all seemed to have its repercussions on our views on the state of global tuna stocks and the way public and private governance of a responsible fishing practice is shaped. Some considered it a miracle that tuna still existed at all, others thought that tuna stocks were much more resilient and robust than could be estimated with our limited data resources. Nonetheless a general consensus transpired that sustainable management of global fisheries had become an inevitable part of the landscape of common governance. Bluefin tuna had played a major part in these anxieties. NGOs such as Oceana, Pew, WWF and Greenpeace had earmarked the fish as an icon for the battle of sustainable fisheries. Documentaries and journalist investigations had exposed the fraudulent plundering practices around tuna fishing. Some scientists had passed a crushing judgement on a powerful fisheries lobby which, right across the globe, managed to capture billions in subsidies to create colossal over-capacity in fishing boats, some even warned for a 'global collapse' of the fish stocks, starting with the large predatory fish, like tuna. This would have had inevitable repercussions for the rest of the marine ecological system. The darkest future perspectives consisted of oceans turning into a warm, salty-acidic soup, filled with vast jellyfish populations and plagues of polyps and algae, like the poisonous cyanobacterium or fireweed that covered fishermen of the Australian west coast in festering blisters.

This nightmarish perspective did not leave me unmoved. After all, I belonged to the privileged generation which had actually just began to discover the rich under-water ocean life, with Jacques Cousteau's red sailor's beanie and Flipper's whistles on television, and now had to watch the destruction of the ocean ecology. Eating tuna became an awkward undertaking. Many people had lost track of how to consume tuna in a responsible way. Only specialists could find their way in the maze of tuna species, fishing locations and gears that indicated a sustainable fishery. Sustainability developed in a broadening concept with many criteria as the health of stocks, the impact on the environment, effective management, social responsibility, CO_2 footprints and illegal fisheries. Even the specialists sometimes had difficulty finding a consensus of what could be considered sustainable. Some consumers got so confused (and guilt-ridden?) that they decided it would be better to refrain from eating or serving any tuna altogether, least of all bluefin, to prevent a total collapse.

But as is often the case with the predicted end of the world, for the time being at least a collapse of stocks did not materialise. The picture was confusing though. Take bluefin tuna. If, in the spring of 2018, you would go to have a look at the *almadrabas* along the south coast of Spain, well after the statistical ground zero for bluefin tuna had been predicted, you would see large tuna weighing 250 kilos or more being hoisted from the fishing boats in the port of Barbate. The Gibraltar Strait was overflowing with increasing numbers of large, plump bluefin tuna. Orcas lay in wait in groups to gorge themselves on their favourite meal. In 2017 the catch quotas for successive years had been drastically raised. Too drastically some thought, but no one could deny that bluefin tuna was back in the water and back on the menu.

Once again, the most cosmopolitan fish in the world had sprung a surprise. This time by not dying out. 'Don't underestimate the resilience of the bluefin tuna stock,' the eminent Japanese tuna expert Peter Makoto Miyake had predicted, at a time when all hope that the fish could be saved from extinction had virtually disappeared. He was proved right. The key question was why. There had clearly been a few good tuna year classes. Some of the local mafia-clans and dictators that specialised in massive illegal tuna fishing had been effectively wiped out of business. Also, the ICCAT tuna management policy, for years a muddle, had paid off. The quota and in particular the minimal catch weight might have helped recovering the stocks. The graphs produced by tuna scientists, confusing as they were, showed an undeniable reversal in the downward trend of populations. The biomass was growing, although no one knew exactly by how much. No wonder: as we have seen, not only the population models by fisheries biologists are rather complex and full of big uncertainties. Data were improving but in the near past a great deal was wrong about the catch statistics which were meant to produce the required input. It was for sure that catches had been underestimated due to a massive illegal and unreported fishery. Complicating enough, the amount of bluefin tuna hidden somewhere in the vast waters of the oceans had probably been underestimated. When the ICCAT management measures kicked in, this resulted in a quicker recovery than many had expected.

Thunnus thynnus seemed to have survived the perfect storm of unbridled upscaling of fleets, a race worth billions of dollars in a borderless, global economy, characterized by brazen corruption, indifference, market power struggles and political horse trading and a seemingly limitless technological development of gear and ships.

But the improving governance did not come about without important setbacks. The biggest crackdown ever on illegal tuna in the Mediterranean bluefin tuna fisheries that popped up in 2018 was a wakeup call that surveillance put in place by ICCAT and the European Union still was all but proactive. The system of policing tuna seemed far from being really water-tight. And shamefully enough Malta, the country that had delivered European Commissioners for Fisheries twice, was once-more the lynch-pin in a blossoming million-euro tuna racket as if nothing had happened in a decade of measures to prevent it.

But everyone who's caught the tuna fever could notice that something has been set in motion. Not only with bluefin tuna, but also within the rest of the extensive tuna family in the world seas that had outpaced the giant tuna during the last century as a fish for global consumption. The realisation had deepened that the development of sustainable tuna fishing had become an inevitable necessity. Tuna became an iconic example in the quest for management and governance to preserve wild fish, not only as a biological specimen in an ecological system, but also as one of the most important nutritious and healthy proteins we can extract from the sea. If we can succeed with tuna, then we might be able to do so with all other fish as well.

That is not only a vague notion, things are actively happening. Right across the globe, through trial and error, we are looking for new organisational forms of governance and partnerships to drive sustainability of catches. Such sustainability is increasingly demanded by the consumer market. Certification is gaining traction; the widespread illegal, unregulated and unreported fisheries are under attack by law makers. Transparency and control of the chain of custody is a new challenge with new rules and techniques such as block chain. The long and complex supply chains of tuna do need put in place a system that guarantees that the fish on your plate is really the one that it is supposed to be. Regional management organisations (until recently struggling where harvesting strategies are concerned) can apparently claim the first discernible success. The EU policy on IUU fishing is beginning to work. China, the biggest emerging fishing power, appears to be interested in making its fisheries sustainable on a policy basis.

The most important point to make is that the bluefin tuna in the Mediterranean has set some kind of precedent. After all, the approach to Atlantic bluefin tuna by ICCAT and the European Union had shown that recovery was possible with sustainability policy. In the summer of 2017 after a lot of wrangling a management plan was agreed for its severely threatened brother in the Pacific Ocean. This plan was welcomed by the NGOs as an important step in the right direction. The Northern Committee of the WCPFC decided to introduce catch restrictions which, by 2034, are intended to restore the population to 20% of its estimated original size. It was about time, too: in 2017 all but 2% of the original Pacific bluefin remained. The year 2034 seemed a long time ahead, but if all parties—especially Japan—actively enforced the agreements, there was light at the end of the tunnel.

Further positive blips on the screen had appeared for other fish, as in the North Sea, where stocks of plaice, sole, herring and cod had recovered to levels that had not been attested since the 1950s. The EU sustainable fisheries policy appears to be bearing fruit, although vigilance is called for. A coalition of NGOs kept a close eye on North Sea policy and rightly so: no one knows what spanner the Brexit will put in the works. The common fishery policy might prove one of the most complicated scrambled eggs to unscramble once Brexit is a fact.

Tuna has become a global thermometer for measuring the efficacy of mechanisms for governance and management of an economy based on sustainable fisheries. During the eighteenth century, Adam Smith examined issues around a fair division of wealth using newly emerging, large-scale industries. In a comparable way, these days the issues of a sustainable economy and social prosperity can be charted using tuna and its role in a complex global value chain. Just imagine: the idea of the 'invisible hand' (never formulated as such by Adam Smith, but always attributed to him) starts from the premise that pursuing one's own individual interest has the unintended consequence of prosperity for society as a whole. The idea of the tragedy of the commons depicts the opposite. The unbridled pursuit of individual benefit, as in the case of the tuna fisherman who tries to catch as much fish as possible, can bring extra personal profit. But as a result, common sustainability of tuna stocks is undermined and this results in lower overall affluence. The question is how to prevent this tragedy in a borderless, global world, where free-market thinking has engendered deregulation since the 1980s and where governments, with their constrained jurisdiction, can exert no more than a limited grip on the situation.

These issues converge nowhere as closely as in tuna: we would like to guarantee the fish a sustainable future during its peregrination stretching thousands of kilometres, even if only to serve our own interest. Tuna teaches us that different governance arrangements have emerged, from state-based, public regulation to market-based, private initiatives [180]. New, innovative means of governance emerge that can lead to sustainable fisheries, with partnerships and new organisational relations between stakeholders in the market. The tuna's rich history as the first industrial fish that was caught and preserved on a large scale, together with its fascinating biological characteristics, makes it once again an exemplary species. In his inaugural address as a professor of the Environmental Policy Group at Wageningen University in the Netherlands Simon Bush formulated it as follows: 'Tuna is not only impressive because it is one of the most biologically dynamic fish in the sea, but also because it is the most globalized.' In this globalisation, we find new ways to tackle the sustainability of tuna, curiously enough, within a complex stratification which seems to rule out any specific solution from the start.

Large-scale global tuna fishing involves now more than 100 countries, organised by five regional management commissions governing tuna. There are six multinationals operating across the globe which dominate the markets and they are all chiefly interested in cheap mass supply of tuna. The US and the EU represent their interests against considerably less powerful small states, in many cases developing countries, which manage the waters in which the tuna is caught. Billions of people consume tuna in all major markets. Many low-income countries are dependent on tuna; in some cases, it even represents their principal revenue and an essential source of protein. There are new players in the market, the NGOs, which, in different ways and from different perspectives, work towards sustainable tuna capture. As we saw in our quest, positioning as an NGO can be pretty tough in the treacherous battlefield of big interests in the hybrid tuna wars. Sustainability has become a new and powerful marketing weapon, with tuna wars being fought that in fact are driven by important commercial interests. As we have seen, this can lead to fierce battles around certifications, including conspiracy of tuna lead firms.

One tuna war we did not research but cannot leave unmentioned: The Battles against Corruption. This is not an easy battle as the sums of money involved can be huge. Take the huge $2 billion fraudulent loan scam in Mozambique that started in 2016. The fraud included a $850 million 'tuna bond' scam around a state-owned tuna fishing company EMOTUM, that bought 24 ships, mostly longliners, that never went out fishing. The fraud heavily affected the Mozambique state debt. At the start of 2019 the US Justice Department and the Mozambique Attorney were investigating Credit Suisse and BNP Paribas for their role in the loans. In total 18 Mozambican suspects were indicted, among them a former minister of finance, public servants and bank officials [29].

Even more damage is done by the widespread corruption that involves officials and politicians in exchange for access to fishing areas, licenses and other arrangements that involved tuna fisheries management institutions. The impact is of concern in particular in the Pacific Island region, where the states depend heavily on the tuna fisheries and the management of its sustainability, but where institutions are very vulnerable for bribes and corruption [128]. This is a problem within the PNA tuna cartel, where the vested interests of the industry at stake are huge and the transparency usually very limited. It has to be feared that only the tip of the iceberg of tuna corruption cases is uncovered. The harm done to the management of sustainable policies might well be systematic and significant. Everyone in the tuna-business has suspicions or even educated guesses of at least some persons that are involved with corruption practices. In the years to come, policies to prioritize the fight against corruption by the law-enforcing authorities will become increasingly important as a condition to make the management of sustainable tuna effective.

Nonetheless: out of these global sustainability policies new initiatives arise as an alternative, in search of new solutions for sustainable fishing and putting an end to the related conflicts. Tuna contributed strongly to the development of the so-called Sustainable Seafood Movement, a social movement that started around the new millennium to reform practices in the supply chain for seafood [123]. New partnerships crop up between the various private and public parties, driving new approaches. Powers are shifting in the end markets with suppliers urging for sustainable fish and by doing so driving market supplies towards a sustainable catch. Big retailers are starting (according to many still too slow) by supplying more sustainable products on their shelves, if only to avoid reputational damage. There are private, market-led sustainability schemes such as MSC, which in turn have a regulatory effect on local authorities. New techniques enable consumers to find out where and how their tuna was caught, which is the next important challenge in making sustainability more transparant and traceable. These techniques are also used to combat illegal, unreported and unregulated catches more effectively. The initiatives can take surprising forms, such as Global Fishing Watch [113] a partnership between Google, NGO Oceana and mapping company SkyTruth, funded in part by Leonardo DiCaprio. The aim is a real-time satellite surveillance system to track down illegal fishing practices and inform the states in which they take place. It won't be long before we will be able to do this with James-Bond-like apps on our mobile phones. Not Skyfall but SkyTruth for tuna.

One of the big questions is: can consumers, by being more aware when buying their cans of tuna and grilling their tuna steaks, ultimately steer the industry in a more sustainable direction? How can we rely on market-related initiatives, in a race to the bottom with an ever expanding range of certifications? How to improve the Theory of Change? [209] What about the fast-growing markets of Asia, Africa and Latin-America, where issues of sustainable fisheries hardly matter to consumers or retailers? Will sustainability certification help to create policies of the companies and states involved in such environment? Can international supervisory bodies improve the quality and impact of their governance in order to tackle sustainable fishing policies and the enormous over-capacity of the fleets with more stringent quotas, tightly controlled licences and effective monitoring? Will we be able to manage two of the most urgent issues: the FAD fisheries and the transparency and traceability of tuna markets, in particular along its long and complex chain of custody?

Tuna is a wicked issue. But one thing has become clear to me: the quest for tuna has become a quest for a world of sustainable solutions, for the tools and opportunities for an ecological transition. This is far from a lost cause; above all it is an imperative issue. After a history of millennia of large-scale tuna fisheries, tuna is at risk of disappearing as a massive source of healthy proteins, leaving a poor shadow of the fish that once reigned in our oceans and our kitchens. The Phoenicians and the Romans, the Moors, and the Sicilians, the Japanese right up to the Pacific island inhabitants, Europeans, Americans, Asians, Australians and Africans, thousands of years of living tuna fishing culture are at stake. We may be the last generation able to prevent the giant tuna's disappearance from the Gibraltar Strait to the Sea of Japan, not to speak of all those other tuna in the oceans. Let us use that opportunity. If we save the tuna, we save part of ourselves.

Cádiz, 2019

References

1. Abulafia D (2014) The great sea. Penguin Books, London
2. Adolf S, Bush S, Vellema S (2016) Reinserting state agency in global value chains: the case of MSC certified skipjack tuna. Fish Res 182:79–87. https://doi.org/10.1016/j.fishres.2015.11.020
3. Aelianus MNG (2011) Aelian's On the nature of animals. Trinity University Press, San Antonio
4. Agnew D, Pearce J, Pramod G et al (2009) Estimating the worldwide extent of illegal fishing. PLoS One 4:e4570. https://doi.org/10.1371/journal.pone.0004570
5. AIDCP-APICD (2001) The new AIDCP dolphin safe tuna certification
6. Alvarez de Toledo L (2007) Las almadrabas de los Guzmanes. Fundacion Casa Medina Sidonia, Sanlucar de Barrameda
7. Appert N (1991) L'art de conserver, pendant plusieurs années, toutes les substances animales et végétales. Jeanne Laffitte, Paris
8. Aqorau T (2010) Pacific countries might start to operate like OPEC. IslandBusiness, 12-1-2010, [online] <http://www.atuna.com/NewsArchive/ViewArticle.asp?ID=7887/>. Accessed 23 July 2015
9. Aristotle A (2015) Aristotle's history of animals. Forgotten Books, London
10. Arnason R (2015) Review of the PNA purse seine vessel day scheme final report. PNA Office
11. Atuna (2013) EII: MSC does not protect dolphins. In: atuna.com. Accessed 6 Sep 2013
12. Atuna (2015) Phil IUU tuna whistle blower shot dead in bed. In: atuna.com. https://atuna.com/index.php/en/2-news/3121-assassinated-anti-iuu-advocate-cooperated-with-wwf?highlight=WyJhbHBham9yYSIsImFscGGqb3JhJhJ3MiXQ==. Accessed 17 Aug 2015
13. Atuna (2015) Authorities: "FAD-Free" declarations are illegal. In: atuna.com. https://atuna.com/index.php/en/2-news/3722-authorities-fad-free-declarations-are-illegal?highlight=WyJuZmEiLCJuZmEncyJd. Accessed 26 Nov 2018
14. Atuna (2015) US big 3 face second price-fixing lawsuit. https://atuna.com/news/us-big-3-face-second-price-fixing-lawsuit?highlight=WyJvbGGpbiIsIm9sZWFufJ3MiXQ==. Accessed 10 Dec 2018
15. Atuna (2017) Korean warship searches for fishermen kidnapped by pirates. https://atuna.com/index.php/en/2-news/7546-korean-warship-searches-for-fishermen-kidnapped-by-pirates. Accessed 3 Apr 2017
16. Atuna (2017) Missing observer mystery still unsolved. https://atunatuna.com/index.php/en/2-news/5550-missing-observer-mystery-still-unsolved. Accessed 6 Jan 2017
17. Atuna (2017) Two Vietnamese fishermen shot dead by Philippine navy. https://atuna.com/index.php/en/2-news/6729-two-vietnamese-tuna-fishermen-shot-dead. Accessed 26 Sep 2017
18. Atuna (2017) More US retailers join price-fixing legal fight. In: Atuna.com. https://atuna.com/news/more-us-retailers-join-price-fixing-legal-fight?highlight=WyJoeS12ZWUiLCJoeS12ZWUncyJd. Accessed 10 Dec 2018

19. Atuna (2017) Walmart lists 50+ tuna execs as alleged conspirators. In: Atuna.com. https://
    atuna.com/news/walmart-lists-50-tuna-execs-as-price-fixing-conspirators?highlight=WyJoe
    S12ZWUiLCJoeS12ZWUncyJd. Accessed 10 Dec 2018
20. Atuna (2017) ISSF canners can now only buy from ISSF companies. In: Atuna.com. https://
    www.atuna.com/news/issf-canners-can-now-only-buy-from-issf-companies?highlight=
    WyJpc3NmIiwiaXNzZidzIiwiaXNzZidzY29uc2VydmF0aW9uIiwiJ2lzc2YiXQ==.
    Accessed 15 Dec 2018
21. Atuna (2018) Indonesia sinks 100+ IUU vessels in 1 day spree. https://atuna.com/index.php/
    en/2-news/8229-indonesia-sinks-100-iuu-vessels-in-1-day-. Accessed 23 Aug 2018
22. Atuna (2018) PNG minister refutes claims over number of missing observers. In: atuna.com.
    https://atuna.com/index.php/en/2-news/7354-png-minister-refutes-claims-over-number-of-
    missing-observers?highlight=WyJiYXNhIiwiYmFzYSdzIl0=. Accessed 20 Feb 2018
23. Atuna   (2018).   https://atuna.com/index.php/en/tuna-info/tuna-species-guide#skipjack.
    Accessed 20 Nov 2018
24. Atuna (2018). https://atuna.com/index.php/en/tuna-info/tuna-species-guide#bigeye. Accessed
    20 Nov 2018
25. Atuna   (2018).   https://atuna.com/index.php/en/tuna-info/tuna-species-guide#albacore.
    Accessed 20 Nov 2018
26. Atuna   (2018).   https://atuna.com/index.php/en/tuna-info/tuna-species-guide#yellowfin.
    Accessed 20 Nov 2018
27. Atuna (2018) Where has ISSF's 12 million dollar investment gone?. In: atuna.com. https://
    atuna.com/index.php/en/2-news/7358-where-has-issf-s-12-million-dollar-investment-gone?
    highlight=WzEyLCIxMidzIiwiaXNzZiIsImlzc2YncyIsImlzc2Ync2NvbnNlcnZhdGlvbiIsIid
    pc3NmIl0=. Accessed 26 Nov 2018
28. Atuna (2018) PNA, from small island nations to smart ocean states. https://atuna.com/news/
    pna-from-small-island-nations-to-smart-ocean-states?highlight=WyJzbWFydCIsIidzb
    WFydCIsIidzbWFydCciLCJpc2xhbmQiLCJpc2xhbmQncyIsImlzbGFuZCcsIiwiJ2lzbGFuZ
    CJd. Accessed 3 Dec 2018
29. Atuna (2019) 18 more arrests in crackdown on $ 850 M tuna bond scandal. In: Atuna.com.
    https://www.atuna.com/news/18-more-arrests-in-crackdown-on-850-m-tuna-bond-scandal?
    highlight=WyJtb3phbWJpccVIiwibW96YW1iaXF1ZSdzIiwibW96YW1iaXF1ZScuHUy
    MDFkIiwibW96YW1iaXF1ZSdzc3RhdGUiXQ==. Accessed 19 Jan 2019
30. Balcombe J (2017) What a fish knows. Oneworld, London
31. Barclay K (2010) History of industrial tuna fishing in the Pacific Islands. Asian Research
    Center/HMAP
32. BBC News (2018) 'Illegal' Libya fishing concerns. In: BBC News. http://www.bbc.co.uk/
    news/science-environment-15597675. Accessed 9 Nov 2018
33. Beijnen J (2017) The closed cycle aquaculture of Atlantic bluefin tuna in Europe: current
    status, market perceptions and future potential
34. Belhabib D, Koutob V, Sall A et al (2014) Fisheries catch misreporting and its implications:
    the case of Senegal. Fish Res 151:1–11. https://doi.org/10.1016/j.fishres.2013.12.006
35. Bennema F (2018) Long-term occurrence of Atlantic bluefin tuna *Thunnus thynnus* in the
    North Sea: contributions of non-fishery data to population studies. Fish Res 199:177–185.
    https://doi.org/10.1016/j.fishres.2017.11.019
36. Bernal D (2014) Garum y Salsamenta. Del origen fenicio a la democratización romana de una
    milenaria tradición salazonera. In Badia i Homs J La salaó de peix a Empúries i a l'Escala.
    Ajuntament de l'Escala, L'Escala
37. Bernal B (2015) Citra Mina: no direct relationship with 'abandoned' fishermen. In: Rappler.
    https://www.rappler.com/nation/85143-tuna-exporter-supplier-abandoned-workers-fiasco.
    Accessed 2 Dec 2018
38. Bernal B (2015) Citra Mina ignored my family's cry for help. In: Rappler. https://www.
    rappler.com/nation/85212-citra-mina-help-fishermen. Accessed 2 Dec 2018

39. Bernal D, Diaz J, Exposito J et al (2018) Atunes y Garum en Baelo Claudia: nuevas investigaciones. Tuna fish & Garum at Baelo Claudia: recent research. Al Qantir 21:73–86
40. Bester TC (2014) Tsukiji: the fish market at the center of the world. University of California Press, Berkeley, CA
41. Bjørndal T, Monroe G (2012) The economics and management of world fisheries. Oxford University Press, Oxford, UK
42. Blánquez Pérez J (2015) Baelo Claudia y la familia Otero. Universidad Autónoma de Madrid, Madrid
43. Block B, Stevens E (eds) (2001) Tuna physiology ecology and evolution. Academic Press, San Diego
44. (2017) Bluefin tuna caught in Hoang Sa set record – news VietNamNet. In: English. vietnamnet.vn. https://english.vietnamnet.vn/fms/society/178935/bluefin-tuna-caught-in-hoang-sa-set-record.html. Accessed 16 Nov 2018
45. Bonanno A, Constance D (1996) Caught in the net. University Press of Kansas, Lawrence, KS
46. Braudel F (1992) The perspective of the world. University of California Press, Oakland, CA
47. Braudel F (1986) La dinámica del capitalismo. Fondo de Cultura Económica, México
48. Braudel F (1992) The structure of everyday life. University of California Press, Oakland, CA
49. Braudel F (1994) A history of civilizations. Int Labor Work Hist. https://doi.org/10.1017/S014754791100007X
50. Braudel F (2003) The wheels of commerce. Theory Soc. https://doi.org/10.1023/A:1024404620759
51. Braudel F (2007) The Mediterranean in the ancient world. Penguin Books, London
52. Brown J (1968) Cosmological Myth and the Tuna of Gibraltar. Trans Proc Am Philol Assoc 99:37. https://doi.org/10.2307/2935832
53. Brownjohn M (2014) PNA and the WCP RFMO Presentation Global Oceans Action Summit, Working Group 3, session 3 PNA-SIDS. The Hague 23rd May 2014
54. Capistrano Z (2016) French environmentalist found dead in Palawan. In: Davao Today. http://davaotoday.com/main/environment/french-environmentalist-found-dead-in-palawan/. Accessed 16 Nov 2018
55. Casasola B, Sáez Romero AM (2016) Fish-salting plants and amphora production in the Bay of Cadiz (Beatica, Hispana). Patterns of settlement from the Punic era to late antiquity. University of Leuven, Leuven
56. CCSBT (2018) About Southern Bluefin Tuna I CCSBT Commission for the conservation of Southern Bluefin tuna. In: Ccsbt.org. https://www.ccsbt.org/en/content/about-southern-bluefin-tuna. Accessed 20 Nov 2018
57. Cervantes Saavedra M, Kelly W (2012) The exemplary novels of Miguel de Cervantes. AUK Classics
58. Cervantes Saavreda M (2003) Don Quixote. Penguin Classics, London
59. Clover C (2006) The end of the line. Blue & Green Tomorrow
60. Collette B (2017) Bluefin tuna science remains vague. Science 358:879–880. https://doi.org/10.1126/science.aar3928
61. Constance D, Bonanno A (1999) Contested terrain of the global fisheries: "Dolphin-Safe" Tuna, The Panama Declaration, and the Marine Stewardship Council. Rural Sociol 64:597–623. https://doi.org/10.1111/j.1549-0831.1999.tb00380.x
62. Corson T (2008) The Zen of fish. Perennial, New York
63. Cort JL, Nottestad L (2007) Fisheries of bluefin tuna (*Thunnus thynnus*) spawners in the northeast Atlantic. Collect Vol Sci Pap ICCAT 60(4):1328–1344
64. Cort JL, Santiago J, Arrizabalaga H et al (2018) Review and update of the catch at age (Caa) for the Spain bay of biscay bluefin tuna fisheries for 1950–2000. Collect Vol Sci Pap ICCAT 74(6):2657–2669
65. Cousteau JY, Dumas F (1953) Le Monde du Silence. Editions de Paris, Paris
66. Cousteau J, Dumas F (2004) The silent world. National Geographic Society, Washington, D.C

67. Cousteau J, Malle L (1956) Le monde du silence. FSJYC Production, Requins Associés, Société Filmad, Titanus
68. Cphpost.dk (2018) The Copenhagen post: Danish news in English. Cphpost.dk. http://cphpost. dk/news/bluefin-tuna-returns-to-danish-waters.html. Accessed 31 Oct 2018
69. Curtis B (1961) The life story of fish. Dover, Mineola, NY
70. Dagne T (2009) Somalia: prospects for a lasting peace. Mediterr Q 20:95–112. https://doi.org/ 10.1215/10474552-2009-007
71. Dagorn L, Holland K, Itano D (2006) Behavior of yellowfin (*Thunnus albacares*) and bigeye (*T. obesus*) tuna in a network of fish aggregating devices (FADs). Mar Biol 151:595–606. https://doi.org/10.1007/s00227-006-0511-1
72. Dagorn L, Holland K, Restrepo V, Moreno G (2012) Is it good or bad to fish with FADs? What are the real impacts of the use of drifting FADs on pelagic marine ecosystems? Fish Fish 14:391–415. https://doi.org/10.1111/j.1467-2979.2012.00478.x
73. (2017) Development off a rebate scheme (RBS) for the Papua New Guinea (PNG) tuna processers. PNG
74. (2018) Distant water fishing fleets and hybrid warfare – Stimson environmental security – medium. In: Medium. https://medium.com/natural-security-forum/distant-water-fishing-fleets-and-hybrid-warfare-7966feac34a0. Accessed 16 Nov 2018
75. Druel E (2018) Fisheries laws need time to perform, not reform. In: ClientEarth. https://www. clientearth.org/fisheries-laws-need-time-perform-not-reform/. Accessed 9 Nov 2018
76. e La Exposición C (2011) Pescar con Arte Fenicios y romanos en el origen de los aparejos andaluces. CUBIERTA SAGENA_3_4_PESCA
77. Earth Island Institute (2010) International tuna monitoring program 2009 annual report
78. EII Philippines (2014) MSC Tuna products
79. Eisenhower D (2006) The military–industrial complex; the farewell address of president Eisenhower. Basements Publications
80. EJF, Oceana, PEW, WWF (2016) The EU IUU regulation: building on success. Brussels
81. Ellis R (2008) Tuna: a love story. Alfred A. Knopf, New York
82. Espejo E (2010) The destruction of the Filipino tuna industry - Asia sentinel. In: Asia Sentinel. https://www.asiasentinel.com/econ-business/the-destruction-of-the-filipino-tuna-industry/. Accessed 2 Dec 2018
83. Espejo E (2014) Gensan no longer PH's tuna capital? In: Rappler. https://www.rappler.com/ business/industries/247-agriculture/68158-general-santos-mindoro-ph-tuna-capital. Accessed 2 Dec 2018
84. Étienne R, Mayet F (1998) Les mercatores de saumure hispanique. Mélanges l'École Française Rome, Antiquité
85. European Commission (2011) In: Ec.europa.eu. https://ec.europa.eu/fisheries/sites/fisheries/ files/docs/body/kiribati_2012_en.pdf. Accessed 3 Dec 2018
86. Europêche (2015) Europêche position on tuna FAD free commercialisation. In: Europeche. chil.me. http://europeche.chil.me/download-doc/122137. Accessed 30 Nov 2018
87. Europeche (2017) EU-fisheries: the essential role of our large scale fishing fleet. In: Europeche.chil.me. http://europeche.chil.me/. Accessed 26 Nov 2018
88. Europol (2018) How the illegal bluefin tuna market made over EUR 12 million a year selling fish in Spain. In: Europol. https://www.europol.europa.eu/newsroom/news/how-illegal-bluefin-tuna-market-made-over-eur-12-million-year-selling-fish-in-spain. Accessed 12 Nov 2018
89. Evans D, Barnabas N, Philipson P et al (2008) Development of a regional strategy to maximize economic benefits from purse seine caught tuna. PSS Associates
90. Faas PCP (1994) Rond de Tafel der Romeinen. Domus, Amsterdam
91. FAO (2010) The state of world fisheries and aquaculture. Food and Agriculture Organization of the United Nations, Rome
92. FAO (2012) The state of world fisheries and aquaculture. Food and Agriculture Organization of the United Nations, Rome

93. FAO (2014) The state of world fisheries and aquaculture. Food and Agriculture Organization of the United Nations, Rome
94. FAO (2016) The state of world fisheries and aquaculture 2016. Food and Agriculture Organization of the United Nations, Rome
95. FAO (2018) Fishery and Aquaculture Statistics Statistiques des pêches et de l' aquaculture Estadísticas de pesca y acuicultura. Food and Agriculture Organization of the United Nations, Rome
96. Fishwick C, Hodal K, Kelly C, Trent S (2018) Slave labour producing prawns for supermarkets in US, UK: your questions answered. In: The Guardian. https://www.theguardian.com/global-development/2014/jun/10/asian-slave-labour-prawns-supermarkets-us-uk-thailand. Accessed 26 Nov 2018
97. FiskerForum.com (2018) Bluefin back in UK waters: FiskerForum.com. FiskerForum.com. http://fiskerforum.dk/en/news/b/bluefin-back-in-uk-waters. Accessed 31 Oct 2018
98. Flipper (2018) (1964 TV series). In: En.wikipedia.org. https://en.wikipedia.org/wiki/Flipper_(1964_TV_series). Accessed 3 Dec 2018
99. (2018) Flipper (1964 TV series). In: En.wikipedia.org. https://en.wikipedia.org/wiki/Flipper_(1964_TV_series). Accessed 3 Dec 2018
100. Foer J (2010) Eating animals. Penguin, London
101. Fontenau A, le Person A (2009) Bluefin fishing by sport fishermen off the Trebeurden Bay, Northern Brittany, during the 1946–1953 period. Collect Vol Sci Pap ICCAT 63:69–78. SCRS/2008/059
102. Fonteneau A (2011) Potential impact of Gillnets fisheries on Indian ocean ecosystems?
103. Fonteneau A, Pallares P, Pianet R (2000) A worldwide review of purse seine fisheries on FADs. Peche thoniere et dispositifs de concentration de poissons
104. Froese R, Proelss A (2012) Evaluation and legal assessment of certified seafood. Mar Policy 36:1284–1289. https://doi.org/10.1016/j.marpol.2012.03.017
105. Fromentin J (2003) The East Atlantic and Mediterranean bluefin tuna stock management: uncertainties and alternatives. Sci Mar 67(S1):51–62. https://doi.org/10.3989/scimar.2003.67s151
106. Fromentin JM (2008) An attempt to evaluate the recent management regulations of the East Atlantic and Mediterranean bluefin tuna stock through a simple simulation model. Collect Vol Sci Pap ICCAT 62:1271–1279
107. Fromentin J (2009) Lessons from the past: investigating historical data from bluefin tuna fisheries. Fish Fish 10:197–216. https://doi.org/10.1111/j.1467-2979.2008.00311.x
108. Fromentin J-M, Powers JE (2005) Atlantic bluefin tuna: population dynamics, ecology, fisheries and management. Fish Fish. https://doi.org/10.1111/j.1467-2979.2005.00197.x
109. García Vargas E, Muñoz Vicente Á (2003) Reconocer la cultura pesquera de la Antiguedaden Andalacia. PH44
110. García Vargas E, Bernal Casasola D, Palacios Macías V et al (2014) Confectio Gari Pompeiani. Procedimiento experimental para la elaboración de salsas de pescado romanas. Spal Revista de Prehistoria y Arqueología de la Universidad de Sevilla:65–82. https://doi.org/10.12795/spal.2014.i23.04
111. Gibbon E (2016) History of the decline and fall of the Roman Empire. André Hoffmann, Dinslaken
112. Gillett R (2007) A short history of industrial fishing in the Pacific Islands. FAO-RAP, Rome
113. Global fishing watch | Sustainability through transparency. (2016) In: Global fishing watch. https://globalfishingwatch.org/. Accessed 14 Dec 2018
114. González JAL, Acevedo JMR (2012) Series cronológicas de capturas del atún rojo en las almadrabas del Golfo de Cádiz (siglos XVI–XXI). Collect Vol Sci Pap – ICCAT 67:139–174
115. Goscinny U (1977) Asterix and the great crossing. Hodder & Stoughton, London
116. Graves R (1989) The Isles of unwisdom. Arrow, London
117. Greenpeace (2008) Game over: European Union sinks tuna agreement, November 24

118. Greenpeace (2014) Monster boats, the scourge of the oceans. Greenpeace's European Ocean Team, Amsterdam
119. Greenpeace (2016) Complaint letter to the president of the University of Washington about research practices of Professor Ray Hilborn, May 11 2016
120. Grotius H, Feenstra R (2009) Mare liberum. Brill, Leiden, pp 1609–2009
121. GSSI (2018) Global sustainable seafood initiative. In: Ourgssi.org. http://www.ourgssi.org/. Accessed 7 Dec 2018
122. Guggenheim D (2006) An inconvenient truth. Paramount
123. Gutiérrez A, Morgan S (2015) The influence of the sustainable seafood movement in the US and UK capture fisheries supply chain and fisheries governance. Front Mar Sci 2:72. https://doi.org/10.3389/fmars.2015.00072
124. Haenlein C (2017) Below the surface how illegal, unreported and unregulated fishing threatens our security. RUSI
125. Hall M, Roman M (2013) Bycatch and non-tuna catch in the tropical tuna purse seine fisheries of the world. FAO Fisheries and Aquaculture. FAO Technical Paper No 568. https://doi.org/10.13140/2.1.1734.4963
126. Hamerschlag K, Venkat K (2011) Meat eaters guide: methodology 2011. Environmental Working Group Cleanmetrics
127. Hamilton A, Lewis A, McCoy M et al (2011) Market and industry dynamics in the global tuna supply chain. FFA, Honiara
128. Hanich Q, Tsamenyi M (2009) Managing fisheries and corruption in the Pacific Islands region. Mar Policy 33:386–392. https://doi.org/10.1016/j.marpol.2008.08.006
129. Hannesson R (2008) The exclusive economic zone and economic development in the Pacific island countries. Mar Policy 32:886–897. https://doi.org/10.1016/j.marpol.2008.01.002
130. Hardin G (1968) The tragedy of the commons. Science 162:1243–1248. https://doi.org/10.1126/science.162.3859.1243
131. Havice E (2010) The structure of tuna access agreements in the Western and Central Pacific Ocean: lessons for vessel day scheme planning. Mar Policy 34:979–987. https://doi.org/10.1016/j.marpol.2010.02.004
132. Havice E (2013) Rights-based management in the Western and Central Pacific Ocean tuna fishery: economic and environmental change under the Vessel Day Scheme. Mar Policy 42:259–267. https://doi.org/10.1016/j.marpol.2013.03.003
133. Henshilwood C, d'Errico F, van Niekerk K et al (2018) An abstract drawing from the 73,000-year-old levels at Blombos Cave, South Africa. Nature 562:115–118. https://doi.org/10.1038/s41586-018-0514-3
134. Hilborn R (2011) Greenpeace (2016) complaint letter to the president of the University of Washington about research practices of Professor Ray Hilborn. Let us eat fish. The New York Times op-ed contribution, April 14th 2011
135. Hilborn R, Pauly D (2013) Does catch reflect abundance? Nature 494:303
136. Hilborn R, Amoroso R, Bogazzi E et al (2017) When does fishing forage species affect their predators? Fish Res 191:211–221. https://doi.org/10.1016/j.fishres.2017.01.008
137. Holden M, Garrod D (1996) The common fisheries policy. Fishing News, Oxford
138. (2018) How fish sauce is made. In: Thaifoodandtravel.com. https://www.thaifoodandtravel.com/features/fishsauce1.html. Accessed 22 Oct 2018
139. Hughes J (2011) The piracy-illegal fishing Nexus in the Western Indian Ocean. Future Directions International
140. Hutchinson R (2013) Spanish Armada. Weidenfeld & Nicolson, London
141. ICCAT (2008) Report of the 2008 Atlantic bluefin tuna stock assessment session. Madrid
142. ICCAT (2008) Report for the biennial period, 2008–2009, part I, vol 2. SCRS, Madrid
143. Infofish (2016) ISSF adopts and amends tuna conservation measures. In: Infofish.org. http://infofish.org/v2/index.php/227-issf-adopts-and-amends-tuna-conservation-measures. Accessed 2 Jan 2019

144. IPNLF (2018) International pole and line foundation. In: Ipnlf.org. http://ipnlf.org/. Accessed 2 Dec 2018
145. ISEAL Alliance (2018) ISEAL is the global membership association for credible sustainability standards. In: Isealalliance.org. https://www.isealalliance.org/. Accessed 7 Dec 2018
146. Issenberg S (2014) The sushi economy. Gotham Books, New York
147. IUCN red list 2011 – in pictures. (2018) In: The Guardian. https://www.theguardian.com/environment/gallery/2011/nov/10/icun-red-list-in-pictures. Accessed 19 Nov 2018
148. IUU Watch (2018) EU carding decisions. In: IUU Watch. http://www.iuuwatch.eu/map-of-eu-carding-decisions/. Accessed 2 Dec 2018
149. Jackson S (2011) Analyzing the PNA draft assessment what it means for a world market in search of sustainable tuna. ISSF
150. Jiménez S (2018) Origen y desarrollo de la industria de conservas de pescado en Andalucía (1879–1936). In: Revistes.ub.edu. http://revistes.ub.edu/index.php/HistoriaIndustrial/article/view/19606. Accessed 16 Oct 2018. Jimenez S Evolucion de la gran empresa almadraberoconservera Andaluza entre 1919 y 1936. Universidade de Santiago de Compostela, Santiago de Compostela
151. Joseph J (1994) The tuna-dolphin controversy in the eastern pacific ocean: biological, economic, and political impacts. Ocean Dev Int Law 25:1–30. https://doi.org/10.1080/00908329409546023
152. Junger S (2010) The perfect storm. Fourth Estate, London
153. Junquera N (2009) Fuego sobre los piratas del 'Alakrana'. In: EL PAÍS. https://elpais.com/diario/2009/11/18/espana/1258498801_850215.html. Accessed 11 Dec 2018
154. Kamen H (2005) Spain, 1469–1714: a society of conflict. Routledge, London
155. Koninklijke Bibliotheek (2018) Adriaen Coenen's Visboek. Kb.nl. https://www.kb.nl/en/themes/middle-ages/adriaen-coenens-visboek. Accessed 30 Oct 2018
156. Kousbroek R (1969) De aaibaarheidsfactor. Atlas Contact, Amsterdam
157. Kurlansky M (1999) Cod. Vintage, London
158. Levine D (2006) Tuna in ancient Greece. American Institute of Wine and Food, New York
159. Lopez J, Moreno G, Sancristobal I, Murua J (2014) Evolution and current state of the technology of echo-sounder buoys used by Spanish tropical tuna purse seiners in the Atlantic, Indian and Pacific Oceans. Fish Res 155:127–137. https://doi.org/10.1016/j.fishres.2014.02.033
160. Lövin I (2013) Commission memo: fisheries partnership agreement (FPA) between the EU and Kiribati. Main elements of the new protocol
161. MacKenzie B, Myers R (2007) The development of the northern European fishery for north Atlantic bluefin tuna *Thunnus thynnus* during 1900–1950. Fish Res 87:229–239. https://doi.org/10.1016/j.fishres.2007.01.013
162. McKee A (1988) From merciless invaders: the defeat of the Spanish Armada. Grafton, London
163. Meere F, Lack M (2008) Assessment of impacts of illegal, unreported and unregulated (IUU) fishing in the Asia-Pacific. Asia-Pacific Economic Cooperation Secretariat, Singapore
164. Mexico Loses 10-Year Battle with US Over Tuna Labeling After New Ruling (2018) In: Mexico News Daily. https://mexiconewsdaily.com/news/mexico-loses-10-year-battle-over-tuna-labeling/. Accessed 19 Feb 2019
165. Mielgo Bregazzi R (2006) The plunder of bluefin tuna in the Mediterranean and east Atlantic in 2004 and 2005. WWF International, Gland
166. Mielgo Bregazzi R (2008) The race for the last bluefin. WWF Mediterranean
167. Mielgo Bregazzi R (2008) Lifting the lid, the 2008 bluefin tuna dossier. WWF, Rome
168. Miles R (2013) Carthage must be destroyed. Penguin, London
169. Miller A, Bush S (2015) Authority without credibility? Competition and conflict between ecolabels in tuna fisheries. J Clean Prod 107:137–145. https://doi.org/10.1016/j.jclepro.2014.02.047
170. Mitsubishi Corporation (2008) Mitsubishi corporation Atlantic bluefin tuna sourcing policy (updated November 2008)

171. Mitsubishi Corporation (2009) Revised position statement on bluefin tuna, January 2009
172. Miyake P (2001) Atlantic bluefin tuna: research and management. European Association of Fisheries Economists
173. Miyake M (2010) Recent developments in the tuna industry. Food and Agriculture Organization of the United Nations, Rome
174. Miyake MP, Miyabe N, Nakano H (2004) Historical trends of tuna catches in the world: FAO Technical Paper 467. FAO, Rome
175. Mol A (2008) Bringing the environmental state back in: partnerships in perspective. In: Glasbergen P, Biermann F, Mol A (eds) Partnerships, governance and sustainable development. Edward Elgar, Cheltenham
176. National Geographic (2018) Watch wicked tuna on National Geographic. In: Watch wicked tuna on FOX. https://www.nationalgeographic.com/tv/wicked-tuna/. Accessed 1 Nov 2018
177. NPR (2017) NPR choice page. In: Npr.org. http://www.npr.org/sections/thesalt/2017/09/01/547903557/countries-pledge-to-recover-dwindling-pacific-bluefin-tuna-population?. Accessed 20 Nov 2018
178. Oceana (2008) Italian driftnets: illegal fishing continues
179. OECD (2012) Rebuilding fisheries. OECD, Paris
180. Oosterveer P, Bush S (2018) Governing sustainable seafood. Routledge, New York
181. OPAGAC (2017) OPAGAC receives the first AENOR responsible tuna fishing (APR) certificates in the world. In: Opagac.org. http://opagac.org/en/opagac-receives-the-first-aenor-apr-certificates-in-the-world/. Accessed 3 Jan 2019
182. Oppian (1722) Halieuticks of the nature of fishes and fishing of the ancients. The Theater, Oxford
183. (1982) Overview – convention & related agreements. In: Un.org. https://www.un.org/Depts/los/convention_agreements/convention_historical_perspective.htm. Accessed 16 Nov 2018
184. Pacific Alliance for Sustainable Tuna (2017) The pacific alliance for sustainable tuna, representing leaders in mexican tuna industry, to earn MSC certification, the highest sustainability standard for wild caught tuna
185. Palmer M (2015) Making sure tuna is dolphin safe. In: HuffPost. https://www.huffingtonpost.com/mark-j-palmer/making-sure-tuna-is-dolph_b_6519742.html. Accessed 7 Dec 2018
186. Parker G (2002) The Dutch revolt. Penguin, London
187. Parker G (2014) Imprudent king. Yale University Press, New Haven
188. Pauly D (2009) Aquacalypse now. The New Republic
189. Pauly D, Zeller D (2016) Catch reconstructions reveal that global marine fisheries catches are higher than reported and declining. Nat Commun 7:10244. https://doi.org/10.1038/ncomms10244
190. Pauly D, Zeller D (2017) Comments on FAOs state of world fisheries and aquaculture (SOFIA 2016). Mar Policy 77:176–181. https://doi.org/10.1016/j.marpol.2017.01.006
191. Peñas Lado E (2016) The common fisheries policy. Wiley, Hoboken, NJ
192. PEW (2012) Estimating the use of drifting Fish Aggregation Devices (FADs) around the globe
193. PEW (2014) The story of Pacific Bluefin tuna. Pew Environment Group
194. PEW (2016) New science puts decline of Pacific Bluefin at 97.4 percent. In: Pewtrusts.org. https://www.pewtrusts.org/en/research-and-analysis/articles/2016/04/25/new-science-puts-decline-of-pacific-bluefin-at-974-percent. Accessed 19 Nov 2018
195. PEW (2018) FAQ: what is a regional fishery management organization? Pewtrusts.org. https://www.pewtrusts.org/en/research-and-analysis/fact-sheets/2012/02/23/faq-what-is-a-regional-fishery-management-organization. Accessed 5 Nov 2018
196. Plinius RH (1983) Natural history. Harvard University Press, Cambridge
197. Polacheck T, Davies C (2008) Considerations of implications of large unreported catches of Southern Bluefin tuna for assessments of tropical tunas, and the need for independent verification of catch and effort statistics. CSIRO Marine and Atmospheric Research, Hobart
198. Ponsich M (1988) Aceite de oliva y salazones de pescado. Universidad Complutense, Madrid

199. Pritchard C (1984) Tuna fishing causes a row between US and Pacific Islands. The Christian Science Monitor
200. Psihoyos L (2009) The Cove. Lions Gate, Santa Monica, CA
201. Ramírez de Haro I (2008) El caso Medina Sidonia. Esfera de los Libros, Madrid
202. Ramsden N (2017) 'Dolphin safe' pioneer Mark Berman passes away. In: undercurrentnews. com. Accessed 3 Aug 2017
203. (2018) Record fish caught in disputed waters. In: Atuna.com. https://atuna.com/index.php/en/2-news/6135-record-fish-caught-in-disputed-waters?highligh%E2%80%A6. Accessed 15 Nov 2018
204. Ricardo Fuentes e Hijos (2018) Ricardo Fuentes e Hijos. https://www.ricardofuentes.com/. Accessed 1 Nov 2018
205. Rittel H, Webber M (1973) Dilemmas in a general theory of planning. Policy Sci 4:155–169. https://doi.org/10.1007/bf01405730
206. RoboTuna (2018) In: RoboTuna. https://robotuna.wordpress.com/. Accessed 29 Oct 2018
207. Roca Rosell A (2018) La ingeniería y el proyecto del Ictíneo de Monturiol, 1857–1868. In: Ub. edu. http://www.ub.edu/geocrit/sn/sn119-96.htm. Accessed 29 Oct 2018
208. Rodríguez-Vidal J, d'Errico F, Pacheco F et al (2014) A rock engraving made by Neanderthals in Gibraltar. Proc Natl Acad Sci 111:13301–13306. https://doi.org/10.1073/pnas.1411529111
209. Roheim C, Bush S, Asche F et al (2018) Evolution and future of the sustainable seafood market. Nat Sustain 1:392–398. https://doi.org/10.1038/s41893-018-0115-z
210. Rossi F (1968) L'Odissea. RAI, Italy/France/Germany/Yugoslavia
211. Rubin A (2007) Rock the boat: the plight of the Southern Bluefin tuna. Michigan State University College of Law
212. Safina C, Klinger D (2008) Collapse of bluefin tuna in the Western Atlantic. Conserv Biol 22:243–246. https://doi.org/10.1111/j.1523-1739.2008.00901.x
213. Sampson G, Sanchirico J, Roheim C et al (2015) Secure sustainable seafood from developing countries. Science 348:504–506. https://doi.org/10.1126/science.aaa4639
214. Sarmiento M, Pérez de Guzmán el Bueno P (1757) De los Atunes y sus transmigraciones. Caixa de Pontevedra, Madrid
215. Schaefer M (1954) Some aspects of the dynamics of populations important to the management of the commercial Marine fisheries. Bull Math Biol 53:253–279. https://doi.org/10.1007/bf02464432
216. Schwartz A, Schwartz A (2018) MIT develops robotic fish to detect environmental pollutants. In: Inhabitat.com. https://inhabitat.com/fish-robots-could-be-used-to-environmental-pollutants/#more-56684. Accessed 29 Oct 2018
217. Scott G, Lopez J (2014) The use of FADs in tuna fisheries. In: Europarl.europa.eu. http://www.europarl.europa.eu/RegData/etudes/note/join/2014/514002/IPOL-PECH_NT(2014)514002_EN.pdf. Accessed 26 Nov 2018
218. Secure Fisheries (2015) Securing Somali fisheries report. One Earth Future
219. Shabecoff P (1990) 3 companies to stop selling tuna netted with dolphins. In: Nytimes.com. https://www.nytimes.com/1990/04/13/us/3-companies-to-stop-selling-tuna-netted-with-dolphins.html. Accessed 8 Dec 2018
220. Sharpless A, Evans S (2013) The perfect protein. Oceana
221. Shepherd JG, Pope JG (2002) Dynamic pool models I: interpreting the past using virtual population analysis. In: Hart P, Reynolds J (eds) Handbook of fish biology and fisheries, vol 2. Blackwell Science, Oxford, UK, pp 127–136
222. Smith A (2010) The wealth of nations. Capstone, Chichester
223. Smith A (2012) American tuna: the rise and fall of an improbable food (California Studies in Food and Culture). University of California Press, Berkeley, CA
224. Smith A (2018) Athenaeus: Deipnosophists – translation. In: Attalus.org. http://www.attalus.org/info/athenaeus.html. Accessed 22 Oct 2018
225. Souter D, Harris C, Banks R et al (2016) Towards the quantification of illegal, unreported and unregulated (IUU) fishing in the Pacific Islands region. MRAG

226. Stefano d V, Heijden v d P (2007) Bluefin tuna fishing and ranching: a difficult management problem. New Medit 6(2):59–64
227. Stewart C (2015) Somalia threatened by illegal fishermen after west chases away pirates. In: The guardian. https://www.theguardian.com/world/2015/oct/31/somalia-fishing-flotillas-pirates-comeback. Accessed 14 Dec 2018
228. Stockton W (1986) U.S. and Mexico seek end to tuna war. In: Nytimes.com. https://www.nytimes.com/1986/05/12/business/us-and-mexico-seek-end-to-tuna-war.html. Accessed 8 Dec 2018
229. Strabo (2015) Geography of Strabo. Forgotten Books, London
230. Stringer CB, Finlayson JC, Barton RNE et al (2008) Neanderthal exploitation of marine mammals in Gibraltar. Proc Natl Acad Sci. https://doi.org/10.1073/pnas.0805474105
231. Sumaila U, Pauly D (2006) Catching more bait: a bottom-up re-estimation of global fisheries. University of British Columbia, Canada
232. Sumaila U, Alder J, Keith H (2006) Global scope and economics of illegal fishing. Mar Policy 30:696–703. https://doi.org/10.1016/j.marpol.2005.11.001
233. TakePart (2018) Bluefin tuna are showing up in the arctic—and that's not good news. TakePart. http://www.takepart.com/article/2014/09/15/bluefin-tuna-arctic. Accessed 31 Oct 2018
234. Teisl M, Roe B, Hicks R (2002) Can eco-labels tune a market? Evidence from dolphin-safe labeling. J Environ Econ Manag 43:339–359. https://doi.org/10.1006/jeem.2000.1186
235. The Black Fish (2012) The black fish free hundreds of endangered bluefin tuna in the Adriatic Sea
236. The Guardian (2010) Environment. In: The Guardian. https://www.theguardian.com/environment/2010/jun/18/sea-shepherd-release-bluefin-tuna-libya 3/. Accessed 18 Jun 2010
237. The Hill (2017) US loses tuna trade battle with Mexico. In: https://thehill.com/policy/international/330549-us-lost-a-trade-war-with-mexico-over-tuna. Accessed 9 Dec 2018
238. The internet classics archive | The Persians by Aeschylus. (2018) In: Classics.mit.edu. http://classics.mit.edu/Aeschylus/persians.html. Accessed 22 Oct 2018
239. The Malta Independent (2018) 5,000 kilos of tuna allegedly smuggled to Malta each week, and exported to EU states. In: Independent.com.mt. http://www.independent.com.mt/articles/2017-12-03/local-news/5-000-kilos-of-tuna-allegedly-smuggled-to-Malta-each-week-and-exported-to-EU-states-6736182205. Accessed 9 Nov 2018
240. The Malta Independent (2018) Europol: bluefin tuna illegally caught in Maltese waters exported to Spain for millions. In: Independent.com.mt. http://www.independent.com.mt/articles/2018-10-19/local-news/Europol-bluefin-tuna-illegally-caught-in-Maltese-waters-exported-to-Spain-for-millions-6736197990. Accessed 12 Nov 2018
241. (2018) The oldest fish hooks and evidence of paleolithic offshore fishing (Circa 21,000 BCE–16,000 BCE): HistoryofInformation.com. In: Historyofinformation.com
242. The PEW Charitable Trust (2016) Netting billions: a global valuation of tuna
243. The World Bank Regional Vice-Presidency for Africa (2013) The pirates of Somalia: ending the threat, rebuilding a nation. The World Bank, Washington, DC
244. Thonier-senneur (2018) La pêche au thon, histoire. Thonier-senneur.net. http://www.thonier-senneur.net/technique3.htm. Accessed 30 Oct 2018
245. Tory S (2017) The mysterious disappearance of Keith Davis. Hakai Magazine Coastal Science and Societies
246. Tuna-Dolphin GATT Case (I and II) (2018) In: En.wikipedia.org. https://en.wikipedia.org/wiki/Tuna-Dolphin_GATT_Case_(I_and_II). Accessed 6 Dec 2018
247. United States (2015) U.S. loses WTO appeal in Mexican tuna dispute. In: U.S. Reuters. https://www.reuters.com/article/usa-mexico-tuna/u-s-loses-wto-appeal-in-mexican-tuna-dispute-idUSL1N13F28S20151120. Accessed 30 Nov 2018
248. Urbina I (2015) Murder at sea: captured on video, but killers go free. In: Nytimes.com. https://www.nytimes.com/2015/07/20/world/middleeast/murder-at-sea-captured-on-video-but-killers-go-free.html. Accessed 16 Nov 2018

249. Waal F (2001) The ape and the sushi master. Allen Lane, London
250. White C (2018) Ray Hilborn study disputes previous findings on forage fish. fisherynation. com. http://fisherynation.com/archives/57400. Accessed 5 Nov 2018
251. Wietecha O (2016) News: Mexico wants to kill dolphins—and the MSC label will let them. In: Savedolphins.eii.org. http://savedolphins.eii.org/news/entry/mexico-wants-to-kill-dolphins-and-the-msc-label-will-let-them. Accessed 11 Dec 2018
252. Wiki (2018) Tuna-dolphin GATT case (I and II). In: En.wikipedia.org. https://en.wikipedia. org/wiki/Tuna-Dolphin_GATT_Case_(I_and_II). Accessed 8 Dec 2018
253. Wikipedia (2018) Battle of the Paracel Islands. In: En.wikipedia.org. https://en.wikipedia.org/ wiki/Battle_of_the_Paracel_Islands. Accessed 16 Nov 2018
254. Wilson K (2018) eBook: looting the seas. In: ICIJ. https://www.icij.org/investigations/looting-the-seas/e-book-looting-seas/. Accessed 9 Nov 2018
255. Worm B, Davis B, Kettemer L et al (2013) Global catches, exploitation rates, and rebuilding options for sharks. Mar Policy 40:194–204. https://doi.org/10.1016/j.marpol.2012.12.034
256. Worms B et al (2006) Impacts of biodiversity loss on ocean ecosystem services. Science 314:787–790. https://doi.org/10.1126/science.1132294
257. Worms B, Hilborn R, Baum J et al (2009) Rebuilding global fisheries. Science 325 (5940):578–585
258. WTO lets Mexico Slap Trade Sanctions on U.S. in Tuna Dispute (2017) In: U.S. https://www. reuters.com/article/us-usa-mexico-tuna-idUSKBN17R1V2. Accessed 9 Dec 2018
259. Wulf A (2016) The invention of nature. John Murray, London
260. WWF (2008) End of the line for tuna commission, time for trade measures, November 24
261. WWF (2018) €12.5 million illegal bluefin tuna trade exposes threat to sustainable fisheries in Europe. In: Wwf.eu. http://www.wwf.eu/?uNewsID=336830. Accessed 12 Nov 2018
262. Wwf.se (2018). https://www.wwf.se/source.php/1718354/PR-Tuna-tagging_english.pdf. Accessed 31 Oct 2018
263. Young N, Irvin W, McLean M (1997) The flipper phenomenon: perspectives on the Panama declaration and the "dolphin safe" label. Ocean Coast Law J 3:1

Index

© Springer Nature Switzerland AG 2019
S. Adolf, *Tuna Wars*, https://doi.org/10.1007/978-3-030-20641-3